OXFORD STATISTICAL SCIENCE SERIES

SERIES EDITORS

OXFORD STATISTICAL SCIENCE SERIES

A. C. Atkinson: *Plots, Transformations, and Regression*
M. Stone: *Coordinate-Free Multivariable Statistics*
W. J. Krzanowski: *Principles of Multivariate Analysis: A User's Perspective*
M. Aitkin, D. Anderson, B. Francis, and J. Hinde: *Statistical Modelling in GLIM*
Peter J. Diggle: *Time Series: A Biostatistical Introduction*
Howell Tong: *Non-linear Time Series: A Dynamical System Approach*
V. P. Godambe: *Estimating Functions*
A. C. Atkinson and A. N. Donev: *Optimum Experimental Design*
U. N. Bhat and I. V. Basawa: *Queueing and Related Models*
J. K. Lindsey: *Models for Repeated Measurements*
N. T. Longford: *Random Coefficient Models*
Philip J. Brown: *Measurement, Regression, and Calibration*
Peter J. Diggle, Kung-Yee Liang, and Scott L. Zeger: *Analysis of Longitudinal Data*
J. I. Ansell and M. J. Phillips: *Practical Methods for Reliability Data Analysis*
J. K. Lindsey: *Modelling Frequency and Count Data*
J. L. Jensen: *Saddlepoint Approximations*
Steffen L. Lauritzen: *Graphical Models*

Analysis of Longitudinal Data

PETER J. DIGGLE

Professor of Statistics
Lancaster University

KUNG-YEE LIANG

and

SCOTT L. ZEGER

Professor of Biostatistics
School of Hygiene and Public Health
Johns Hopkins University, Maryland

CLARENDON PRESS • OXFORD

Oxford University Press, Walton Street, Oxford OX2 6DP
Oxford New York
Athens Auckland Bangkok Bombay
Calcutta Cape Town Dar es Salaam Delhi
Florence Hong Kong Istanbul Karachi
Kuala Lumpur Madras Madrid Melbourne
Mexico City Nairobi Paris Singapore
Taipei Tokyo Toronto
and associated companies in
Berlin Ibadan

Oxford is a trade mark of Oxford University Press

Published in the United States by
Oxford University Press Inc., New York

First published 1994
Reprinted 1994, 1995 (with corrections), 1996

A catalogue record for this book is available from the British Library

Library of Congress Cataloging in Publication Data
Diggle, Peter.
Analysis of longitudinal data / Peter J. Diggle, Kung-Yee Liang, and Scott L. Zeger.
p. cm.—(Oxford statistical science series ; 13)
Includes bibliographical references and index.
1. Multivariate analysis. 2. Time-series analysis.
3. Longitudinal method. 4. Life sciences—Statistical methods.
I. Liang, Kung-Yee. II. Zeger, Scott L. III. Title. IV. Series.
QA278.D545 1994 519.5'35—dc20 94–1859
ISBN 0 19 852284 3

Printed in Great Britain on acid-free paper by
Bookcraft (Bath) Ltd, Midsomer Norton, Avon

To Mandy, Yung-Kuang, and Joanne

Preface

This book describes statistical models and methods for the analysis of longitudinal data, with a strong emphasis on applications in the biological and health sciences. The technical level of the book is roughly that of a final year undergraduate or first year postgraduate course in statistics. However, we have tried to write in such a way that readers with a lower level of technical knowledge, but experience of dealing with longitudinal data from an applied point of view, will be able to appreciate and evaluate the main ideas. Also, we hope that readers with interests across a wide spectrum of application areas will find the ideas relevant and interesting.

In classical univariate statistics, a basic assumption is that each of a number of subjects, or experimental units, gives rise to a single measurement on some relevant variable, termed the *response*. In multivariate statistics, the single measurement on each subject is replaced by a vector of measurements. For example, in a univariate medical study we might measure the blood pressure of each subject, whereas in a multivariate study we might measure blood pressure, heart-rate, temperature, and so on. In longitudinal studies, each subject again gives rise to a vector of measurements, but these now represent the same physical quantity measured at a sequence of observation times. Thus, for example, we might measure a subject's blood pressure on each of five successive days.

Longitudinal data therefore combine elements of multivariate and time series data. However, they differ from classical multivariate data in that the time series aspect of the data typically imparts a much more highly structured pattern of interdependence among measurements than for standard multivariate data sets; and they differ from classical time series data in consisting of a large number of short series, one from each subject, rather than a single, long series.

The book divides naturally into three blocks. The first three chapters provide an introduction to the subject, and cover basic issues of design and exploratory analysis. Chapters 4, 5, 6, and 11 develop linear models and associated statistical methods for data sets in which the response variable is a continuous measurement. Chapters 7, 8, 9, and 10 are concerned with generalized linear models for discrete response variables. Appendix A gives a brief review of the statistical background assumed.

We have chosen not to discuss software explicitly in the book. Many commercially available packages, for example SAS, BMDP, or GENSTAT, include some facilities for longitudinal data analysis. However, none of the currently available packages contains enough facilities to cope with the full range of longitudinal data analysis problems which we cover in the book. For our own analyses, we have used the S system (Becker *et al.*, 1988; Chambers and Hastie, 1992) with additional user-defined functions for longitudinal data analysis.

Many friends and colleagues have helped us to write this book. Patty Hubbard typed the book. Larry Park and Mary Joy Argo facilitated its production. Larry Magder, Daniel Tsou, Patrick Heagerty, Bev Mellen-Harrison, Beth Melton, John Hanfelt, Sterling Hilton, Larry Moulton, Nick Lange and Joanne Katz gave assistance with computing, preparation of diagrams, and reading a first draft. We gratefully acknowledge support from a Merck Development Grant to Johns Hopkins University. Our colleagues in the Department of Mathematics at Lancaster and the Department of Biostatistics at Johns Hopkins have helped to provide a stimulating work environment in which our ideas could flourish. We thank all of these people, whilst retaining full responsibility for any deficiencies in the final product!

Lancaster
March 1994

P.J.D.
K.Y.L.
S.L.Z.

Contents

1	**Introduction**	**1**
	1.1 Longitudinal studies	1
	1.2 Examples	3
	1.3 Notation	16
	1.4 Merits of longitudinal studies	17
	1.5 Approaches to longitudinal data analysis	18
	1.6 Organization of subsequent chapters	21
2	**Design considerations**	**23**
	2.1 Introduction	23
	2.2 Bias	23
	2.3 Efficiency	25
	2.4 Sample size calculations	27
	2.4.1 Continuous responses	29
	2.4.2 Binary responses	31
	2.5 Further reading	32
3	**Exploring longitudinal data**	**33**
	3.1 Introduction	33
	3.2 Graphical presentation of longitudinal data	34
	3.3 Fitting smooth curves to longitudinal data	41
	3.4 Exploring correlation structure	47
	3.5 Further reading	52
4	**General linear models for longitudinal data**	**55**
	4.1 Motivation	55
	4.2 The general linear model with correlated errors	55
	4.2.1 The uniform correlation model	56
	4.2.2 The exponential correlation model	57
	4.3 Weighted least-squares estimation	58
	4.4 Maximum likelihood estimation under Gaussian assumptions	63
	4.5 Restricted maximum likelihood estimation	64
	4.6 Robust estimation of standard errors	68
5	**Parametric models for covariance structure**	**78**
	5.1 Introduction	78

5.2 Models 79
5.2.1 Pure serial correlation 81
5.2.2 Serial correlation plus measurement error 87
5.2.3 Random intercept plus serial correlation plus measurement error 87
5.2.4 Random effects plus measurement error 88
5.3 Model-fitting 90
5.3.1 Formulation 91
5.3.2 Estimation 92
5.3.3 Inference 93
5.3.4 Diagnostics 95
5.4 Examples 95
5.5 Non-parametric modelling of the mean response 106
5.6 Estimation of individual trajectories 113
5.7 Further reading 115

6 Analysis of variance methods 117
6.1 Preliminaries 117
6.2 Time-by-time ANOVA 118
6.3 Derived variables 122
6.4 Repeated measures 126
6.5 Conclusions 129

7 Generalized linear models for longitudinal data 131
7.1 Marginal models 131
7.2 Random effects models 133
7.3 Transition (Markov) models 135
7.4 Contrasting approaches 137
7.5 Inferences 142

8 Marginal models 146
8.1 Introduction 146
8.2 Binary responses 147
8.2.1 The log-linear model 147
8.2.2 Log-linear models for marginal means 148
8.2.3 Generalized estimating equations 151
8.3 Examples 153
8.4 Counted responses 162
8.4.1 Parametric modelling for count data 162
8.4.2 Generalized estimating equation approach 165
8.4.3 An example 165
8.5 Further reading 167

9 Random effects models 169
9.1 Introduction 169

9.2 Estimation for generalized linear mixed models 171
9.2.1 Conditional likelihood 171
9.2.2 Maximum likelihood estimation 172
9.3 Logistic regression for binary responses 175
9.3.1 Conditional likelihood approach 175
9.3.2 Random effects models for binary data 178
9.3.3 Examples of logistic models with Gaussian random effects 180
9.4 Counted responses 183
9.4.1 Conditional likelihood method 183
9.4.2 Random effects models for counts 185
9.4.3 Poisson–Gaussian random effects models 187
9.5 Further reading 188

10 Transition models 190
10.1 General 190
10.2 Fitting transition models 192
10.3 Transition models for categorical data 194
10.3.1 Indonesian children's study example 197
10.3.2 Ordered categorical data 201
10.4 Log-linear transition models for count data 203
10.5 Further reading 207

11 Missing values in longitudinal data 208
11.1 Classification of missing values for longitudinal data 208
11.2 Testing for completely random dropouts 211
11.3 Modelling the dropout process 216
11.4 Discussion 221

A Statistical Background 235
A.1 Introduction 235
A.2 The linear model and the method of least squares 235
A.3 Multivariate Gaussian theory 237
A.4 Likelihood inference 238
A.5 Generalized linear models 241
A.5.1 Logistic regression 241
A.5.2 Poisson regression 242
A.5.3 The general class 243
A.6 Quasi-likelihood 244

Index 247

1 Introduction

1.1 Longitudinal studies

The defining characteristic of a longitudinal study is that individuals are measured repeatedly through time. Longitudinal studies are in contrast to cross-sectional studies, in which a single outcome is measured for each individual. While it is often possible to address the same scientific questions with a longitudinal or cross-sectional study, the major advantage of the former is its capacity to separate what in the context of population studies are called *cohort* and *age* effects. The idea is illustrated in Fig. 1.1. In Fig. 1.1(a), reading ability is plotted against age for a hypothetical cross-sectional study of children. Reading ability appears to be poorer among older children; little else can be said. In Fig. 1.1(b), we suppose the same data were obtained in a longitudinal study in which each individual was measured twice. Now it is clear that while younger children began at a higher reading level, everyone improved with time. Such a pattern might have resulted from introducing elementary education into a poor rural community beginning with the younger children. If the data set were as in Fig. 1.1(c), a different explanation would be required. The cross-sectional and longitudinal patterns now tell the same unusual story – that reading ability deteriorates with age.

The point of this example is that longitudinal studies (Figs. 1.1(b) and 1.1(c)) can distinguish changes over time within individuals (ageing effects) from differences among people in their baseline levels (cohort effects). Cross-sectional studies cannot. This concept is developed in more detail in Section 1.4.

In some studies a third time scale, the *period* or calendar date of a measurement, is also important. Any two of age, period, and cohort determine the third. For example, an individual's age and birth cohort at a given measurement determine the date. Analyses which must consider all three scales require external assumptions which unfortunately are difficult to validate. See Mason and Feinberg (1985) for details.

Longitudinal data can be collected either prospectively, following subjects forward in time, or retrospectively, by extracting multiple measurements on each person from historical records. The statistical methods dis-

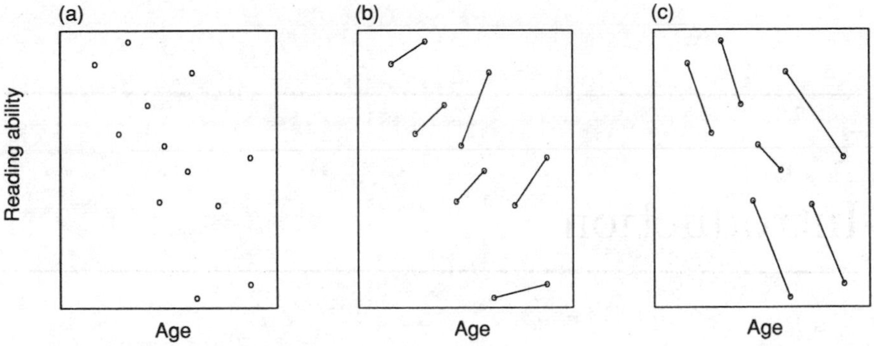

Fig. 1.1. Hypothetical data on the relationship between reading ability and age.

cussed in this book apply to both situations. Longitudinal data are more commonly collected prospectively since the quality of repeated measurements collected from past records or from a person's recollection may be inferior (Goldfarb, 1960).

Clinical trials are prospective studies which often have time to a clinical outcome as the principal response. The dependence of this univariate measure on treatment and other factors is the subject of *survival analysis* (Cox, 1972). This book does not discuss survival problems. The interested reader is referred to Cox and Oakes (1984) or Kalbfleisch and Prentice (1980).

The defining feature of a longitudinal data set is repeated observations on individuals enabling direct study of change. Longitudinal data require special statistical methods because the set of observations on one subject tends to be intercorrelated. This correlation must be taken into account to draw valid scientific inferences.

The issue of accounting for correlation also arises when analysing a single long *time series* of measurements. Diggle (1990) discusses time series analysis in the biological sciences. Analysis of longitudinal data tends to be simpler because subjects can usually be assumed independent. Valid inferences can be made by borrowing strength across people. That is, the consistency of a pattern across subjects is the basis for substantive conclusions. For this reason, inferences from longitudinal studies can be made more robust to model assumptions than those from time series data, particularly to assumptions about the nature of the correlation.

Sociologists and economists often refer to longitudinal studies as *panel studies.* Although many of the statistical methods which we discuss in this book will be applicable to the analysis of data from the social sciences, our emphasis is firmly motivated by our experience of longitudinal data

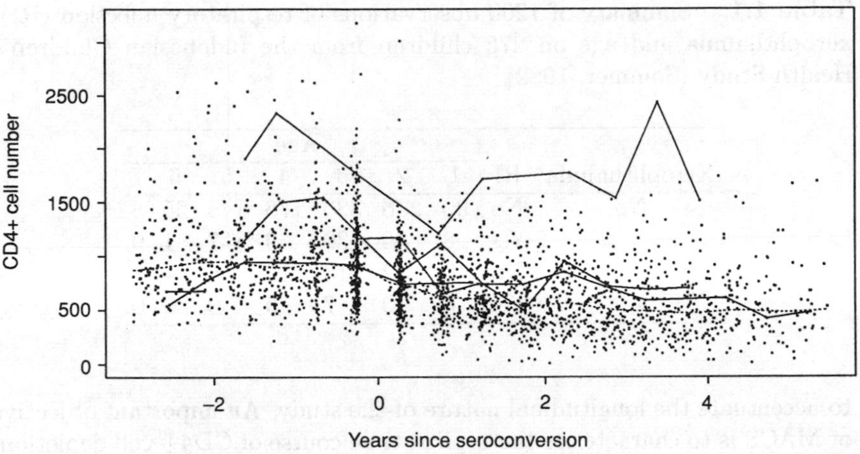

Fig. 1.2. Relationship between CD4+ cell numbers and time since seroconversion due to infection with the HIV virus. · : individual counts; —— : sequences of measurements for randomly selected subjects; .... : lowess curve.

problems arising in the biological health sciences. A somewhat different emphasis would have been appropriate for many social science applications, and is reflected in books written with such applications in mind. See, for example, Goldstein (1979), Plewis (1985), or Heckman and Singer (1985).

1.2 Examples

We now introduce six longitudinal data sets that will be used throughout the book for illustration. The examples have been chosen from the biological and health sciences to represent a range of challenges for analysis. After presenting each data set, common and distinguishing features are discussed.

Example 1.1. CD4+ cell numbers

The human immune deficiency virus (HIV) causes AIDS by reducing a person's ability to fight infection. HIV attacks an immune cell called the CD4+ cell which orchestrates the body's immunoresponse to infectious agents. An uninfected individual has around 1100 cells per millilitre of blood. CD4+ cells decrease in number with time from infection so that an infected person's CD4+ cell number can be used to monitor disease progression. Figure 1.2 displays 2376 values of CD4+ cell number plotted against time since seroconversion (time when HIV becomes detectable) for 369 infected men enrolled in the Multicenter AIDS Cohort Study or MACS (Kaslow *et al.*, 1987).

In Fig. 1.2, repeated measurements for some individuals are connected

Table 1.1. Summary of 1200 observations of respiratory infection (RI), xerophthalmia and age on 275 children from the Indonesian Children's Health Study (Sommer, 1982)

		Age						
Xerophthalmia	RI	1	2	3	4	5	6	7
No	No	90	236	330	176	143	65	5
	Yes	8	36	39	9	7	1	0
Yes	No	0	2	18	15	8	4	1
	Yes	0	0	7	0	0	0	0

to accentuate the longitudinal nature of the study. An important objective of MACS is to characterize the typical time course of CD4+ cell depletion. This helps to clarify the interaction of HIV with the immune system and can assist when counselling infected men. A non-parametric smooth curve (Zeger and Diggle, 1994) has been added to the figure to highlight the average trend. Note that the average CD4+ cell number is constant until the time of seroconversion and then decreases, more quickly at first.

Objectives of a longitudinal analysis of these data are to:

- estimate the average time course of CD4+ cell depletion;
- estimate the time course for individual men taking account of the measurement error in CD4+ cell determinations;
- characterize the degree of heterogeneity across men in the rate of progression;
- identify factors which predict CD4+ cell changes.

Example 1.2. Indonesian children's health study
Alfred Sommer and colleagues conducted a study (which we will refer to as the Indonesian Children's Health Study or ICHS) in West Java, Indonesia to determine the causes and effects of vitamin A deficiency in pre-school children (Sommer, 1982). Over 3000 children were medically examined quarterly for up to six visits to assess whether they suffered from respiratory or diarrheal infection and xerophthalmia, an ocular manifestation of vitamin A deficiency. Weight and height were also measured. We will focus on the question of whether vitamin A deficient children are at increased risk of respiratory infection, one of the leading causes of morbidity and mortality in children from the developing world. Such a relationship is plausible because vitamin A is required for the integrity of epithelial cells, the first line of defence against infection in the respiratory tract. It has public health significance because vitamin A deficiency can be eliminated

by diet modification or if necessary by food fortification for only a few pence per child per year.

The data on 275 children are summarized in Table 1.1. The objectives of analysis are to estimate the increase in risk of respiratory infection for children who are vitamin A deficient while controlling for other demographic factors, and to estimate the degree of heterogeneity in the risk of disease among children.

Example 1.3. Growth of Sitka spruce

Dr Peter Lucas of the Biological Sciences Division at Lancaster University provided these data on the growth of Sitka spruce trees. The study objective is to assess the effect of ozone pollution on tree growth. As ozone pollution is common in urban areas, the impact of increased ozone concentrations on tree growth is of considerable interest. The response variable is log tree size, where size is conventionally measured by the product of tree height and diameter squared. The data for 79 trees over two growing seasons are listed in Table 1.2, and displayed graphically in Fig. 1.3. A total of 54 trees were grown with ozone exposure at 70 ppb; 25 were grown under control conditions. The objective is to compare the growth patterns of trees under the two conditions.

Example 1.4. Protein content of milk

In this example, milk was collected weekly from 79 Australian cows and analysed for its protein content. The cows were maintained on one of three diets: barley, a mixture of barley and lupins, or lupins alone. The data were provided by Ms Alison Frensham, and are listed in Table 1.3. Figure 1.4 displays the three subsets of the data corresponding to each of the three diets. The repeated measurements on each animal are joined to accentuate the longitudinal nature of the data set. The objective of the study is to determine how diet affects the protein in milk. It appears from the figure that barley gives higher values than the mixture, which in turn gives higher values than lupins alone. A plot of the average traces for each group (Diggle, 1990) confirms this pattern. One problem with simple inferences, however, is that in this example, time is measured in weeks since calving, and the experiment was terminated 19 weeks after the earliest calving. Thus, about half of the 79 sequences of milk protein measurements are incomplete. Calving date may well be associated, directly or indirectly, with the physiological processes that also determine protein content. If this is the case, the missing observations should not be ignored in inference. This issue is taken up in Chapter 11.

Table 1.2. Measurements of log-size for 79 Sitka spruce trees grown in normal or ozone-enriched environments. Within each year, the data are organized in four blocks, corresponding to four controlled environment chambers. The first two chambers, containing 27 trees each, have an ozone-enriched atmosphere, the remaining two, containing 12 and 13 trees respectively, have a normal (control) atmosphere

Time in days since 1 January 1988												
152	174	201	227	258	469	496	528	556	579	613	639	674
4.51	4.98	5.41	5.9	6.15	6.16	6.18	6.48	6.65	6.87	6.95	6.99	7.04
4.24	4.2	4.68	4.92	4.96	5.2	5.22	5.39	5.65	5.71	5.78	5.82	5.85
3.98	4.36	4.79	4.99	5.03	5.87	5.88	6.04	6.34	6.49	6.58	6.65	6.61
4.36	4.77	5.1	5.3	5.36	5.53	5.56	5.68	5.93	6.21	6.26	6.2	6.19
4.34	4.95	5.42	5.97	6.28	6.5	6.5	6.79	6.83	7.1	7.17	7.21	7.16
4.59	5.08	5.36	5.76	6.0	6.33	6.34	6.39	6.78	6.91	6.99	7.01	7.05
4.41	4.56	4.95	5.23	5.33	6.13	6.14	6.36	6.57	6.78	6.82	6.81	6.86
4.24	4.64	4.95	5.38	5.48	5.61	5.63	5.82	6.18	6.42	6.48	6.47	6.46
4.82	5.17	5.76	6.12	6.24	6.48	6.5	6.77	7.14	7.26	7.3	6.91	7.28
3.84	4.17	4.67	4.67	4.8	4.94	4.94	5.05	5.33	5.53	5.56	5.57	5.6
4.07	4.31	4.9	5.1	5.1	5.26	5.26	5.38	5.66	5.81	5.84	5.93	5.89
4.28	4.8	5.27	5.55	5.65	5.76	5.77	5.98	6.18	6.39	6.43	6.44	6.41
4.47	4.89	5.23	5.55	5.74	5.99	6.01	6.08	6.39	6.45	6.57	6.57	6.58
4.46	4.84	5.11	5.34	5.46	5.47	5.49	5.7	5.93	6.06	6.15	6.12	6.12
4.6	4.08	4.17	4.35	4.59	4.65	4.69	5.01	5.21	5.38	5.58	5.46	5.5
3.73	4.15	4.61	4.87	4.93	5.24	5.25	5.25	5.45	5.65	5.65	5.76	5.83
4.67	4.88	5.18	5.34	5.49	6.44	6.44	6.61	6.74	7.06	7.11	7.04	7.11
2.96	3.47	3.76	3.89	4.3	4.15	4.15	4.41	4.72	4.76	4.93	4.98	5.07
3.24	3.93	4.76	4.62	4.64	4.63	4.64	4.77	5.08	5.27	5.3	5.43	5.2
4.36	4.77	5.02	5.26	5.45	5.44	5.44	5.49	5.73	5.77	6.01	5.96	5.96
4.04	4.64	4.86	5.09	5.25	5.25	5.27	5.5	5.65	5.69	5.97	5.97	5.89
3.53	4.25	4.68	4.97	5.18	5.64	5.64	5.53	5.74	5.78	5.94	6.18	5.99
4.22	4.69	5.07	5.37	5.58	5.76	5.8	6.11	6.37	6.35	6.58	6.55	6.55
2.79	3.1	3.3	3.38	3.55	3.61	3.65	3.93	4.18	4.13	4.36	4.43	4.39
3.3	3.9	4.34	4.96	5.4	5.46	5.49	5.77	6.03	6.07	6.2	6.26	6.28
3.34	3.81	4.21	4.54	4.86	4.93	4.96	5.15	5.48	5.49	5.7	5.74	5.74
3.76	4.36	4.7	5.44	5.32	5.65	5.67	5.63	6.04	6.02	6.05	6.03	5.91
4.49	4.76	5.15	5.37	5.56	5.73	5.73	5.8	5.97	6.1	6.16	6.22	6.13
4.88	5.14	5.52	6.08	6.17	6.32	6.33	6.37	6.68	6.83	6.94	6.93	6.95
4.88	5.32	5.63	5.75	5.94	6.09	6.09	6.14	6.51	6.61	6.68	6.64	6.74
3.8	4.16	4.45	4.89	5.05	5.06	5.06	5.13	5.32	5.46	5.46	5.42	5.49
4.46	4.62	5.0	5.4	5.49	5.68	5.72	5.95	6.13	6.32	6.33	6.33	6.3
4.29	4.82	5.32	5.46	5.5	5.54	5.54	5.54	5.6	5.57	5.55	5.55	5.55
4.06	4.58	4.81	5.12	5.27	5.48	5.51	5.58	5.93	6.36	6.17	6.07	6.13
5.16	5.43	5.71	6.08	6.21	6.37	6.38	6.41	6.64	6.82	6.89	7.11	7.14
3.81	4.12	4.42	4.62	4.6	4.74	4.76	4.94	5.1	5.21	5.23	5.22	5.23
5.09	5.62	5.9	6.36	6.49	6.72	6.72	6.74	6.87	6.87	6.87	6.83	6.97
4.13	4.71	5.27	5.56	5.72	6.06	6.06	6.21	6.44	6.66	6.71	6.65	6.64
4.85	5.36	5.52	5.96	6.13	6.22	6.24	6.41	6.58	6.78	6.83	6.82	6.8
4.11	4.62	4.95	5.28	5.43	5.8	5.8	5.87	6.2	6.44	6.44	6.4	6.44

Table 1.2.— *continued*

Time in days since 1 January 1988												
152	174	201	227	258	469	496	528	556	579	613	639	674
4.95	5.39	5.82	6.42	6.48	6.61	6.61	6.66	6.73	6.83	6.9	6.83	6.63
4.36	4.65	5.04	5.38	5.47	5.48	5.48	5.47	5.84	5.97	5.93	5.95	6.01
4.05	4.65	5.09	5.44	5.6	5.79	5.8	6.07	6.14	6.3	6.32	6.34	6.42
3.76	4.27	4.59	5.1	5.25	5.41	5.44	5.48	5.93	5.97	6.08	6.29	6.24
2.84	3.25	3.69	4.16	4.21	4.3	4.3	4.45	4.59	4.74	4.84	4.64	4.64
4.33	4.8	5.09	5.42	5.61	5.85	5.88	6.01	6.22	6.45	6.55	6.55	6.55
3.99	4.55	4.91	5.26	5.3	5.69	5.69	5.9	5.89	5.98	6.05	6.25	6.25
3.5	3.75	3.97	4.71	4.85	5.01	5.02	5.27	5.45	5.59	5.67	5.83	5.86
3.31	3.45	4.16	4.48	4.54	4.72	4.74	4.93	5.07	5.26	5.26	5.35	5.35
3.03	3.55	3.97	4.4	4.58	4.47	4.47	4.66	4.8	5.1	5.08	5.12	5.12
3.27	3.83	4.44	4.8	4.89	5.08	5.09	5.34	5.63	5.81	5.93	5.94	5.94
3.56	4.18	4.7	5.27	5.28	5.5	5.5	5.77	5.98	6.05	6.19	6.14	6.14
3.39	3.73	3.92	4.11	4.15	4.49	4.52	4.82	5.18	5.26	5.32	5.28	5.28
3.72	4.16	4.55	5.03	5.02	5.16	5.16	5.28	5.52	5.7	5.7	5.67	5.67
4.53	5.05	5.18	5.41	5.42	5.71	5.71	5.96	6.17	6.34	6.44	6.46	6.43
4.97	5.32	5.83	6.29	6.45	6.61	6.61	6.79	7.13	7.24	7.32	7.29	7.35
4.37	4.81	5.03	5.19	5.4	5.57	5.59	5.82	6.03	6.17	6.32	6.25	6.29
4.58	4.99	5.37	5.68	5.93	6.14	6.17	6.43	6.56	6.69	6.81	6.82	6.76
4.0	4.5	4.92	5.44	5.87	6.02	6.04	6.11	6.47	6.61	6.66	6.7	6.65
4.73	5.05	5.33	5.92	6.01	6.26	6.26	6.36	6.49	6.63	6.92	6.92	6.92
5.15	5.63	6.11	6.39	6.61	6.82	6.82	6.95	7.11	7.44	7.53	7.46	7.56
4.1	4.46	4.84	5.29	5.48	5.68	5.68	5.85	5.99	6.08	6.25	6.15	6.18
3.22	3.85	4.47	4.85	5.11	5.28	5.28	5.45	5.74	5.95	6.05	6.07	6.02
2.23	2.89	3.16	3.4	3.52	3.89	3.93	4.22	4.51	4.65	4.7	4.73	4.68
3.65	4.36	4.76	5.18	5.44	5.7	5.7	5.89	6.09	6.39	6.57	6.36	6.44
3.4	3.92	4.5	4.97	5.14	5.34	5.35	5.61	5.83	6.03	6.09	5.98	5.95
5.16	5.49	5.74	6.05	6.21	6.37	6.37	6.52	6.65	6.86	6.87	6.88	6.84
4.04	4.52	5.15	5.59	5.87	5.96	5.96	6.17	6.37	6.53	6.6	6.52	6.59
4.52	4.91	5.04	5.71	5.97	6.11	6.12	6.24	6.44	6.54	6.65	6.63	6.64
4.56	5.12	5.4	5.69	5.89	6.16	6.17	6.13	6.44	6.72	6.81	6.87	6.8
4.9	5.35	5.71	6.12	6.25	6.39	6.39	6.52	6.86	7.05	7.09	6.9	6.88
4.83	5.1	5.43	5.59	6.04	6.21	6.21	6.46	6.45	6.59	6.7	6.63	6.66
5.46	5.79	6.12	6.41	6.63	6.73	6.73	6.77	6.68	6.75	6.75	6.62	6.6
4.17	4.67	5.16	5.56	5.75	6.0	6.02	6.14	6.28	6.55	6.66	6.63	6.63
3.35	4.05	4.51	5.22	5.44	5.79	5.82	6.05	6.29	6.22	6.39	6.47	6.42
3.33	3.82	4.38	4.99	5.17	5.4	5.4	5.73	5.85	5.75	5.99	6.1	6.15
3.41	3.68	4.03	4.28	4.54	4.52	4.57	5.01	5.13	5.11	5.3	5.46	5.35
4.5	4.8	5.28	5.83	6.16	6.33	6.34	6.56	6.63	6.75	6.89	6.96	6.94
2.99	3.61	4.48	4.91	5.06	5.23	5.25	5.56	5.95	5.98	6.21	6.28	6.34

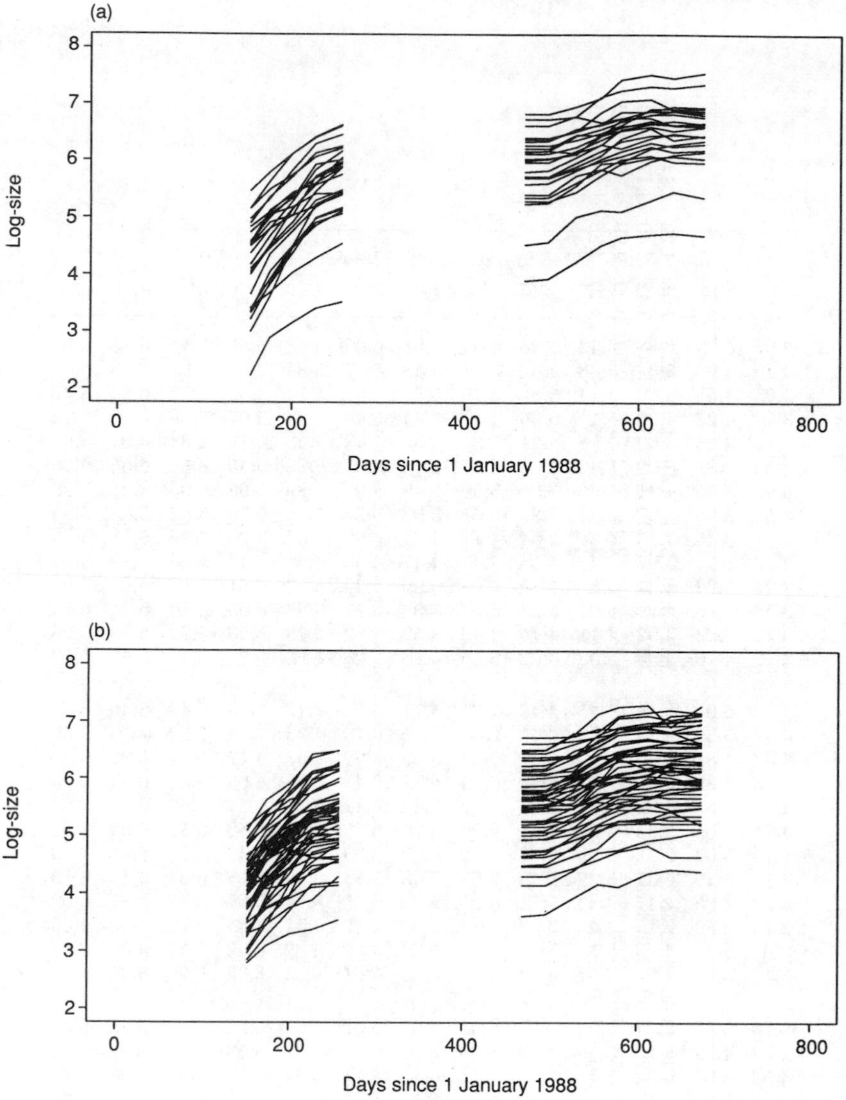

Fig. 1.3. Log-size of 79 Sitka spruce over two growing seasons: (a) control; (b) ozone-treated

Table 1.3. Protein content of milk samples (Ms. A. Frensham). Percentage protein content of milk samples taken at weekly intervals from each of 79 cows on one of three diets: cows 1–25, barley; cows 26–52, barley+lupins; cows 53–79, lupins. 9.99 signifies missing

(a) Barley diet

3.63	3.57	3.47	3.65	3.89	3.73	3.77	3.90	3.78	3.82	3.83	3.71	4.10	4.02	4.13	4.08	4.22	4.44	4.30
3.24	3.25	3.29	3.09	3.38	3.33	3.00	3.16	3.34	3.32	3.31	3.27	3.41	3.45	3.12	3.42	3.40	3.17	3.00
3.98	3.60	3.43	3.30	3.29	3.25	2.93	3.20	3.27	3.22	2.93	2.92	2.82	2.64	9.99	9.99	9.99	9.99	9.99
3.66	3.50	3.05	2.90	2.72	3.11	3.05	2.80	3.20	3.18	3.14	3.18	3.24	3.37	3.30	3.40	3.35	3.28	9.99
4.34	3.76	3.68	3.51	3.45	3.53	3.60	3.77	3.90	3.87	3.61	3.85	3.94	3.87	3.60	3.06	3.47	3.50	3.42
4.36	3.71	3.42	3.95	4.06	3.73	3.92	3.99	3.70	3.88	3.71	3.62	3.74	3.42	9.99	9.99	9.99	9.99	9.99
4.17	3.60	3.52	3.10	3.78	3.42	3.66	3.64	3.83	3.73	3.72	3.65	3.50	3.32	2.95	3.34	3.51	3.17	9.99
4.40	3.86	3.56	3.32	3.64	3.57	3.47	3.97	9.99	3.78	3.98	3.90	4.05	4.06	4.05	3.92	3.65	3.60	3.74
3.40	3.42	3.51	3.39	3.35	3.13	3.21	3.50	3.55	3.28	3.75	3.55	3.53	3.52	3.77	3.77	3.74	4.00	3.87
3.75	3.89	3.65	3.42	3.32	3.27	3.34	3.35	3.09	3.65	3.53	3.50	3.63	3.91	3.73	3.71	4.18	3.97	4.06
4.20	3.59	3.55	3.27	3.19	3.60	3.50	3.55	3.60	3.75	3.75	3.75	3.89	3.87	3.60	3.68	3.68	3.56	3.34
4.02	3.76	3.60	3.53	3.95	3.26	3.73	3.96	9.99	3.70	9.99	3.45	3.50	3.13	9.99	9.99	9.99	9.99	9.99
4.02	3.90	3.73	3.55	3.71	3.40	3.49	3.74	3.61	3.42	3.46	3.40	3.38	3.13	9.99	9.99	9.99	9.99	9.99
3.90	3.33	3.25	3.22	3.35	3.24	3.16	3.33	3.12	2.93	2.84	3.07	3.02	2.75	9.99	9.99	9.99	9.99	9.99
3.81	4.00	3.57	3.47	3.52	3.63	3.45	3.50	3.71	3.55	3.13	3.04	3.31	3.22	2.92	9.99	9.99	9.99	9.99
3.62	3.22	3.62	3.02	3.28	3.15	3.52	3.22	3.45	3.51	3.38	3.00	2.94	3.52	3.48	3.02	9.99	9.99	9.99
3.66	3.66	3.28	3.10	2.66	3.00	3.15	3.01	3.50	3.29	3.16	3.33	3.50	3.46	3.48	3.98	3.70	3.36	3.55
4.44	3.85	3.55	3.22	3.40	3.28	3.42	3.35	3.01	3.55	3.70	3.73	3.65	3.78	3.82	3.75	3.95	3.85	3.72
4.23	3.75	3.82	3.60	4.09	3.84	3.62	3.36	3.65	3.41	3.15	3.68	3.54	3.75	3.72	4.05	3.60	3.88	3.98
3.82	9.99	3.27	3.33	3.25	2.97	3.57	3.43	3.50	3.58	3.70	3.55	3.58	3.70	3.60	3.42	3.33	3.53	3.40
3.53	3.10	3.90	3.48	3.35	3.35	3.65	3.56	3.27	3.61	3.66	3.47	3.34	3.32	3.22	3.18	9.99	9.99	9.99
4.47	3.86	3.34	3.49	3.74	3.24	3.71	3.46	3.88	3.60	4.00	3.83	3.80	4.12	3.98	3.77	3.52	3.50	3.42
3.93	3.79	3.68	3.58	3.76	3.66	3.57	3.85	3.75	3.37	3.00	3.24	3.44	3.23	9.99	9.99	9.99	9.99	9.99
3.27	3.84	3.46	3.44	3.40	3.50	3.63	3.47	3.32	3.47	3.40	3.27	3.74	3.76	3.68	3.68	3.93	3.80	3.52
3.32	3.61	3.25	3.48	3.58	3.47	3.60	3.51	3.74	3.50	3.08	2.77	3.22	3.35	3.14	9.99	9.99	9.99	9.99

Continued overleaf

Table 1.3.—*continued*

(b) Mixed diet

3.38	3.38	3.10	3.09	3.15	2.77	3.40	3.25	3.35	3.57	3.80	3.70	3.44	3.51	3.78	3.52	3.26	3.52	3.25
3.80	3.51	3.19	3.11	3.35	3.46	3.20	3.52	3.66	3.87	3.85	3.90	4.06	3.93	3.40	3.85	3.93	3.52	3.62
4.17	3.71	3.32	3.10	3.07	3.11	3.29	3.00	3.32	3.25	3.50	3.31	3.51	3.54	3.45	3.59	3.38	3.44	3.18
4.59	3.86	3.62	3.60	3.65	3.75	3.71	3.73	3.62	3.63	3.40	3.44	3.45	3.20	9.99	9.99	9.99	9.99	9.99
4.07	3.45	3.56	3.10	3.92	3.20	3.28	3.36	3.28	3.25	2.99	3.10	3.05	2.70	9.99	9.99	9.99	9.99	9.99
4.32	3.37	3.47	3.46	3.31	3.57	3.38	3.01	3.66	3.46	3.21	3.76	3.44	3.47	3.45	3.89	3.65	3.59	3.80
3.56	3.14	3.60	3.36	3.37	3.44	3.57	3.55	3.57	3.68	3.58	3.65	3.29	3.53	3.26	2.89	9.99	9.99	9.99
3.67	3.33	3.20	2.72	2.95	3.07	2.90	3.56	3.19	3.35	3.26	3.40	3.18	3.43	3.57	3.49	3.59	3.27	3.48
4.15	3.55	3.27	3.27	3.65	3.58	3.55	3.47	3.48	3.55	3.30	3.32	3.22	2.97	9.99	9.99	9.99	9.99	9.99
3.51	3.90	2.75	3.37	3.51	3.64	3.55	3.60	3.70	3.60	3.65	3.55	3.30	3.15	2.90	9.99	9.99	9.99	9.99
4.20	3.31	3.30	3.32	3.29	3.17	3.04	3.09	3.14	3.04	3.46	3.12	3.30	3.09	3.45	3.47	3.30	3.45	3.42
4.12	3.56	3.58	3.47	3.20	3.50	3.40	3.56	3.02	3.25	3.44	3.52	3.72	3.55	3.65	3.30	3.35	3.20	2.84
3.52	3.64	3.10	3.19	2.95	2.75	3.15	3.20	3.03	3.53	3.42	3.46	3.33	3.53	3.59	4.42	3.99	3.88	3.80
4.08	3.62	3.55	3.43	3.13	3.28	3.13	3.28	3.02	3.48	3.34	3.25	3.38	3.68	3.40	3.54	3.66	3.52	3.40
4.02	3.45	3.28	3.31	3.37	3.26	3.25	3.48	3.23	3.03	2.80	3.17	3.04	3.05	9.99	9.99	9.99	9.99	9.99
3.18	3.12	2.87	3.04	2.88	3.08	3.15	3.22	3.20	3.46	3.33	3.26	3.40	3.36	3.97	3.20	3.10	3.19	3.10
4.11	4.09	3.54	3.15	3.57	3.23	3.10	3.30	3.60	3.74	3.55	3.66	3.68	3.81	3.97	3.78	3.41	3.15	3.32
3.27	3.10	3.14	2.80	3.34	3.11	3.17	3.15	3.52	3.35	3.30	3.55	9.99	3.11	2.94	3.25	3.05	2.95	9.99
3.27	3.40	3.50	3.38	3.29	3.60	3.50	3.33	3.66	3.68	3.20	3.10	3.32	3.26	3.08	9.99	9.99	9.99	9.99
3.97	3.60	3.50	3.50	3.58	3.58	3.57	3.65	3.71	3.37	3.15	3.20	3.32	3.01	9.99	9.99	9.99	9.99	9.99
3.31	3.42	3.24	3.38	3.39	3.70	3.47	3.49	3.64	3.50	3.27	2.95	3.04	3.21	3.04	9.99	9.99	9.99	9.99
4.12	3.25	3.14	3.00	3.20	3.55	3.45	3.63	3.50	3.90	3.30	3.70	3.92	3.87	3.25	3.12	3.04	3.45	3.38
3.92	3.55	3.74	3.46	3.45	3.48	3.26	3.60	3.50	2.86	3.65	3.17	3.18	3.05	9.99	9.99	9.99	9.99	9.99
3.78	3.40	3.17	3.28	3.27	3.69	3.24	3.30	3.24	2.91	3.00	2.85	3.41	3.30	3.22	3.55	3.62	3.59	3.45
4.00	3.74	3.48	3.75	3.78	3.77	3.70	3.91	3.94	3.94	3.10	3.10	3.30	3.27	9.99	9.99	9.99	9.99	9.99
4.37	4.06	3.77	3.40	3.30	3.27	3.05	3.10	3.72	3.25	3.55	3.70	3.65	3.80	3.78	4.08	4.10	3.57	9.99
3.79	4.07	3.35	3.46	3.20	3.99	3.54	9.99	3.25	3.30	3.18	3.21	3.75	3.67	3.77	3.76	3.75	3.92	3.50

Table 1.3.—*continued*

(c) Lupins diet

3.69	3.38	3.00	3.50	3.09	3.30	3.07	3.22	2.97	3.60	3.42	3.59	3.77	3.74	3.70	3.78	3.78	3.77	3.53
4.20	3.35	3.37	3.07	2.82	3.05	3.12	2.85	3.20	3.38	3.25	3.26	3.30	3.17	3.40	3.41	3.28	3.42	3.25
3.31	3.04	2.80	3.17	2.92	2.84	2.87	3.13	2.87	3.05	3.00	3.26	3.22	3.31	3.20	3.18	3.09	3.30	3.12
3.13	3.34	3.34	3.25	2.79	3.00	3.11	3.27	3.59	3.28	3.49	3.60	3.63	3.60	3.60	3.10	3.38	3.34	3.03
3.73	3.61	3.82	3.61	3.45	3.36	3.02	3.37	3.44	3.00	3.42	3.18	3.37	3.45	3.64	3.68	3.56	3.63	3.46
4.32	3.70	3.62	3.50	3.57	3.28	3.38	3.40	3.33	3.45	3.22	3.28	3.24	2.93	9.99	9.99	9.99	9.99	9.99
3.04	2.89	2.78	2.84	2.61	3.00	3.08	3.12	2.80	3.25	3.13	3.15	3.04	2.95	3.08	2.76	2.89	2.75	2.66
3.84	3.51	3.39	2.88	3.41	3.49	3.30	3.37	3.47	3.30	3.21	3.64	3.55	3.43	3.17	3.23	3.15	2.95	9.99
3.98	3.30	3.02	2.99	2.84	2.92	3.08	2.85	3.17	3.15	3.15	2.87	3.07	3.05	3.10	3.21	3.13	3.17	2.78
4.18	4.12	3.84	3.65	3.65	3.54	3.52	3.41	3.44	3.32	3.23	2.70	3.03	2.88	2.68	9.99	9.99	9.99	9.99
4.20	3.40	3.55	3.45	3.45	3.30	3.25	3.43	3.38	3.07	2.94	3.17	3.18	3.00	9.99	9.99	9.99	9.99	9.99
4.10	3.59	3.52	3.43	9.99	3.33	3.57	3.90	3.74	3.51	3.29	3.31	3.31	3.22	9.99	9.99	9.99	9.99	9.99
3.25	3.75	3.42	3.38	3.52	2.72	3.38	3.75	3.90	3.76	4.22	3.14	3.30	3.35	3.17	9.99	9.99	9.99	9.99
3.34	3.00	3.10	3.31	3.30	3.26	3.00	3.12	3.00	2.45	2.55	2.97	2.79	2.88	9.99	9.99	9.99	9.99	9.99
3.50	3.35	3.05	2.92	2.83	3.15	3.03	2.95	3.35	3.11	2.57	2.67	3.04	3.05	2.87	9.99	9.99	9.99	9.99
4.13	3.12	2.96	2.86	2.74	3.24	3.06	2.86	2.90	3.20	3.02	2.87	3.20	3.14	2.84	2.75	2.96	2.90	2.71
3.21	3.52	3.43	3.45	3.48	4.30	9.99	9.99	3.96	9.99	3.57	3.29	3.52	3.35	3.20	9.99	9.99	9.99	9.99
3.90	3.19	3.53	3.45	3.31	3.45	2.83	3.50	3.60	3.76	3.60	3.18	3.00	3.29	3.15	2.96	9.99	9.99	9.99
3.50	3.12	3.33	3.22	3.20	2.82	2.74	2.77	3.02	2.80	3.08	3.04	3.17	3.26	3.51	3.30	3.26	3.70	3.40
4.10	3.42	3.58	3.11	3.06	3.10	3.13	3.01	3.05	3.05	2.88	3.52	3.58	3.36	3.40	3.58	3.19	3.29	3.65
2.69	3.39	3.20	3.06	3.46	3.14	3.05	2.94	2.88	3.00	3.25	3.38	3.20	3.13	3.27	3.42	3.04	3.20	3.52
4.30	3.37	3.47	3.37	3.10	3.64	9.99	3.55	3.60	3.55	3.45	3.59	3.75	3.75	3.18	2.92	3.32	3.31	3.18
4.06	3.85	3.93	4.00	4.20	3.90	3.65	4.13	3.84	3.46	3.22	3.33	3.37	3.20	9.99	9.99	9.99	9.99	9.99
3.88	3.81	3.44	3.14	3.10	3.35	3.27	3.10	3.00	3.12	3.15	2.90	3.51	3.21	3.40	3.49	3.60	3.45	3.50
3.95	3.22	3.38	3.20	3.40	3.34	3.20	3.50	3.35	3.25	2.97	3.14	3.20	3.04	9.99	9.99	9.99	9.99	9.99
3.67	3.26	3.35	3.28	2.94	3.30	3.27	3.61	3.60	3.78	3.55	3.55	4.00	3.87	3.71	3.47	3.15	3.35	3.09
4.27	3.95	3.85	3.85	3.95	3.45	3.70	3.95	3.92	3.34	3.45	3.19	3.67	3.25	9.99	9.99	9.99	9.99	9.99

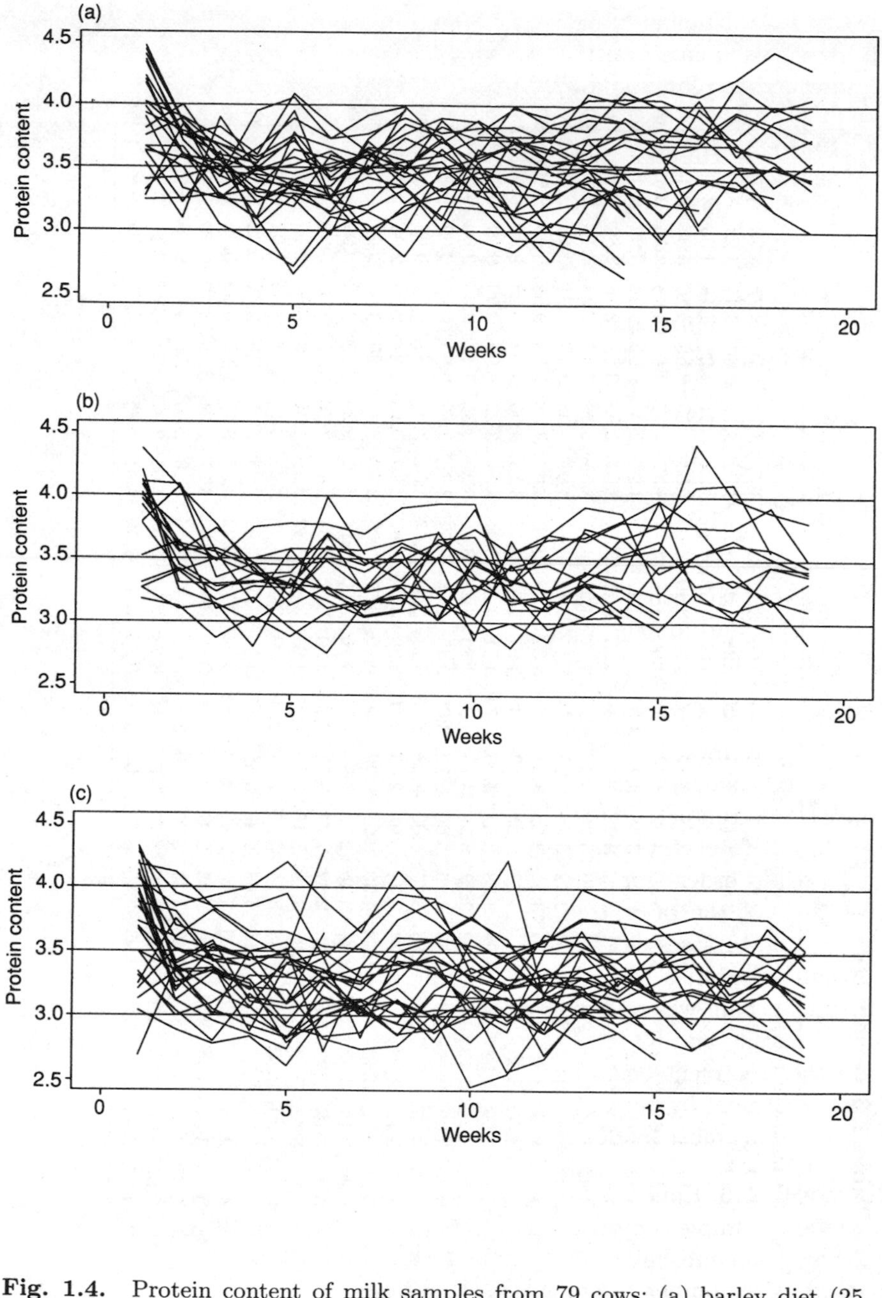

Fig. 1.4. Protein content of milk samples from 79 cows: (a) barley diet (25 cows); (b) mixed diet (27 cows); (c) lupins diet (27 cows).

Table 1.4. Number of patients for each treatment and response sequence in three period crossover trial of analgesic treatment for pain from primary dysmenorrhea (Jones and Kenward, 1987)

Treatment	Response sequence in periods 1,2,3 (0=no relief; 1=relief)								
sequence*	000	100	010	001	110	101	011	111	Total
012	0	0	2	2	1	0	9	1	15
021	2	1	0	0	0	0	9	4	16
102	0	1	1	1	0	8	3	1	15
120	0	1	1	1	8	0	0	1	12
201	3	0	0	0	1	7	2	1	14
210	1	5	0	0	4	3	1	0	14
Total	6	8	4	4	14	18	24	8	86

*Treatment: 0=placebo; 1=low; 2=high analgesic

Note how the multitude of lines in Fig. 1.4 confuses the group comparison. On the other hand, the lines are useful to show the variability across time and among individuals. Chapter 3 discusses compromise displays which more effectively capture patterns in longitudinal data.

Example 1.5. Crossover trial

Jones and Kenward (1987) report a data set from a three period crossover trial of an analgesic drug for relieving pain from primary dysmenorrhea (menstrual cramps). Three levels of the analgesic (control, low, and high) were given to each of 86 women. Women were randomized to one of the six possible orders for administering the three treatment levels so that the effect of the prior treatment on the current response or *carry-over effect* could be assessed. Table 1.4 is a cross-tabulation of the eight possible outcome categories with the six orderings. Ignoring for now the order of treatment, pain was relieved for 22 women (26%) on placebo, 61 (71%) on low analgesic, and 69 (80%) on high analgesic. This pattern is consistent with the treatment being beneficial. However, there may be carry-over or other treatment by period interactions which can also explain the observed pattern. This must be determined in a longitudinal data analysis.

Example 1.6. Epileptic seizures

The final example comprises data from a clinical trial of 59 epileptics, analysed by Thall and Vail (1990) and by Breslow and Clayton (1993). For each patient, the number of epileptic seizures was recorded during a baseline period of eight weeks. Patients were then randomized to treatment with the anti-epileptic drug progabide, or to placebo in addition to standard chemotherapy. The number of seizures was then recorded in four consec-

Table 1.5. Four successive two-week seizure counts for each of 59 epileptics. Covariates are adjuvant treatment (0=placebo, 1=progabide), eight week baseline seizure counts, and age (in years)

Y_1	Y_2	Y_3	Y_4	Trt.	Base	Age	Y_1	Y_2	Y_3	Y_4	Trt.	Base	Age
5	3	3	3	0	11	31	0	4	3	0	1	19	20
3	5	3	3	0	11	30	3	6	1	3	1	10	20
2	4	0	5	0	6	25	2	6	7	4	1	19	18
4	4	1	4	0	8	36	4	3	1	3	1	24	24
7	18	9	21	0	66	22	22	17	19	16	1	31	30
5	2	8	7	0	27	29	5	4	7	4	1	14	35
6	4	0	2	0	12	31	2	4	0	4	1	11	57
40	20	23	12	0	52	42	3	7	7	7	1	67	20
5	6	6	5	0	23	37	4	18	2	5	1	41	22
14	13	6	0	0	10	28	2	1	1	0	1	7	28
26	12	6	22	0	52	36	0	2	4	0	1	22	23
12	6	8	5	0	33	24	5	4	0	3	1	13	40
4	4	6	2	0	18	23	11	14	25	15	1	46	43
7	9	12	14	0	42	36	10	5	3	8	1	36	21
16	24	10	9	0	87	26	19	7	6	7	1	38	35
11	0	0	5	0	50	26	1	1	2	4	1	7	25
0	0	3	3	0	18	28	6	10	8	8	1	36	26
37	29	28	29	0	111	31	2	1	0	0	1	11	25
3	5	2	5	0	18	32	102	65	72	63	1	151	22
3	0	6	7	0	20	21	4	3	2	4	1	22	32
3	4	3	4	0	12	29	8	6	5	7	1	42	25
3	4	3	4	0	9	21	1	3	1	5	1	32	35
2	3	3	5	0	17	32	18	11	28	13	1	56	21
8	12	2	8	0	28	25	6	3	4	0	1	24	41
18	24	76	25	0	55	30	3	5	4	3	1	16	32
2	1	2	1	0	9	40	1	23	19	8	1	22	26
3	1	4	2	0	10	19	2	3	0	1	1	25	21
13	15	13	12	0	47	22	0	0	0	0	1	13	36
11	14	9	8	1	76	18	1	4	3	2	1	12	37
8	7	9	4	1	38	32							

utive two-week intervals. These data are reprinted in Table 1.5, and are graphically displayed using boxplots (Tukey, 1977) in Fig. 1.5. The medical question is whether or not the progabide reduces the rate of epileptic seizures. Figure 1.5 is suggestive of a small reduction in the average number except possibly at week two. Inferences must take into account the very strong variation among people in the baseline level of seizures, which appears to persist across time. In this case, the natural heterogeneity in rates will facilitate the detection of a treatment effect as will be discussed in later chapters.

In all six examples, there are repeated observations on each experimental unit. The units can reasonably be assumed independent of one another, but the multiple responses within each unit are likely to be correlated. The scientific objectives of each study can be formulated as regression problems whose purpose is to describe the dependence of the response on explanatory

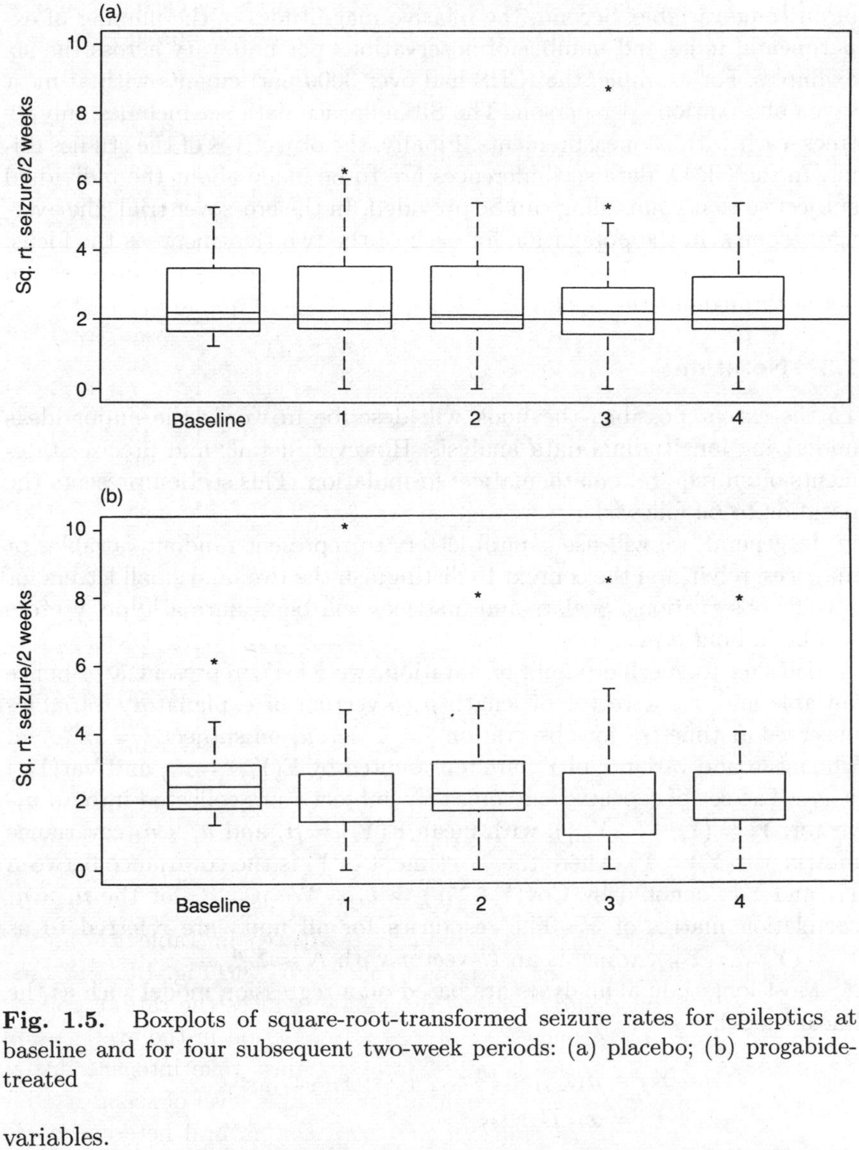

Fig. 1.5. Boxplots of square-root-transformed seizure rates for epileptics at baseline and for four subsequent two-week periods: (a) placebo; (b) progabide-treated

variables.

There are important differences among the examples as well. The responses in Examples 1.1 (CD4+ cells), 1.3 (tree size), and 1.4 (protein content) are continuous variables which, perhaps after transformation, can be adequately described by linear statistical models. However, the response is binary in Examples 1.2 (respiratory disease) and 1.5 (presence of pain); and is a count in Example 1.6 (number of seizures). Linear models will not suffice here. The choice of statistical model must depend on the type of outcome variable. Second, the relative magnitudes of the number of experimental units and number of observations per unit vary across the six examples. For example, the ICHS had over 3000 participants with at most seven observations per person. The Sitka spruce data set includes only 79 trees, each with 23 measurements. Finally, the objectives of the studies differ. In the CD4+ data set, inferences are to be made about the individual subject so that counselling can be provided. In the crossover trial, the average response in the population for each of the two treatments is the focus. These differences influence the specific approach to analysis as discussed in detail throughout the book.

1.3 Notation

To the extent possible, the book will describe in words the major ideas underlying longitudinal data analysis. However, details and precise statements often require a mathematical formulation. This section presents the notation to be followed.

In general, we will use capital letters to represent random variables or matrices, relying on the context to distinguish the two, and small letters for specific observations. Scalars and matrices will be in normal type, vectors will be in bold type.

Turning to specific points of notation, we let Y_{ij} represent a response variable and $\boldsymbol{x}_{ij}$ a vector of length p (p-vector) of explanatory variables observed at time t_{ij}, for observation $j = 1, \ldots, n_i$ on subject $i = 1, \ldots, m$. The mean and variance of Y_{ij} are represented by $\mathrm{E}(Y_{ij}) = \mu_{ij}$ and $\mathrm{Var}(Y_{ij}) = v_{ij}$. The set of repeated outcomes for subject i are collected into an n_i-vector, $\boldsymbol{Y}_i = (Y_{i1}, \ldots, Y_{in_i})$, with mean $\mathrm{E}(\boldsymbol{Y}_i) = \boldsymbol{\mu}_i$ and $n_i \times n_i$ covariance matrix $\mathrm{Var}(\boldsymbol{Y}_i) = V_i$, where the jk element of V_i is the covariance between Y_{ij} and Y_{ik}, denoted by $\mathrm{Cov}(Y_{ij}, Y_{ik}) = v_{ijk}$. We use R_i for the $n_i \times n_i$ correlation matrix of $\boldsymbol{Y}_i$. The responses for all units are referred to as $\boldsymbol{Y} = (\boldsymbol{Y}_1, \ldots, \boldsymbol{Y}_m)$, which is an N-vector with $N = \Sigma_{i=1}^m n_i$.

Most longitudinal analyses are based on a regression model such as the linear model,

$$\begin{aligned} Y_{ij} &= \beta_1 x_{ij1} + \beta_2 x_{ij2} + \cdots + \beta_p x_{ijp} + \epsilon_{ij} \\ &= \boldsymbol{x}_{ij}{}'\boldsymbol{\beta} + \epsilon_{ij} \end{aligned}$$

where $\boldsymbol{\beta} = (\beta_1, \ldots, \beta_p)$ is a p-vector of unknown regression coefficients and ϵ_{ij} is a zero-mean random variable representing the deviation of the response from the model prediction, $\boldsymbol{x}_{ij}{}'\boldsymbol{\beta}$. Typically, $x_{ij1} = 1$ for all i and all j, and β_1 is then the intercept term in the linear model.

In matrix notation, the regression equation for the *ith* subject takes the form

$$\boldsymbol{Y}_i = X_i \boldsymbol{\beta} + \boldsymbol{\epsilon}_i$$

where X_i is a $n_i \times p$ matrix with $\boldsymbol{x}_{ij}$ in the jth row and $\boldsymbol{\epsilon}_i = (\epsilon_{i1}, \ldots, \epsilon_{in_i})$.

Note that in longitudinal studies, the natural experimental unit is not the individual measurement Y_{ij}, but the sequence, $\boldsymbol{Y}_i$, of measurements on an individual subject. For example, when we talk about replication we refer to the number of subjects, not the number of individual measurements. We shall use the following terms interchangeably, according to context: subject, experimental unit, person, animal, individual.

1.4 Merits of longitudinal studies

As mentioned above, the prime advantage of a longitudinal study is its effectiveness for studying change. The artificial reading example of Fig. 1.1 can easily be generalized to the class of linear regression models using the new notation. The distinction between cross-sectional and longitudinal inference is made clearer by consideration of the simple linear regression without intercept. The general case follows naturally by thinking of the explanatory variable as a vector rather than a scalar.

In a cross-sectional study ($n_i = 1$) we are restricted to the model

$$Y_{i1} = \beta_C x_{i1} + \epsilon_{i1}, \quad i = 1, \ldots, m \tag{1.4.1}$$

where β_C represents the difference in average Y across two sub-populations which differ by one unit in x. With repeated observations, the linear model can be extended to the form

$$Y_{ij} = \beta_C x_{i1} + \beta_L (x_{ij} - x_{i1}) + \epsilon_{ij}, \quad \begin{matrix} j = 1, \ldots, n_i \\ i = 1, \ldots, m \end{matrix} \tag{1.4.2}$$

(Ware *et al.*, 1990). Note that when $j = 1$, (1.4.2) reduces to (1.4.1) so β_C has the same cross-sectional interpretation. However, we can now also estimate β_L whose interpretation is made clear by subtracting (1.4.1) from (1.4.2) to obtain

$$(Y_{ij} - Y_{i1}) = \beta_L (x_{ij} - x_{i1}) + \epsilon_{ij} - \epsilon_{i1}.$$

That is, β_L represents the expected change in Y over time per unit change in x for a given subject.

In Fig. 1.1(b), β_C and β_L have opposite signs; in Fig. 1.1(c), they have the same sign. To estimate how individuals change with time from a cross-sectional study, we must assume $\beta_C = \beta_L$. With a longitudinal study, this strong assumption is unnecessary since both can be estimated.

Even when $\beta_C = \beta_L$, longitudinal studies tend to be more powerful than cross-sectional studies. The basis of inference about β_C is a comparison of individuals with a particular value of x to others with a different value. In contrast, the parameter β_L is estimated by comparing a person's response at two times, assuming x changes with time. In a longitudinal study, each person can be thought of as serving as his or her own control. For most outcomes, there is considerable variability across individuals due to the influence of unmeasured characteristics such as genetic make-up, environmental exposures, personal habits, and so on. These tend to persist over time. Their influence is cancelled in the estimation of β_L; they obscure the estimation of β_C.

Another merit of the longitudinal study is its ability to distinguish the degree of variation in Y across time for one person from the variation in Y among people. This partitioning of the variation in Y is important for the following reason. Much of statistical analysis can be viewed as estimating unobserved quantities. For example, in the CD4+ problem we want to estimate a man's immune status as reflected in his CD4+ level. With cross-sectional data, one man's estimate must draw upon data from others to overcome measurement error. But averaging across people ignores the natural differences in CD4+ level among persons. With repeated values, we can borrow strength across time for the person of interest as well as across people. If there is little variability among people, one man's estimate can rely on data for others as in the cross-sectional case. However, if the variation across people is large, we might prefer to use only data for the individual. Given longitudinal data, we can acknowledge the naturally occurring differences among subjects when estimating a person's current value or predicting his future one.

1.5 Approaches to longitudinal data analysis

With one observation on each experimental unit, we are confined to modelling the population average of Y, called the *marginal* mean response; there is no other choice. With repeated measurements, there are several different approaches that can be adopted. A simple and often effective strategy is to:

(1) reduce the repeated values into one or two summaries;

(2) analyse each summary variable as a function of covariates, $\boldsymbol{x}_i$.

For example, with the Sitka spruce data of Example 1.3, the linear growth rate of each tree can be estimated and the rates compared across the ozone groups. This so-called *two-stage* or *derived variable* analysis, which dates at least from Wishart (1938), works when $\boldsymbol{x}_{ij} = \boldsymbol{x}_i$ for all i and j since the summary value which results from stage (1) can only be regressed on $\boldsymbol{x}_i$ in stage (2). This approach is less useful if important explanatory variables change over time.

In lieu of reducing the repeated responses to summary statistics, we can model the individual Y_{ij} in terms of $\boldsymbol{x}_{ij}$. This book will discuss three distinct strategies.

The first is to model the marginal mean as in a cross-sectional study. For example, in the Indonesian Children's Health Study (ICHS), the frequency of respiratory disease in children who are and are not vitamin A deficient would be compared. Or in the CD4+ example, the average CD4+ level would be characterized as a function of time. Since repeated values are not likely to be independent, this *marginal analysis* must also include assumptions about the form of the correlation. For example, in the linear model we can assume $\mathrm{E}(\boldsymbol{Y}_i) = X_i\boldsymbol{\beta}$, and $\mathrm{Var}(\boldsymbol{Y}_i) = V_i(\boldsymbol{\alpha})$ where $\boldsymbol{\beta}$ and $\boldsymbol{\alpha}$ must be estimated. The marginal model approach has the advantage of separately modelling the mean and covariance. As will be discussed below, valid inferences about $\boldsymbol{\beta}$ can sometimes be made even when an incorrect form for $V(\boldsymbol{\alpha})$ is assumed.

A second approach, the random effects model, assumes that correlation arises among repeated responses because the regression coefficients vary across individuals. Here, we model the conditional expectation of Y_{ij} given the person-specific coefficients, $\boldsymbol{\beta}_i$, by

$$\mathrm{E}(Y_{ij} \mid \boldsymbol{\beta}_i) = \boldsymbol{x}_{ij}'\boldsymbol{\beta}_i. \tag{1.5.1}$$

Because there is too little data on a single person to estimate $\boldsymbol{\beta}_i$ from $(\boldsymbol{Y}_i, X_i)$ alone, we further assume that the $\boldsymbol{\beta}_i$s are independent realizations from some distribution with mean $\boldsymbol{\beta}$. If we write $\boldsymbol{\beta}_i = \boldsymbol{\beta} + \boldsymbol{U}_i$ where $\boldsymbol{\beta}$ is fixed and $\boldsymbol{U}_i$ is a zero-mean random variable, then the basic heterogeneity assumption can be restated in terms of the latent variables, $\boldsymbol{U}_i$. That is, there are unobserved factors represented by the $\boldsymbol{U}_i$s that are common to all responses for a given person but which vary across people, thus inducing the correlation. In the Indonesian Children's Health Study (ICHS), it is reasonable to assume that the propensity of respiratory infection naturally varies across children irrespective of their vitamain A status due to genetic and environmental factors which cannot easily be measured. Random effects models are particularly useful when inferences are to be made about individuals as in the CD4+ example.

The final approach, which we will refer to as a *transition model* (Ware *et al.*, 1988) focuses on the conditional expectation of Y_{ij} given past outcomes, $Y_{ij-1}, \ldots, Y_{i1}$. Here the data analyst specifies a regression model for the conditional expectation, $\mathrm{E}(Y_{ij} \mid Y_{ij-1}, \ldots, Y_{i1}, \boldsymbol{x}_{ij})$, as an explicit function of $\boldsymbol{x}_{ij}$ and of the past responses. An example for equally spaced binary data is the logistic regression

$$\log \frac{\Pr(Y_{ij} = 1 \mid Y_{ij-1}, \ldots, Y_{i1}, \boldsymbol{x}_{ij})}{1 - \Pr(Y_{ij} = 1 \mid Y_{ij-1}, \ldots, Y_{i1}, \boldsymbol{x}_{ij})} = \boldsymbol{x}_{ij}'\boldsymbol{\beta} + \alpha Y_{ij-1}. \tag{1.5.2}$$

Transition models like (1.5.2) combine the assumptions about the dependence of Y on x and the correlation among repeated Ys into a single equation. As an example, the chance of respiratory infection for a child in the

ICHS might depend on whether she was vitamin A deficient but also on whether she had an infection at the prior visit.

In each of the three approaches, we model both the dependence of the response on the explanatory variables and the autocorrelation among the responses. With cross-sectional data, only the dependence of Y on $\boldsymbol{x}$ need be specified; there is no correlation. There are two consequences of ignoring the correlation when it exists in longitudinal data:

(1) incorrect inferences about regression coefficients, $\boldsymbol{\beta}$;

(2) estimates of $\boldsymbol{\beta}$ which are inefficient, that is, less precise than possible.

To illustrate, suppose $Y_{ij} = \beta_0 + \beta_1 t_j + \epsilon_{ij}$, $t_j = -3, -2, \ldots, 2, 3$ where the errors, ϵ_{ij}, follow the first-order autoregressive model, $\epsilon_{ij} = \alpha\epsilon_{ij-1} + Z_{ij}$ and the Z_{ij}s are independent, mean-zero, Gaussian variates. Suppose further that we ignore the correlation and use ordinary least squares (OLS) to obtain a slope estimate, $\hat{\boldsymbol{\beta}}_{OLS}$, and an estimate of its variance, $\hat{V}_{OLS}$. Let $\hat{\boldsymbol{\beta}}$ be the optimal estimator of $\boldsymbol{\beta}$ obtained by taking the correlation into account. Figure 1.6 shows the true variances of $\hat{\boldsymbol{\beta}}_{OLS}$ and $\hat{\boldsymbol{\beta}}$ as well as the stated variance from OLS, $\hat{V}_{OLS}$, as a function of the correlation, α, between observations one time unit apart. Note first that $\hat{V}_{OLS}$ can be grossly incorrect when the correlation is substantial. Ignoring correlation leads to invalid assessment of the evidence about trend. Second, $\hat{\boldsymbol{\beta}}$ is less variable than $\hat{\boldsymbol{\beta}}_{OLS}$. That is, better use is made of the available information about the trend by accounting for correlation. The two costs listed above are true for most regression problems and correlation patterns encountered in practice.

Longitudinal data analysis problems can be partitioned into two groups:

(1) those where the regression of Y on $\boldsymbol{x}$ is the scientific focus and the number of experimental units (m) is much greater than the number of observations per unit (n);

(2) problems where the correlation is of prime interest or where m is small.

Using the MACS data in Example 1.1 to estimate the average number of CD4+ cells as a function of time since seroconversion falls into group 1. The nature of the correlation among repeated observations is immaterial to this purpose and m is large relative to n. Estimation of one individual's CD4+ curve is of type 2 since the correlation structure must be correctly modelled. Drawing correct inferences about the effect of ozone on Sitka spruce growth (Example 1.3) is also of type 2 since m and n are of similar size.

The classification scheme above is imprecise, but is nevertheless useful as a rough guide to strategies for analysis. With objectives of type 1, the data

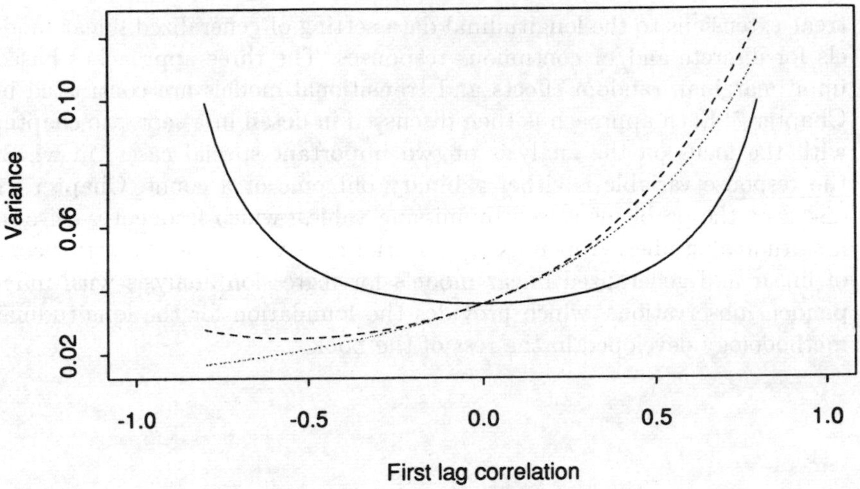

Fig. 1.6. Actual and estimated variances of ordinary least squares (OLS) estimates, and actual variance of optimally weighted least squares estimates, as functions of the correlation between successive measurements: —— : reported by OLS – – – : actual for OLS .... : best possible

analyst must invest the majority of time in correctly modelling the mean, that is, including all necessary explanatory variables and their interactions and identifying the best functional form for each predictor. Less time need be spent in modelling the correlation. As we will show in Chapter 4, if m is large relative to n a robust variance estimate can be used to draw valid inferences about regression parameters even when the correlation is misspecified. In group 2, however, both the mean and covariance models must be approximately correct to obtain valid inferences.

1.6 Organization of subsequent chapters

This chapter has given a brief overview of the ideas underlying longitudinal data analysis. Chapter 2 discusses design issues such as choosing the number of persons (experimental units) and the number of repeated observations to attain a given power. Analytic methods begin in Chapter 3, which focuses on exploratory tools for displaying the data and summarizing the mean and correlation patterns. The linear model for longitudinal data is treated in Chapters 4 to 6. In Chapter 4, the emphasis is on problems of type 1, where relatively less effort will be devoted to careful modelling of the covariance structure. Details necessary for modelling covariances are given in Chapter 5. Chapter 6 discusses traditional analysis of variance models for repeated measures, a special case of the linear model. Chapters 7 to 10 treat extensions to the longitudinal data setting of generalized linear models for discrete and/or continuous responses. The three approaches based upon marginal, random effects and transitional models are contrasted in Chapter 7. Each approach is then discussed in detail in a separate chapter with the focus on the analysis of two important special cases, in which the response variable is either a binary outcome or a count. Chapter 11 discusses the problems raised by missing values, which frequently arise in longitudinal studies. Appendix A is a brief review of the statistical theory of linear and generalized linear models for regression analysis with independent observations, which provides the foundation for the longitudinal methodology developed in the rest of the book.

2

Design considerations

2.1 Introduction

In this chapter we contrast cross-sectional and longitudinal studies from the design perspective. As discussed in Chapter 1, the major advantage of a longitudinal study is its ability to distinguish the cross-sectional and longitudinal relationships between the explanatory variables and the response. The price associated with this advantage is the potential higher cost incurred by repeated sampling of individuals. At the design stage, we can benefit from guestimates of the potential size of bias from a cross-sectional study and the increase in precision of inferences from a longitudinal study. This information can then be weighed either formally or informally against the additional cost of repeated sampling. Then if a longitudinal study is selected, we must choose the number of persons and number of measurements on each.

In Sections 2.2 and 2.3 we quantify the potential bias of a cross-sectional study as well as its efficiency relative to a longitudinal study. Section 2.4 presents sample size calculations when the response variables are either continuous or dichotomous.

2.2 Bias

As the reading example in Section 1.1 demonstrates, it is possible for the association between an explanatory variable, x, and response, Y, determined from a cross-sectional study to be very different from the association measured in a longitudinal study.

We begin our study of bias by formalizing this idea using the model from Section 1.4. Consider a response variable that both changes over time and varies among subjects. Examples include age, blood pressure, or weight. Adopting the notation given in Section 1.3, we begin with a model of the form

$$Y_{ij} = \beta_0 + \beta x_{ij} + \epsilon_{ij}, \; j = 1, \ldots, n; \; i = 1, \ldots, m. \qquad (2.2.1)$$

Re-expressing (2.2.1) as

$$Y_{ij} = \beta_0 + \beta x_{i1} + \beta(x_{ij} - x_{i1}) + \epsilon_{ij}, \qquad (2.2.2)$$

we note that this model assumes implicitly that the cross-sectional effect due to x_{i1} is the same as the longitudinal effect represented by $x_{ij} - x_{i1}$ on the right hand side. This assumption is rather a strong one and doomed to fail in many studies. The model can be modified by allowing each person to have their own intercept, β_{0i}, that is, by replacing $\beta_0 + \beta x_{i1}$ with β_{0i} so that

$$Y_{ij} = \beta_{0i} + \beta(x_{ij} - x_{i1}) + \epsilon_{ij}. \tag{2.2.3}$$

Both (2.2.1) and (2.2.3) represent extreme cases for modelling the cross-sectional variation in the response variable at baseline. In the former case, one assumes that the cross-sectional association with x_{i1} is the same as the longitudinal effect; in the latter case, the baseline level is allowed to be different for every person. An intermediate and often more effective way to modify (2.2.1) is to assume a model of the form

$$Y_{ij} = \beta_0 + \beta_C x_{i1} + \beta_L(x_{ij} - x_{i1}) + \epsilon_{ij} \tag{2.2.4}$$

as suggested in Section 1.4. The inclusion of x_{i1} in the model with separate coefficient β_C allows both cross-sectional and longitudinal effects to be examined separately. We can also use this form to test whether or not the cross-sectional and longitudinal effects of particular explanatory variables are the same, that is, whether $\beta_C = \beta_L$. From a second perspective, one may simply view x_{i1} as a confounding variable whose absence may bias our estimate of the true longitudinal effect.

We now examine the bias of the least-squares estimate of β derived from the model (2.2.1). The estimate is

$$\hat{\beta} = \sum_{i=1}^{m}\sum_{j=1}^{n}(x_{ij} - \bar{x})(y_{ij} - \bar{y}) / \sum_{i=1}^{m}\sum_{j=1}^{n}(x_{ij} - \bar{x})^2,$$

where $\bar{x} = \Sigma_{ij} x_{ij}/(nm)$ and $\bar{y} = \Sigma_{ij} y_{ij}/(nm)$. When the true model is (2.2.4), simple algebra leads to

$$\mathrm{E}(\hat{\beta}) = \beta_L + \frac{\sum_{i=1}^{m} n(x_{i1} - \bar{x}_1)(\bar{x}_i - \bar{x})}{\sum_{i=1}^{m}\sum_{j=1}^{n}(x_{ij} - \bar{x})^2}(\beta_C - \beta_L),$$

where $\bar{x}_i = \Sigma_j x_{ij}/n$ and $\bar{x}_1 = \Sigma_i x_{i1}/m$. Thus the cross-sectional estimate $\hat{\beta}$, which assumes $\beta_L = \beta_C$, is a biased estimate of β_L and is unbiased only if either $\beta_L = \beta_C$ or the variables $\{x_{i1}\}$ and $\{\bar{x}_i\}$ are orthogonal to each other. The latter result is of no surprise if one re-expresses (2.2.4) as

$$Y_{ij} = \beta_0 + \beta_L x_{ij} + x_{i1}(\beta_C - \beta_L) + \epsilon_{ij}. \tag{2.2.5}$$

In this re-expression, (2.2.1) is seen as a special case of (2.2.4) where the variable x_{i1} has been omitted from the model.

The direction of the bias in $\hat{\beta}$ as an estimate for the longitudinal effect, β_L, depends upon the correlation between x_{i1} and $\bar{x}_i$, as is clear from the expression for $E(\hat{\beta})$ above. The main message to be conveyed here is that when dealing with covariates which change over time and vary across subjects at the baseline, it is good practice to use model (2.2.4) instead of (2.2.1). The former model allows one to separate the longitudinal effect, which describes individual change, from the cross-sectional effect where comparison on Y is made between individuals with different xs.

The equation (2.2.5) above has a significant design implication when x represents the age at each visit in which case β_L describes the progression of the response variable over time. The orthogonality between x_{i1} and x_{ij} for each j is achieved if $x_{ij} - x_{i1} = \Delta_j$, independent of i. That is, the least-squares estimator, $\hat{\beta}$, will be an unbiased estimate of β_L if the spacings in x between two consecutive visits are the same for all subjects. We note that this kind of design is not uncommon in many clinical trials and in some observational studies.

2.3 Efficiency

The previous section addressed the issue of bias in cross-sectional studies of change. Even when $\beta_C = \beta_L$, longitudinal studies tend to be more powerful, as we now examine in detail. Assuming the model specified in (2.2.1), the variance of $\hat{\beta}_C$, which uses the data from the first visit only, is

$$\text{Var}(\hat{\beta}_C) = \sigma^2 / \sum_{i=1}^{m} (x_{i1} - \bar{x}_1)^2$$

where $\sigma^2 = \text{Var}(\epsilon_{ij})$. On the other hand, the variance of $\hat{\beta}_L$, which uses all the data, is the lower right entry of

$$\sigma^2 \{ \sum_{i=1}^{m} (X_i' R_i^{-1} X_i) \}^{-1},$$

where R_i is the $n_i \times n_i$ correlation matrix for $\mathbf{Y}_i = (Y_{i1}, \ldots, Y_{in_i})$ and

$$X_i' = \begin{pmatrix} 1 & 1 & \ldots 1 \\ x_{i1} & x_{i2} \ldots & x_{in_i} \end{pmatrix}.$$

We can address the question of how much more can be learned by taking repeated measurements on each person by comparing the variances of $\hat{\beta}_L$ and $\hat{\beta}_C$ in (2.2.5) when $\beta_C = \beta_L$. We use $e = \text{Var}(\hat{\beta}_L)/\text{Var}(\hat{\beta}_C)$ as the specific measure of efficiency. The smaller the value of e the greater is the information gained by taking additional measurements on each person.

As expected, the value of e depends on the true structure of the correlation matrix. We consider two correlation matrices commonly occurring in longitudinal studies. For simplicity, we continue to assume $n_i = n$ for all i. In case 1, we assume the uniform correlation matrix, $R_{jk} = 1$ if $j = k$ and $R_{jk} = \rho$ for any $j \neq k$. The e function defined above then reduces to

$$e = \frac{\{1 + (n-1)\rho\}(1-\rho)}{n(1+\delta)\{1 - p + n\rho\delta/(1+\delta)\}}$$

where

$$\delta = \frac{\text{the averaged within-subject variation in } x}{\text{the between-subjects variation in } x \text{ at visit 1}} = \frac{\sum_{i,j}(x_{ij} - \bar{x}_i)^2/\{m(n-1)\}}{\sum_i(\bar{x}_i - \bar{x})^2/(m-1)}.$$

Figure 2.1 gives plots of e against δ for some selected values of ρ and n. Except when δ is small and the common correlation ρ is high, there is much to be gained by conducting longitudinal studies even when the number of repeated observations is as small as two.

In Case 2, we consider the situation when the true correlation matrix has the form $R_{jk} = \rho^{|j-k|}$. This is the correlation structure of a first order autoregressive process discussed in Chapter 5. In this case

$$e = \frac{1-\rho^2}{(1-\rho)\{n - (n-2)\rho\} + \delta\gamma/(n+1)},$$

where $\gamma = n(n+1) - 2(n-3)(n+1)\rho + (n-3)(n-2)\rho^2$. The message regarding efficiency gain is similar as can be seen in Fig. 2.2.

We note that in either case, e is a decreasing function of δ. One implication for design is that we may obtain a more efficient (less variable) estimate of β by increasing the within-subject variation in x when possible. Suppose x_{ij} is the age of person i at visit j so that β represents the time rate of change of the response variable. With three observations for each person, it is better, all other factors being equal, to arrange the visits at years 0, 1, and 3 rather than 0, 1, and 2. In the former design, the within-subject variation in x is $\{(0-4/3)^2 + (1-4/3)^2 + (3-4/3)^2\}/3 = 1.45$ as opposed to $\{(0-1)^2 + (1-1)^2 + (2-1)^2\}/3 = 0.67$ in the latter case. The efficiency for the former design depends on the type of correlation structure, the magnitude of ρ and the between-subject variation in x at visit 1 (the denominator of δ). For example, with exchangeable correlation structure with $\rho = 0.5$ and the between-subject variation at visit one equal to 10, the variance of $\hat{\beta}$ for the (0,1,3) design is only 78 per cent of that for the (0,1,2) design. In other words, delaying the last visit from year two to year three is equivalent to increasing the number of persons in the study by a factor of $0.78^{-1} = 1.28$.

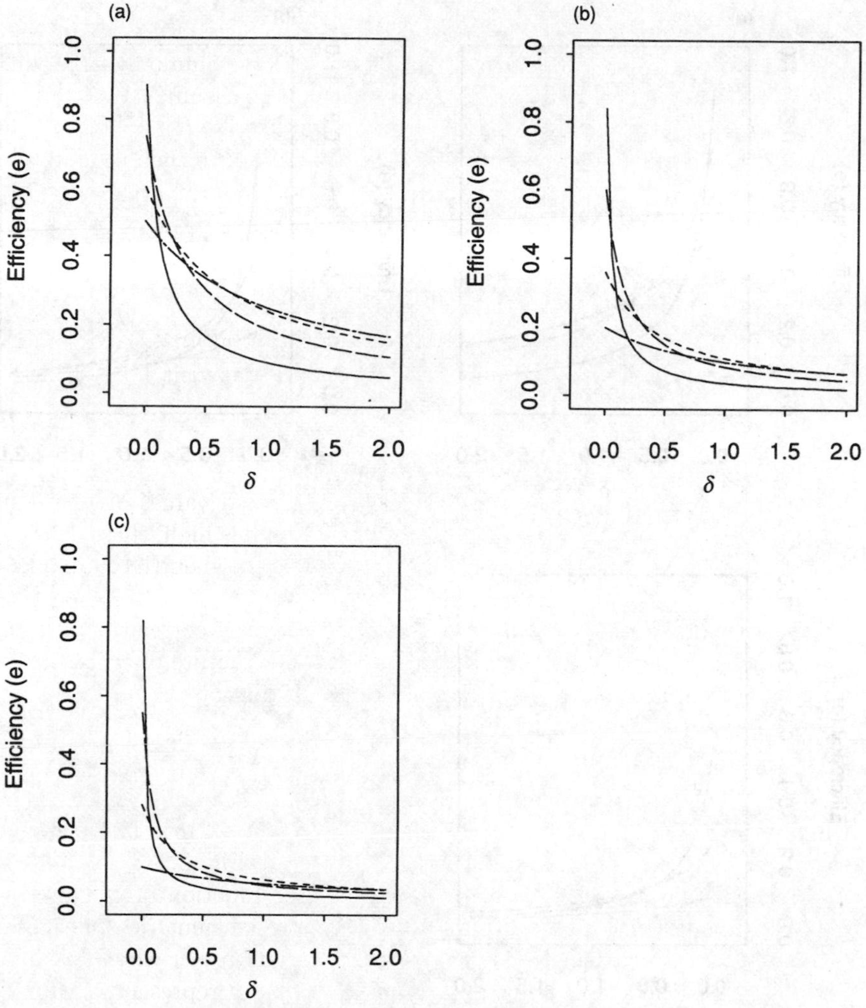

Fig. 2.1. Relationship between relative efficiency (e) of cross-sectional and longitudinal estimation and δ, the nature of variation in x, with uniform correlation structure and n observations per subject: (a) $n = 2$; (b) $n = 5$; (c) $n = 10$. —— : $\rho = 0.8$; – – – : $\rho = 0.5$; - - - - : $\rho = 0.2$; – · – · – : $\rho = 0.0$.

2.4 Sample size calculations

As with cross-sectional studies, investigators conducting longitudinal studies need to know in advance approximately the number of subjects required to achieve a specified statistical power. In any study, investigators must provide the following quantities to determine the required sample sizes.

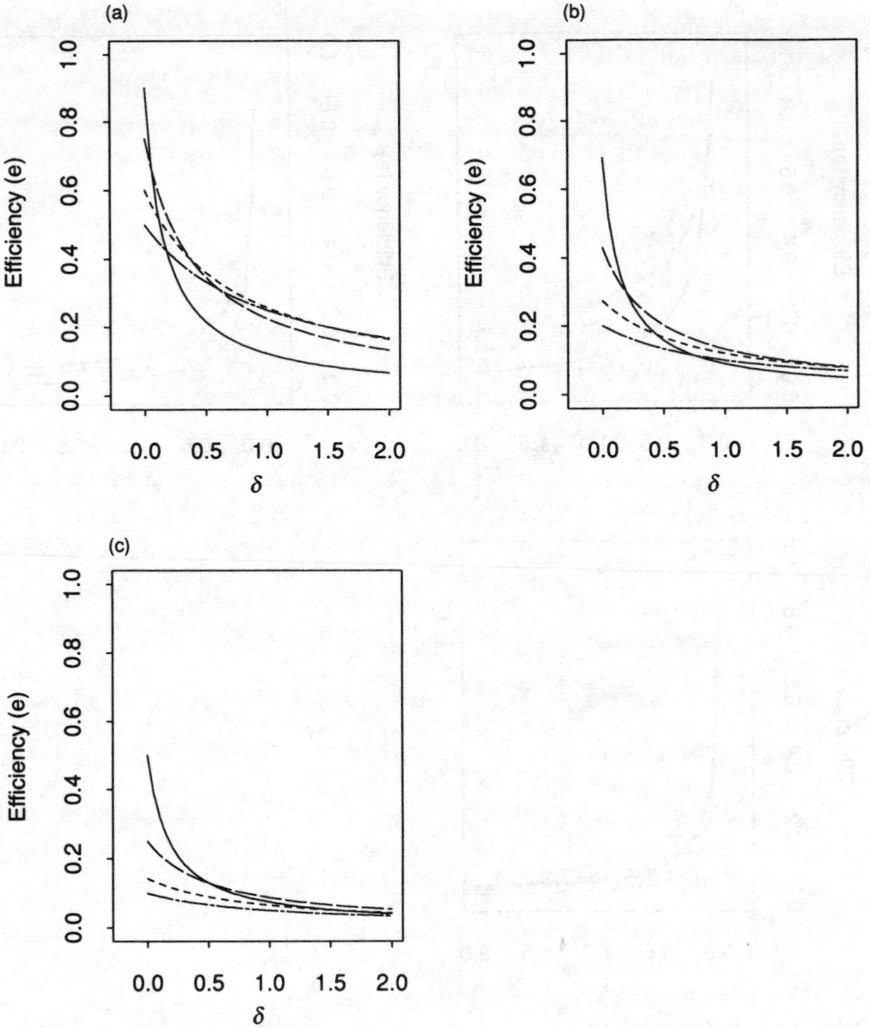

Fig. 2.2. Relationship between relative efficiency (e) of cross-sectional and longitudinal estimation and δ, the nature of variation in x, with exponential correlation structure and n observations per subject: (a) $n = 2$; (b) $n = 5$; (c) $n = 10$. —— : $\rho = 0.8$; – – – : $\rho = 0.5$; - - - - : $\rho = 0.2$; – · – · – : $\rho = 0.0$.

1. **Type I error rate** (α): this quantity is the probability that the study will reject the null hypothesis when it is correct; for example, it would correspond to the probability of declaring a significant difference between treatment and control groups when the treatment is useless. A typical choice of α is 0.05.

2. **Smallest meaningful difference to be detected** (d): the investigators typically want their study to reject the null hypothesis with high probability when the parameter of interest deviates from its value under the null hypothesis by an amount d or more, where the value of d is chosen to be of practical significance. We call this value the smallest meaningful difference.

3. **Power** (P): the power of a statistical test is the probability that the study rejects the null hypothesis when it is incorrect. Of course, this must depend on *how* incorrect is the null hypothesis, as the power must approach the type I error rate, α, as the deviation from the null hypothesis approaches zero. A typical requirement might be that the power is at least 0.9 when the deviation from the null hypothesis is at least d, the smallest meaningful difference.

4. **Measurement variation** (σ^2): for a continuous response variable, Y_{ij}, the quantity $\sigma^2 = \text{Var}(Y_{ij})$ measures the unexplained variability in the response. The value of σ^2 is sometimes available to a reasonable approximation through either pilot studies or similar studies previously reported in the literature. Otherwise, the statistician must elicit from the investigator a plausible guess at its value.

In longitudinal studies, the following additional quantities are needed:

5. **Number of repeated observations per person** (n): the number, n, of observations per person may be constrained by practical considerations, or may need to be balanced against the sample size. For a given total cost, the investigator may be free to choose between a small value of n and a large sample size, or vice versa.

6. **Correlation among the repeated observations**: as with the measurement variance, the pattern of correlation among repeated observations can sometimes be estimated from previous studies. When this is not possible, a reasonable guess must be made.

In the remainder of this section, we give sample size formulas separately for continuous and binary responses. For simplicity, we present only formulas for two-group comparisons, and assume an exchangeable correlation structure. However, the principles apply to general study designs and correlation structures.

2.4.1 *Continuous responses*

We consider the simple problem of comparing two groups, A and B. We assume that in group A, the response depends on a single explanatory variable as follows:

$$Y_{ij} = \beta_{0A} + \beta_{1A}x_{ij} + \epsilon_{ij}, \quad j = 1, \ldots, n; \quad i = 1, \ldots, m.$$

In group B, the same equation holds but with different coefficients, β_{0B} and β_{1B}. Both groups have the same number of subjects, m; each person has n repeated observations. We assume that $\text{Var}(\epsilon_{ij}) = \sigma^2$ and $\text{Corr}(Y_{ij}, Y_{ik}) = \rho$ for all $j \neq k$. In addition, we assume that each person has the same set of explanatory variables so that $x_{ij} = x_j$. A typical example of such x_j is the duration between the first and the jth visit, in which case β_{1A} and β_{1B} are the rates of change in Y for groups A and B respectively.

Let z_p denote the pth quantile of a standard Gaussian distribution and $d = \beta_{1B} - \beta_{1A}$ be the meaningful difference of interest. With n fixed and known, the number of subjects per group that are needed to achieve type I error rate α and power P, is

$$m = \frac{2(z_\alpha + z_Q)^2\sigma^2(1-\rho)}{ns_x^2 d^2}$$

where $Q = 1 - P$ and $s_x^2 = \Sigma_j(x_j - \bar{x})^2/n$, the within-subject variance of the x_j.

To illustrate, consider a hypothetical clinical trial on the effect of a new treatment in reducing blood pressure. Three visits, including the baseline visit, are planned at years 0, 2 and 5. Thus, $n = 3$ and $s_x^2 = 4.22$. For type I error rate $\alpha = 0.05$, power $P = 0.8$ and smallest meaningful difference $d = 0.5$ mmHg/year, the table below gives the number of subjects needed for both treated and control groups for some selected values of ρ and σ^2.

	σ^2		
ρ	100	200	300
0.2	313	625	937
0.5	195	391	586
0.8	79	157	235

Note that for each value of σ^2, the required sample size decreases as the correlation, ρ, increases.

A second important type of problem is to estimate the time-averaged difference in a response between two groups. The appropriate statistical model is written

$$Y_{ij} = \beta_0 + \beta_1 x + \epsilon_{ij}, \quad j = 1, \ldots, n; \quad i = 1, \ldots, 2m$$

where x is the treatment assignment indicator variable. Let d be the meaningful difference between the average response for groups A and B. The number of subjects needed per group is

$$\begin{aligned} m &= 2(z_\alpha + z_Q)^2\sigma^2\{1 + (n-1)\rho\}/nd^2 \\ &= 2(z_\alpha + z_Q)^2\{1 + (n-1)\rho\}/(n\Delta^2) \end{aligned}$$

where $\Delta = d/\sigma$ is the smallest meaningful difference in standard deviation units. For $\alpha = 0.05, P = 0.8$, and $n = 3$, the table below gives the sample size needed per group for some selected values of ρ and Δ.

	Δ			
ρ	20%	30%	40%	50%
0.2	146	65	37	24
0.5	208	93	52	34
0.8	270	120	68	44

Note that in this case, for each value of σ^2, the required sample size increases with the correlation, ρ.

It may seem counter-intuitive at first glance that the sample size formula is decreasing in ρ in the first case, but increasing in the second case. The explanation is that in the first case, the parameter of interest, β, is the rate of the change in Y. As a result, the contribution from each subject to the estimation of β is a linear contrast of the Ys whose variance is decreasing in ρ. However, in the second case, the parameter of interest is the expected average of the Ys for individuals in a group and the variance of the corresponding estimate is increasing in ρ.

2.4.2 *Binary responses*

For a binary response variable, we consider the situation similar to the last case in Section 2.4.1. That is, we assume

$$\Pr(Y_{ij} = 1) = \begin{cases} p_A \text{ in group A} \\ p_B \text{ in group B} \end{cases} \quad j = 1, \ldots, n; \; i = 1, \ldots, m.$$

Assuming also that $\text{Corr}(Y_{ij}, Y_{ik}) = \rho$ for all $j \neq k$ and that d is the smallest meaningful difference between the response probabilities for groups A and B, the number of subjects needed per group is

$$m = \frac{\left[z_\alpha\{2\bar{p}\bar{q}(1 + (n-1)\rho)\}^{1/2} + z_Q\{(1 + (n-1)\rho)(p_A q_A + p_B q_B)\}^{1/2}\right]^2}{nd^2}$$

where $\bar{p} = (p_A + p_B)/2$ and $\bar{q} = 1 - \bar{p}$. For $\alpha = 0.05, P = 0.8, n = 3$, and $p_A = 0.5$, the table below gives the sample size needed per group for some selected ρ and $d = p_B - p_A$. As in the analogous problem with a continuous

	d		
ρ	0.3	0.2	0.1
0.2	15	35	143
0.5	21	49	204
0.8	27	64	265

response variable, the required sample size increases with the correlation, ρ.

In summary, this section has shown that correlation between repeated observations can affect required sample sizes differently, depending on the problem. A simple rule of thumb is that positive correlation increases the variance, and hence the required sample size for a given power, when estimating group averages or differences between averages for more than one group. In contrast, positive correlation decreases the required sample size when estimating a change over time within individuals, or differences between groups in the average change.

2.5 Further reading

In this chapter, we have discussed some of the most basic design considerations which arise with longitudinal studies. In particular, we have assumed that each subject is allocated to a single treatment group for the duration of the study.

There is an extensive literature on the more sophisticated design issues which arise in the related area of crossover trials, in which each subject receives a sequence of different treatments, with the order of presentation of the treatments varying between subjects. Much of this work emphasizes the construction of designs which enable valid inferences to be made in the presence of carry-over effects, whereby the response to a given treatment may include a residual dependence on the previous treatment.

Two of the earliest references to crossover designs are Williams (1949) and Patterson (1951). Later developments are discussed in Hedayat and Afsarinejad (1975, 1978), Afsarinejad (1983), and Bishop and Jones (1984). The books by Jones and Kenward (1989) and by Senn (1992) give general introductions to the design and analysis of crossover trials.

3
Exploring longitudinal data

3.1 Introduction

Longitudinal data analysis, like other statistical methods, has two components which operate side by side: exploratory and confirmatory analysis. Exploratory data analysis (EDA) is detective work. It comprises techniques to visualize patterns in data. Confirmatory analysis is judicial work, weighing evidence in data for or against hypotheses. This chapter discusses methods for exploring longitudinal data.

John W. Tukey, the father of EDA, said, 'EDA can never be the whole story, but nothing else can serve as the foundation stone – as the first step' (Tukey, 1977). Data analysis must begin by making displays that expose the patterns relevant to the scientific question. The best methods are capable of uncovering patterns which are unexpected. Like a good detective, the data analyst must keep an open mind ready to discover new clues.

There is no single prescription for making effective graphical displays of longitudinal data, but the following are a few simple guidelines:

- show as much of the relevant raw data as possible rather than only data summaries;
- highlight aggregate patterns of potential scientific interest;
- identify both cross-sectional and longitudinal patterns as distinguished in Section 1.1;
- make easy the identification of unusual individuals or unusual observations.

Most longitudinal analyses address the relationship of a response with explanatory variables, often including time. Hence, a scatterplot of the response against an explanatory variable is a basic display. Section 3.2 discusses scatterplots designed to follow the guidelines above. Special attention is given to large data sets where care is required to avoid graphs that are excessively busy. Section 3.3 discusses smoothing techniques that highlight the typical response as a function of an explanatory variable without reliance on specific parametric models. Smoothing splines, kernel estimators, and a robust method, *lowess*, are reviewed. Section 3.4 discusses methods to explore the association among repeated observations for an individual. When observations are made at equally spaced times, association is conveniently measured in terms of correlation. With unequally spaced observations, the variogram is often more effective. Throughout this chapter, we focus on continuous responses. Data displays for discrete data are illustrated in later chapters.

3.2 Graphical presentation of longitudinal data

With longitudinal data, an obvious first graph is the scatterplot of the response variable against time.

Example 3.1. Weights of pigs

Table 3.1 lists some data provided by Dr Philip McCloud (Monash University, Melbourne) on the weights of 48 pigs measured in nine successive weeks. Figure 3.1 displays the data graphically. Lines connect the repeated observations for each animal. This simple graph makes apparent a number of important patterns. First, all animals are gaining weight. Second, the pigs which are largest at the beginning of the observation period tend to be largest throughout. This phenomena is called 'tracking'. Third, the spread among the 48 animals is substantially smaller at the beginning of the study than at the end. This pattern of increasing variance over time could be explained in terms of variation in the growth rates of the individual animals.

Figure 3.1 is an adequate display for exploring these growth data, although it is hard to pick out individual response profiles. We can do slightly better by adding a second display obtained from the first by standardizing each observation. This is achieved by subtracting the mean, $\bar{y}_j$, and dividing by the standard deviation, s_j, of the 48 observations at time j, and replacing each y_{ij} by the standardized quantity $y^*_{ij} = (y_{ij} - \bar{y}_j)/s_j$. The resulting plot is shown in Fig. 3.2. Its effect is as if we were running a magnifying glass along the overall mean response profile, adjusting the magnification as we go so as to maintain a roughly constant amount of variation. As a result, the plot is able to highlight the degree of 'tracking', whereby animals tend to maintain their relative size over time. With large data sets, connected line graphs become unduly cluttered, and an alternative strategy for the basic time plot may be needed.

Table 3.1. Bodyweights of 48 pigs in 9 successive weeks of follow-up.

Week								
1	2	3	4	5	6	7	8	9
24.0	32.0	39.0	42.5	48.0	54.5	61.0	65.0	72.0
22.5	30.5	40.5	45.0	51.0	58.5	64.0	72.0	78.0
22.5	28.0	36.5	41.0	47.5	55.0	61.0	68.0	76.0
24.0	31.5	39.5	44.5	51.0	56.0	59.5	64.0	67.0
24.5	31.5	37.0	42.5	48.0	54.0	58.0	63.0	65.5
23.0	30.0	35.5	41.0	48.0	51.5	56.5	63.5	69.5
22.5	28.5	36.0	43.5	47.0	53.5	59.5	67.5	73.5
23.5	30.5	38.0	41.0	48.5	55.0	59.5	66.5	73.0
20.0	27.5	33.0	39.0	43.5	49.0	54.5	59.5	65.0
25.5	32.5	39.5	47.0	53.0	58.5	63.0	69.5	76.0
24.5	31.0	40.5	46.0	51.5	57.0	62.5	69.5	76.0
24.0	29.0	39.0	44.0	50.5	57.0	61.5	68.0	73.5
23.5	30.5	36.5	42.0	47.0	55.0	59.0	65.5	73.0
21.5	30.5	37.0	42.5	48.0	52.5	58.5	63.0	69.5
25.0	32.0	38.5	44.0	51.0	59.0	66.0	75.5	86.0
21.5	28.5	34.0	39.5	45.0	51.0	58.0	64.5	72.5
31.0	38.0	48.0	54.0	60.0	62.0	66.5	75.5	84.0
27.5	32.5	36.0	43.0	49.5	52.5	56.0	61.0	64.0
30.0	37.0	45.0	51.0	58.0	63.0	67.5	74.5	81.0
26.0	32.0	40.5	45.5	52.5	55.5	62.5	69.5	74.0
26.0	32.5	39.5	44.0	48.0	54.5	58.0	66.0	73.0
28.5	35.5	41.5	47.5	54.0	59.5	63.5	71.0	78.5
26.5	34.5	42.0	48.5	55.5	62.0	68.0	76.5	85.0
27.5	33.5	41.0	45.0	50.5	56.0	62.5	71.0	78.0
22.5	27.0	33.5	38.5	41.0	49.0	56.0	64.0	68.0
22.0	26.5	32.5	38.5	43.5	50.5	56.5	63.5	68.5
23.5	29.0	35.5	40.0	45.0	50.0	56.5	63.0	67.5
22.5	29.5	36.5	42.0	45.0	55.0	61.0	68.0	72.0
27.5	34.5	42.0	47.5	53.0	63.0	72.0	79.0	85.5
23.5	28.0	33.0	37.0	38.5	48.0	52.5	62.0	64.5
24.5	30.0	38.5	42.0	47.5	54.0	62.5	71.5	77.0
24.5	31.5	40.5	46.5	51.5	61.5	68.5	77.5	84.5
24.5	32.0	39.0	45.0	51.0	55.5	61.5	69.0	75.5
24.0	32.5	40.0	48.0	54.5	61.5	68.0	74.5	81.0
24.0	31.5	38.5	44.0	51.5	57.5	64.0	72.5	79.0
24.5	32.5	39.5	44.5	52.5	56.5	62.0	67.5	72.5
24.5	32.0	38.5	44.0	50.0	56.0	63.5	69.5	76.0
25.5	33.0	41.5	47.0	55.5	60.5	66.5	77.0	82.0
25.5	32.0	39.0	45.5	51.0	57.5	63.5	72.0	78.5
25.0	31.0	36.5	43.0	50.5	55.0	62.5	69.0	75.5
26.5	30.5	33.0	39.0	43.5	49.5	56.5	61.0	65.0
24.0	32.0	39.0	44.5	50.0	56.0	63.0	67.5	74.0
24.5	31.0	37.5	43.5	48.0	56.0	62.5	66.5	70.5
27.0	34.5	42.0	48.5	53.0	60.0	67.0	73.0	76.0
31.0	39.0	47.5	51.0	57.0	64.0	71.0	77.0	80.5
27.0	33.5	40.0	46.5	53.0	60.0	66.5	72.5	80.0
29.5	37.0	46.0	52.5	60.0	67.5	76.0	81.5	88.0
28.5	36.0	42.5	49.0	55.0	63.5	72.0	78.5	85.5

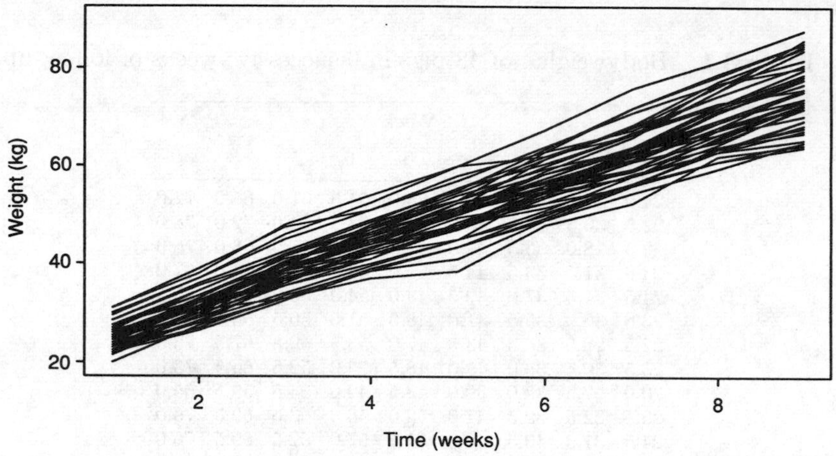

Fig. 3.1. Data on the weights of 48 pigs over a nine-week period.

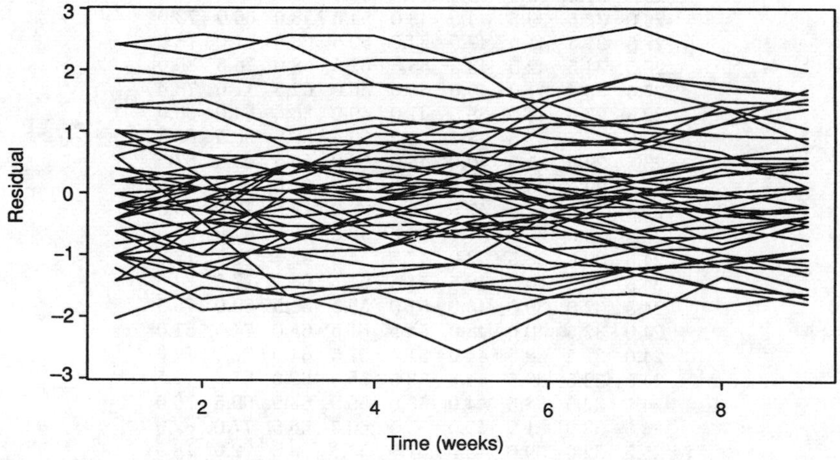

Fig. 3.2. Standardised residuals from data on the weights of pigs.

Example 3.2. CD4+ cell numbers

Figure 3.3 displays the CD4+ cell numbers for a subset of 100 seroconverters from the Multicenter AIDS Cohort Study (MACS) against the time variable, years since seroconversion. The plot includes only those men with at least seven observations that cover the date of seroconversion. We can see that the CD4+ cell numbers are relatively constant at about 1000 cells

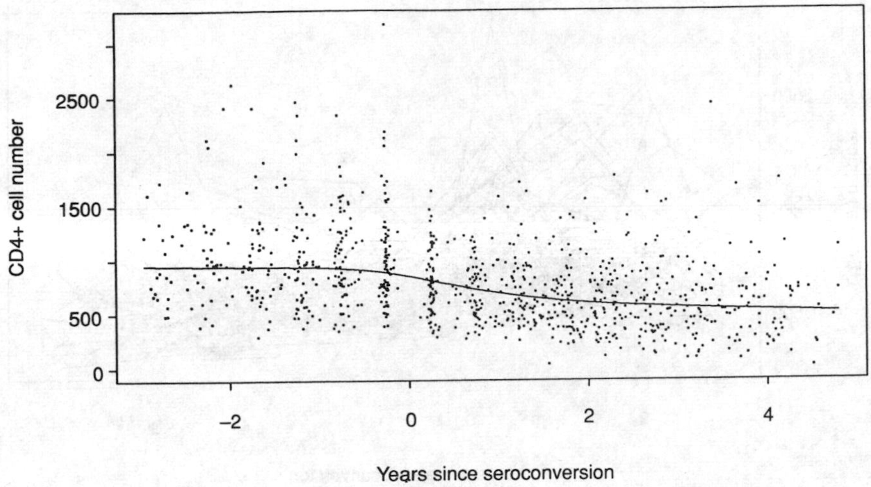

Fig. 3.3. CD4+ counts against time since seroconversion, with lowess smooth curve. · : data; —— : lowess curve.

until the time of seroconversion and then decrease afterwards. This mean pattern is highlighted by the non-parametric curve also shown in Fig. 3.3. The curve was calculated using lowess (Cleveland, 1979); the degree of smoothness was chosen subjectively. The curve indicates that the rate of CD4+ cell loss may be more rapid immediately after seroconversion.

The repeated observations for an individual have not been connected in Fig. 3.3, so that it is not possible to discern whether the pattern of decreasing cells is common to most individuals. That is, we cannot distinguish cross-sectional from longitudinal trends in the data. The reader is referred to Fig. 1.1 for a discussion of cross-sectional and longitudinal patterns.

Figure 3.4 displays the CD4+ cell data for all men, now with each person's repeated observations connected. Unfortunately, this plot is extremely busy, reducing its usefulness except perhaps from the perspective of an ink manufacturer.

The issue is clearly drawn by these two examples. On the one hand, we want to connect repeated measurements to display changes through time for individuals. On the other, presenting every person's curve creates little more than confusion in large data sets. The problem can be even more severe than in the CD4+ example. See Jones and Boadi-Boteng (1991) for a further illustration.

The ideal solution would be to display each individual's data with a very thin grey line and to use darker lines for the typical pattern as well as for a subset of persons. This is consistent with Tufte's (1990) 'micro/macro'

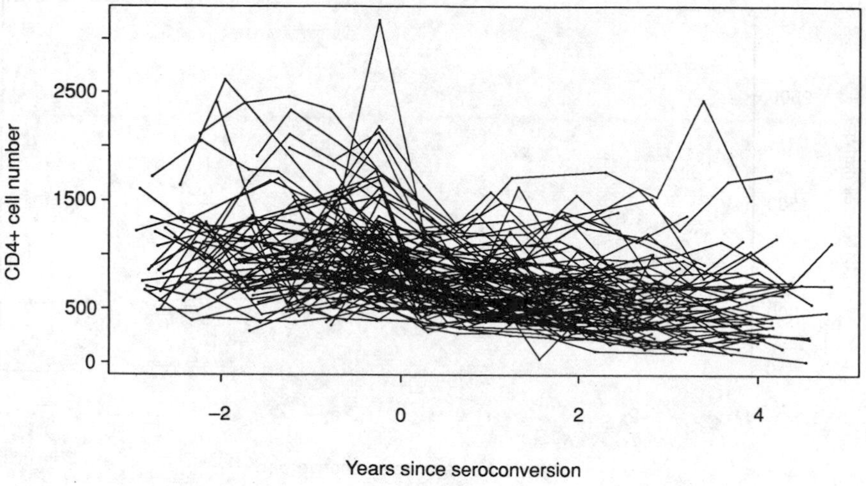

Fig. 3.4. CD4+ counts against time since seroconversion, with sequences of data on each subject shown as connected line segments.

design strategy whereby communication of statistical information is enhanced by adding detail in the background. Unfortunately, such a plot is difficult on the type of graphics devices which are available to many users. An alternative solution is to connect the repeated observations for only a judicious selection of individuals. Jones and Rice (1992) have proposed this idea for presenting a large number of curves. The simplest way to choose a subset is at random, as is done for seven people in Fig. 3.5.

There are two problems with random selection. First, it is possible to obtain a non-representative group by chance, especially when a relatively small number of curves are to be highlighted. Second, this display is unlikely to uncover outlying individuals. We therefore prefer a second approach, in which individual curves are ordered with respect to some characteristic that is relevant to the model of interest. We then connect data for individuals with selected quantiles for this ordering statistic. Ordering statistics can be chosen to measure: the average level; variability within an individual; trend; or correlation between successive values. Resistant statistics are preferred for ordering, so that one or a few outlying observation do not determine an individual's summary score. The median, median absolute deviation, and biweight trend (Mosteller and Tukey, 1977) are examples.

When the data naturally divide into treatment groups, such as in Example 1.3, a separate plot can be made for each group, or one plot can be produced with a separate summary curve for each group and distinct plotting symbols for the data from each group.

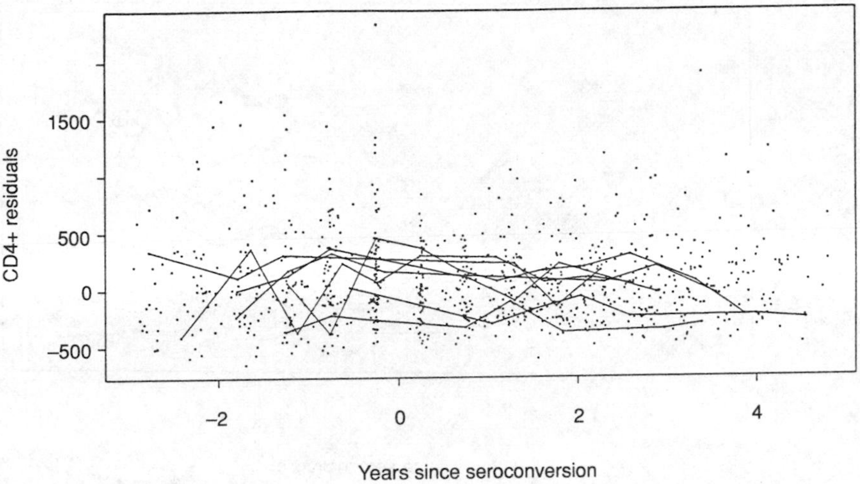

Fig. 3.5. CD4+ residuals against time since seroconversion, with sequences of data from randomly selected subjects shown as connected line segments.

For equally spaced observation times, Jones and Rice (1992) apply principal components analysis to identify ordering statistics. Principal components do not exploit the longitudinal nature of the data, nor do they adapt readily to situations with unequal numbers of observations or different sets of observation times across subjects.

Example 3.2.(continued)
Figure 3.3 displays the average number of CD4+ cells for the MACS seroconverters. Figure 3.6 shows the residuals from this average curve with the repeated values for nine individuals connected.

These persons had *median* residual value at the extrema or the 5th, 10th, 25th, 50th, 75th, 90th, or 95th percentile. We have used residuals rather than the raw data as this sometimes helps to uncover more subtle patterns in individual curves. Note that the data for each individual tend to track at different levels of CD4+ cell numbers, but not nearly to the same extent as in Fig. 3.2. The CD4+ data have considerably more variation across time within a person.

Thus far, we have considered displays of the response against time. In many longitudinal problems, the primary focus is the relationship between the response and an explanatory variable other than time. For example, in the MACS data set, an interesting question is whether CD4+ cell number depends on the depressed mood of an individual. There is some prior belief

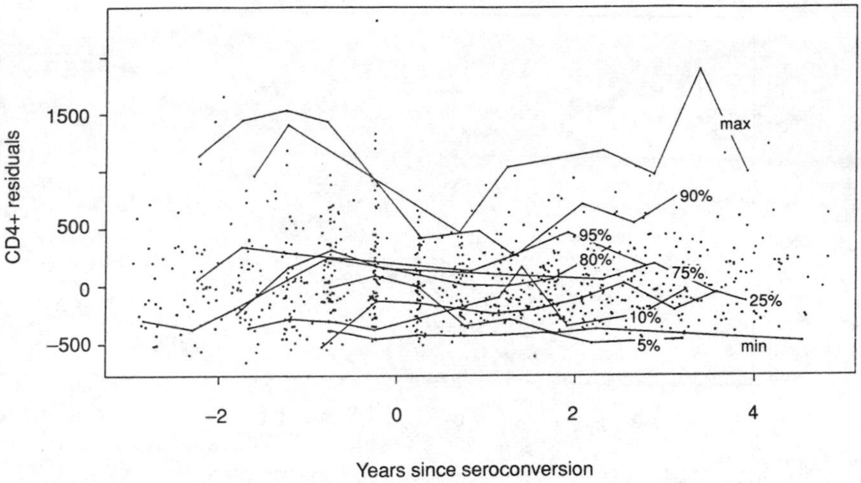

Fig. 3.6. CD4+ residuals against time since seroconversion, with sequences of data from systematically selected subjects shown as connected line segments.

that depressive symptoms are negatively correlated with the capacity for immune response. The MACS collected a measure of depressive symptoms called CESD; a higher score indicates greater depressive symptoms.

Example 3.3. Relationship between CD4+ cell numbers and CESD score

Figure 3.7 plots the residuals with time trends removed for CD4+ cell numbers against similar residuals for CESD scores. Also shown is a lowess curve, which is barely discernible from the horizontal line, $y = 0$. Thus, there is very little evidence for an association between depressive symptoms (CESD score) and immune response (CD4+ numbers), although such evidence as there is points to a negative association, a larger CESD score being associated with a lower CD4+ count.

An interesting question is whether the evidence for any such relationship would derive largely from differences across people, that is, from cross-sectional information, or from changes across time within a person. As previously discussed in Chapter 1, cross-sectional information is more likely biased by unobserved factors. In this example, 70% of the variation in CESD is within individuals, so that both cross-sectional and longitudinal information are present.

We would like to separate the cross-sectional and longitudinal evidence about the association between the response and explanatory variable in EDA. The model in equation (1.4.2)

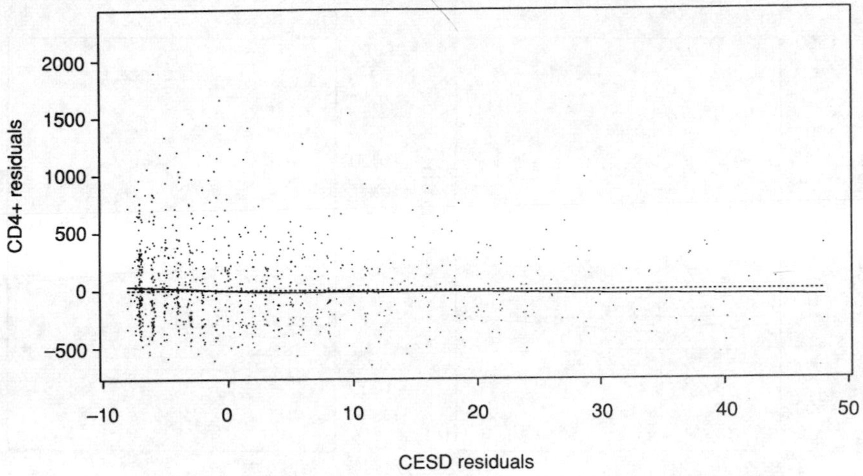

Fig. 3.7. CD4+ residuals against CESD residuals · : data; —— : lowess curve; - - - - : $y = 0$.

$$Y_{ij} = \beta_C x_{i1} + \beta_L(x_{ij} - x_{i1}) + \epsilon_{ij},\ j = 1, \ldots, n_i;\ i = 1, \ldots, m,$$

expresses the outcomes in terms of the baseline value and the change in the explanatory variable. The model implies two facts:

- $Y_{i1} = \beta_C x_{i1} + \epsilon_{i1}$;
- $Y_{ij} - Y_{i1} = \beta_L(x_{ij} - x_{i1}) + \epsilon_{ij} - \epsilon_{i1},\ j = 2, \ldots, n_i$.

This suggests making two scatterplots:

- y_{i1} against x_{i1}; and
- $y_{ij} - y_{i1}$ against $x_{ij} - x_{i1}$.

Example 3.3. (continued)
Figure 3.8 shows both scatterplots for the CD4+ and CESD data. Note that there is little evidence of strong relationship in either the cross-sectional (a) or in the longitudinal (b) display.

3.3 Fitting smooth curves to longitudinal data

A curve-fitting method, lowess, was used in the previous section to highlight the average change in CD4+ cell numbers over time. Lowess is one

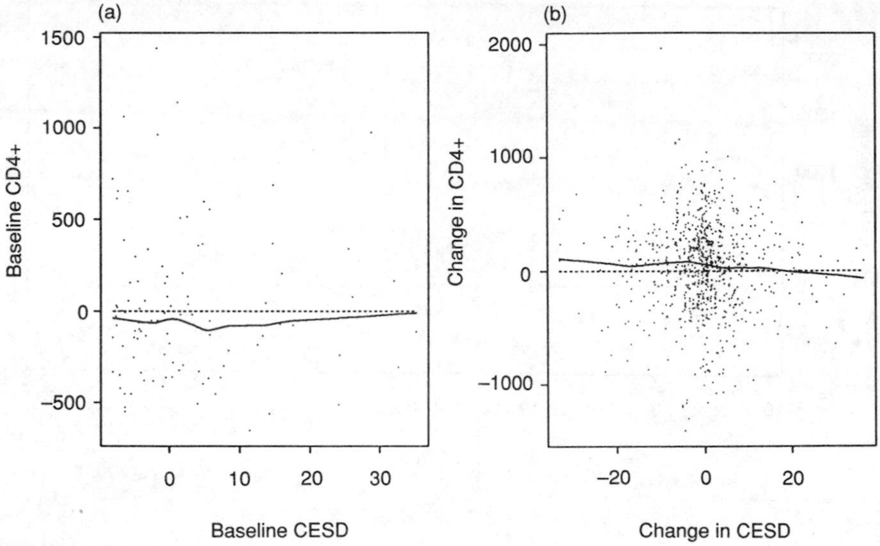

Fig. 3.8. Relationship between CD4+ residuals and CESD residuals. (a) cross-sectional (baseline residuals); (b) longitudinal (change in residuals); · · : data; —— : lowess curve; - - - - : $y = 0$.

of a number of non-parametric regression methods that can be used to estimate the mean response profile as a function of time. Other methods include kernel and spline estimation. Figure 3.9 is a scatterplot of CD4+ cell numbers against time using only a single observation, chosen at random, for each person. Smooth curves have been fit by kernel estimation, smoothing splines, and lowess. All three methods give qualitatively similar results.

We now briefly review each of these smoothing techniques. Details can be found in Hastie and Tibshirani (1990) and the original references therein.

To simplify the discussion of smoothing, we assume there is a single observation on each individual, denoted y_i, observed at time t_i. The general problem is to estimate from data of the following form

$$(t_i, y_i),\ i = 1, \ldots, m,$$

an unknown mean response curve $\mu(t)$ in the underlying model

$$Y_i = \mu(t_i) + \epsilon_i \tag{3.3.1}$$

where the ϵ_i are independent, mean zero errors.

The first method, kernel estimation, is visualized in Fig. 3.10. At the left side of the figure, a *window* is centred at t_1. Then $\hat{\mu}(t_1)$, the estimate

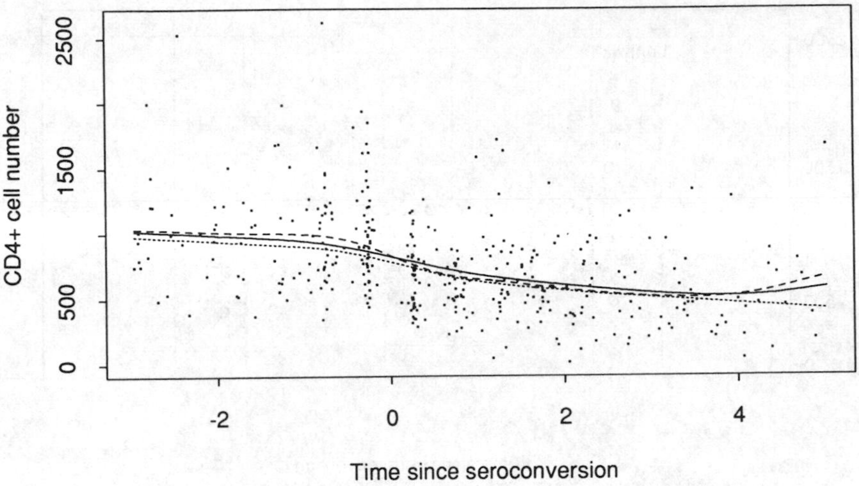

Fig. 3.9. Comparison between kernel, spline and lowess smooths of a sub-set of the CD4+ counts (one count per subject) · : data; —— : kernel; - - - : spline; .... : lowess.

of the mean response at time t_1, is just the average Y value of all points (highlighted by circles) which are visible in the window. Two other windows are shown in the centre and on the right of the plot for which we similarly obtain $\hat{\mu}(t_2)$ and $\hat{\mu}(t_3)$. To obtain an estimate of the smooth curve at every time, we simply slide a window from the extreme left across the data to the extreme right, calculating the average of the points within the window at every time.

If the window is so narrow as to contain only one observation at any time, the resulting estimate will interpolate the data. If it is so wide as to include all data at every time, the resulting estimate is a constant line equal to the average Y value. There is a continuum of curves, one for each possible window width. The wider the window, the smoother the resulting curve.

Taking a straight average of the points within each window is referred to as 'using a boxcar window' because $\hat{\mu}(t)$ is a weighted average of the y_i's with weights equal to zero or one. The weights form the shape of a boxcar (rail van) when plotted against time. An alternative, slightly better strategy is to use a weighting fuction that changes smoothly with time and gives more weight to the observations close to t. A common weight function is the Gaussian kernel, $K(u) = \exp(-0.5u^2)$. The kernel estimate is defined as

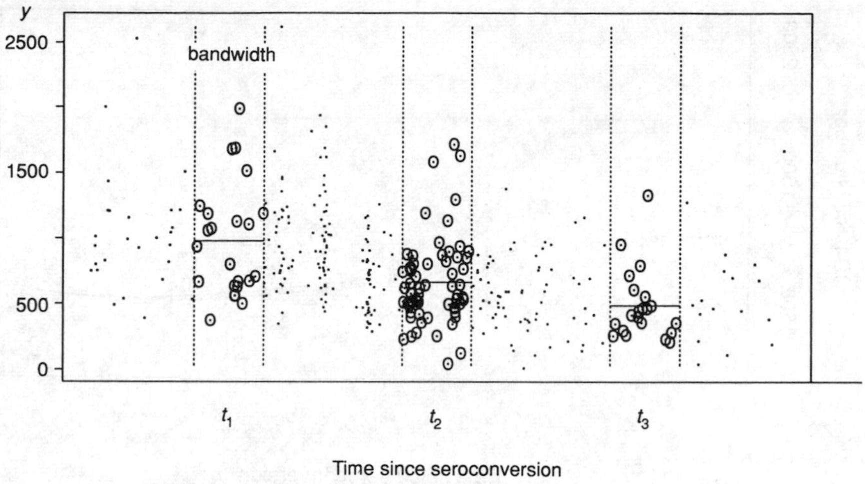

Fig. 3.10. Visualisation of the kernel smoother (see text for detailed explanation).

$$\hat{\mu}(t) = \sum_{i=1}^{m} w(t, t_i, h) y_i \Big/ \sum_{i=1}^{m} w(t, t_i, h) \tag{3.3.2}$$

where $w(t, t_i, h) = K\{(t - t_i)/h\}$ and h is the *bandwidth* of the kernel. Larger values of h produce smoother curves.

Example 3.4. CD4+ cell numbers

Figure 3.11 shows the CD4+ data with kernel estimates of the conditional CD4+ mean as a function of time for two choices of the bandwidth. Note how the estimated curve is considerably smoother for the larger bandwidth.

The second non-parametric curve in common use is the smoothing spline (Silverman, 1985). The name *spline* refers to a draftsman's tool which is a flexible rod used to interpolate smoothly points in a design drawing. A cubic smoothing spline is the function, $s(t)$, which minimizes the criterion

$$J(\lambda) = \sum_{i=1}^{m} \{y_i - s(t_i)\}^2 + \lambda \int \{s''(t)\}^2 \, dt \tag{3.3.3}$$

where $s''(t)$ is the second derivative of $s(t)$. The first term in the equation above quantifies the fidelity of the function, $s(t)$, to the observations, y_i; the integral is a measure of curvature, or roughness of the function, sometimes called a *roughness penalty*. The constant λ determines the degree of smoothness of the spline. When λ is small, little weight is given to the roughness penalty and the spline will be less smooth.

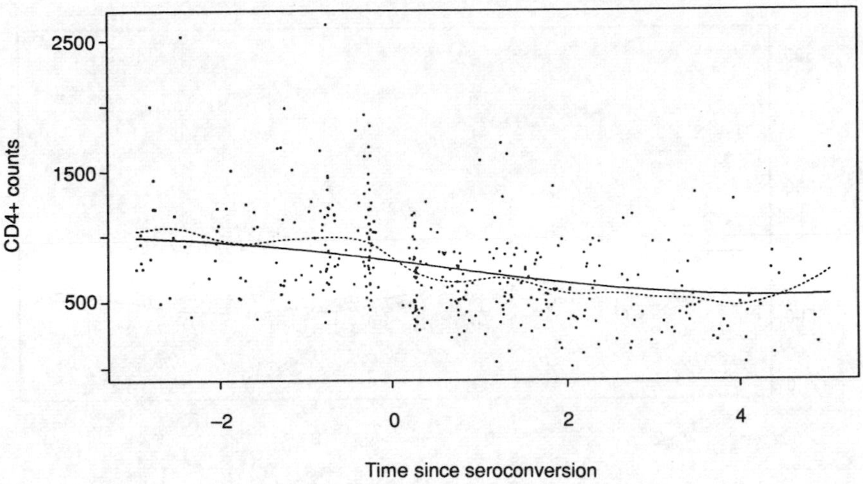

Fig. 3.11. Comparison between two kernel smoothers —— : large bandwidth - - - - : small bandwidth

The smoothing spline which minimizes the criterion above is a twice-differentiable piecewise cubic polynomial. It can be calculated from the observations $(t_i, y_i), i = 1, \ldots, m$ by solving relatively simple linear equations.

Silverman (1984) has shown that a cubic smoothing spline can be closely approximated by a kernel estimate with local bandwidth that is proportional to $g(t)^{-.25}$ where $g(t)$ is the density of points in the vicinity of time t. While splines have adaptive bandwidths, the power -0.25 indicates that the effective bandwidths do not change very quickly with the density of observations.

The final method, lowess (Cleveland, 1979), is a natural extension of kernel methods made 'robust', that is less sensitive, to outlying Y values. The lowess curve estimate at time t_i starts by centring a window there as in Fig. 3.10. Rather than calculating a weighted mean of the points in the window, a weighted least-squares line is fitted. As before, more weight is given to observations close to the middle of the window. Once the line has been fitted, the residual (vertical) distance from the line to each point in the window is determined. The outliers, that is, points with large residuals, are then downweighted and the line is re-estimated. This process is iterated a few times. The net result is a fitted line that is insensitive to observations with outlying Y values. The value of the lowess curve at t_i is just the predicted value for the line. The entire lowess curve is obtained by repeating this process at the desired times.

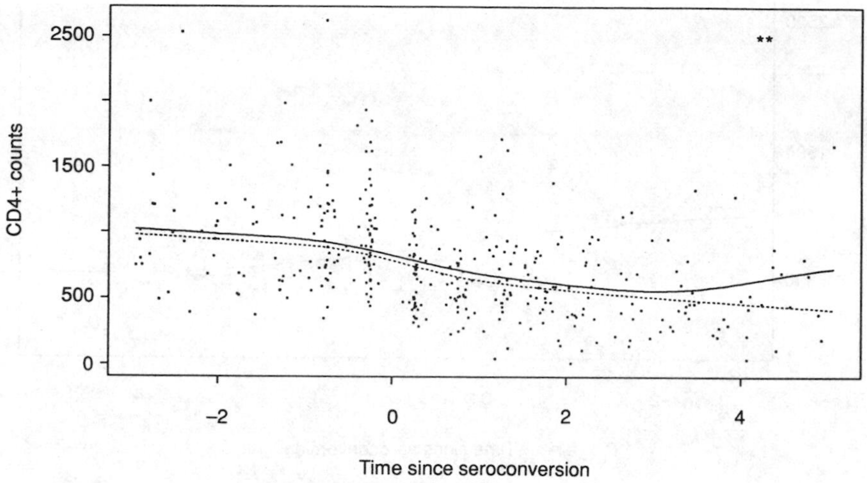

Fig. 3.12. Comparison between lowess and kernel smoothers, for data with two artificial outliers (shown as $\star$) —— : kernel - - - - : lowess.

Example 3.4. (continued)
Figure 3.12 shows kernel and lowess estimates of similar smoothness fitted to the CD4+ data where we have altered two of the points to create outliers. Note that the kernel estimator is affected more than lowess by the outliers. When exploring longitudinal data, it is good practice to use robust smoothing methods as this avoids excessive dependence of results on a few observations.

With each of the non-parametric curve estimation techniques, there is a bandwidth parameter which controls the smoothness of the fitted curve. In choosing the bandwidth, there is a classic trade-off between bias and variance. The wider the bandwidth, the smaller the variance of the estimated curve at any fixed time since more observations have been averaged. But estimates which are smoother than the functions they are approximating are biased. The objective in curve estimation is to choose from the data a smoothness parameter that balances bias and variance.

A generally accepted aggregate measure that combines bias and variance is the *average predictive squared error* defined by

$$PSE(\lambda) = \frac{1}{m} \sum_{i=1}^{m} \mathrm{E}\{Y_i^* - \hat{\mu}(t_i; \lambda)\}^2 \tag{3.3.4}$$

where Y_i^* is a new observation at t_i. We cannot compare the observed y_i with the corresponding value, $\hat{\mu}(t_i)$, in the criterion above because the optimal curve would interpolate the points. PSE can be estimated by *cross-validation* where we compare the observed y_i to the predicted curve $\hat{\mu}^{-i}(t_i)$ obtained by leaving out the *ith* observation. The cross-validation criterion is given by

$$CV(\lambda) = \frac{1}{m}\sum_{i=1}^{m}\{y_i - \hat{\mu}^{-i}(t_i;\lambda)\}^2. \qquad (3.3.5)$$

The expected value of $CV(\lambda)$ is approximately equal to the average predictive squared error. See Hastie and Tibshirani (1990) for further discussion.

3.4 Exploring correlation structure

In this section, graphical displays for exploring the degree of association in a longitudinal data set are considered. To remove the effects of explanatory variables, we first regress the response, y_{ij}, on the explanatory variables, $\boldsymbol{x}_{ij}$, to obtain residuals, $r_{ij} = y_{ij} - \boldsymbol{x}_{ij}'\hat{\beta}$. With data collected at a fixed number of equally spaced time points, correlation can be studied using a scatterplot matrix in which r_{ij} is plotted against r_{ik} for all $j < k = 1, \ldots, n$.

Example 3.5. CD4+ cell numbers

To illustrate the idea of a scatterplot matrix, we have rounded the CD4+ observation times to the nearest year, so that there are a maximum of seven observations between -2 and 4 for each individual. Figure 3.13 shows each of the 7 choose 2 scatterplots of responses from a person at different times.

Notice from the main diagonal of the scatterplot matrix that there is substantial positive correlation between repeated observations on the same individual that are one year apart. The degree of correlation decreases as the observations are moved farther from one another in time, which corresponds to moving farther from the diagonal. Notice also that the correlation is reasonably consistent along a diagonal in the matrix. This indicates that the correlation depends more strongly on the time between observations than on their absolute times.

If the residuals have constant mean and variance and if $\mathrm{Corr}(y_{ij}, y_{ik})$ depends only on $|t_{ij} - t_{ik}|$, the process Y_{ij} is said to be *weakly stationary* (Box and Jenkins, 1970). This is a reasonable first approximation for the CD4+ data.

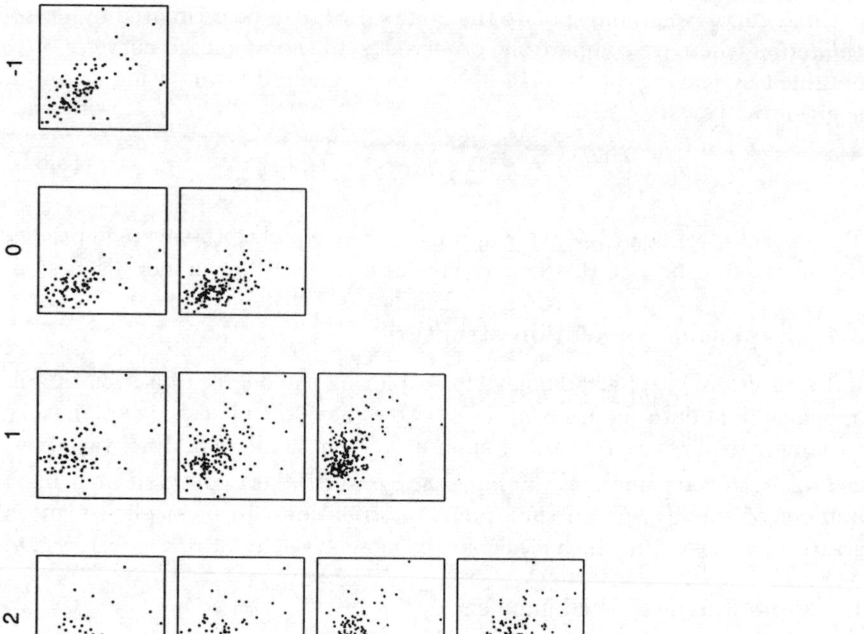

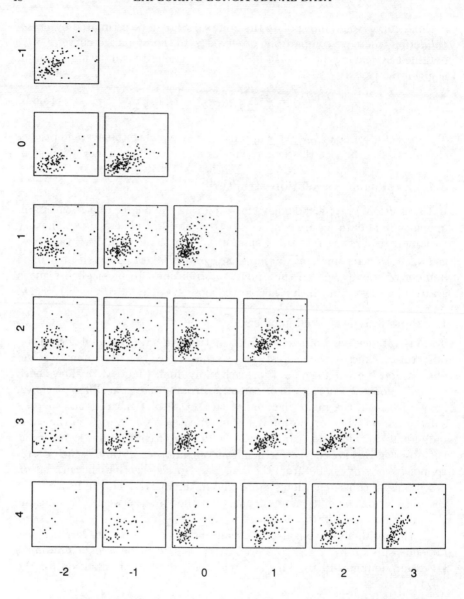

Fig. 3.13. Scatterplot matrix of CD4+ residuals. Axis labels are years relative to seroconversion.

When each scatterplot in the matrix appears like a sample from the bivariate Gaussian (normal) distribution, we can summarize the association with a *correlation matrix*, comprised of a correlation coefficient for each plot.

Example 3.5. (continued)
The estimated correlation matrix for the CD4+ data is presented in Table 3.2. The correlations show some tendency to decrease with increasing time lag, but remain substantial at all lags. The estimated correlation of 0.89 at lag 6 is quite likely misleading, as it is calculated from only 9 pairs of measurements.

Table 3.2. Estimated autocorrelation matrix for CD4+ residuals. Entries are $\text{Corr}(Y_{ij}, Y_{ik}), 1 = t_{ij} < t_{ik} \leq 7$ years.

		t_{ij}					
		1	2	3	4	5	6
	2	0.66					
	3	0.56	0.49				
t_{ik}	4	0.41	0.47	0.51			
	5	0.29	0.39	0.51	0.68		
	6	0.48	0.52	0.51	0.65	0.75	
	7	0.89	0.48	0.44	0.61	0.70	0.75

Assuming stationarity, a single correlation estimate can be obtained for each distinct value of the time separation or *lag*, $| t_{ij} - t_{ik} |$. This corresponds to pooling observation pairs along the diagonals of the scatterplot matrix. This *autocorrelation function* for the CD4+ data takes the values presented in Table 3.3.

Table 3.3. Estimated autocorrelation function for CD4+ residuals. Entries are $\text{Corr}(Y_{ij}, Y_{ij-u}), u = 1, \ldots, 6$.

u:	1	2	3	4	5	6
$\hat{\rho}(u)$:	0.60	0.54	0.46	0.42	0.47	0.89

It is obviously desirable to indicate the uncertainty in an estimated correlation coefficient when examining the correlation matrix or autocorrelation function. The simplest rule of thumb is that under the null condition of no correlation, a correlation coefficient has standard error which is roughly $1/\sqrt{N}$ where N is the number of independent pairs of observations in the

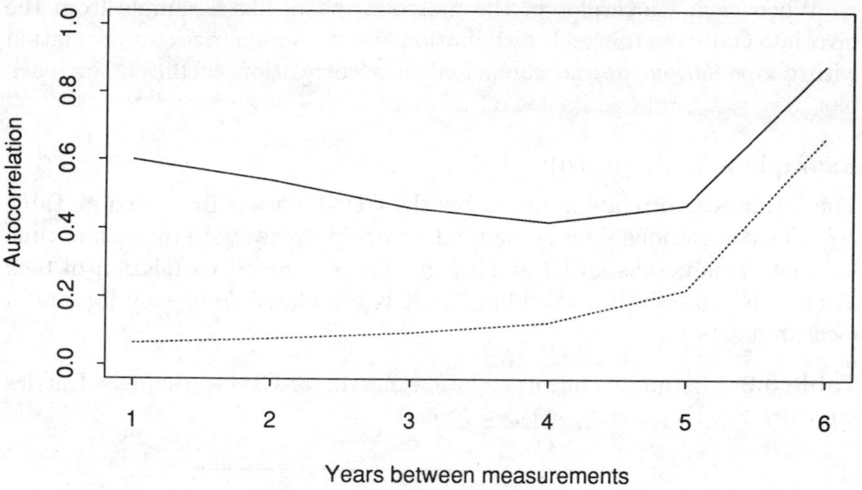

Fig. 3.14. Sample autocorrelation function of CD4+ residuals, and upper 95% tolerance limits assuming zero autocorrelation. —— : sample autocorrelation function - - - - : tolerance limits.

calculation. Using this rule of thumb, a plot of the autocorrelation function can be enhanced with tolerance limits for a true autocorrelation of zero. These take the form of pairs of values $\pm 2/\sqrt{N_u}$, where N_u is the number of pairs of observations at lag u.

Example 3.5. (continued)

Figure 3.14 shows the estimated autocorrelation function for the CD4+ cell numbers, together with its associated tolerance limits for zero autocorrelation. The size of the tolerance limit at lag 6 is an effective counter to spurious over-interpretation of the large estimated autocorrelation. All that can be said is that the autocorrelation at lag 6 is significantly greater than zero.

Calculating confidence intervals for non-zero autocorrelations is more complex. See Box and Jenkins (1970) for a detailed discussion.

In subsequent chapters, the autocorrelation function will be one tool for identifying sensible models for the correlation in a longitudinal data set. The empirical function described above will be contrasted with the theoretical correlations for a candidate model. Hence, the EDA displays are later useful for model criticism as well.

The autocorrelation function is most effective for studying equally spaced data that are roughly stationary. Autocorrelations are more difficult to estimate with irregularly spaced data unless we round observation times as was done above for the CD4+ data. An alternative function that describes the association among repeated values and is easily estimated with irregular observation times is the *variogram* (Diggle, 1990). For a stochastic process $Y(t)$, the variogram is defined as

$$\gamma(u) = \tfrac{1}{2}\mathrm{E}\left[\{Y(t) - Y(t-u)\}^2\right], \ u \geq 0. \qquad (3.4.1)$$

If $Y(t)$ is stationary, the variogram is directly related to the autocorrelation function, $\rho(u)$, by

$$\gamma(u) = \sigma^2 \{1 - \rho(u)\}$$

where σ^2 is the variance of $Y(t)$. However, the variogram is also well-defined for a limited class of non-stationary processes for which the *increments*, $Y(t) - Y(t-u)$, are stationary.

The origins of the variogram as a descriptive tool go back at least to Jowett (1955). It has since been used extensively in the body of methodology known as *geostatistics*, which is concerned primarily with the estimation of spatial variation (Journel and Huijbregts, 1978). In the longitudinal context, the empirical counterpart of the variogram, which we call the *sample variogram*, is calculated from observed half-squared-differences between pairs of residuals,

$$v_{ijk} = \tfrac{1}{2}(r_{ij} - r_{ik})^2,$$

and the corresponding time-differences

$$u_{ijk} = t_{ij} - t_{ik}.$$

If the times t_{ij} are not totally irregular, there will be more than one observation at each value of u. We then let $\hat{\gamma}(u)$ be the average of all of the v_{ijk} corresponding to that particular value of u. With highly irregular sampling times, the variogram can be estimated from the data $(u_{ijk}, v_{ijk}), j < k = 1, \ldots, n_i; i = 1, \ldots, m$ by fitting a non-parametric curve. The process variance, σ^2, is estimated as the average of all half-squared-differences $\frac{1}{2}(y_{ij} - y_{lk})^2$ with $i \neq l$. The autocorrelation function at any lag u can then be estimated from the sample variogram by the formula

$$\hat{\rho}(u) = 1 - \hat{\gamma}(u)/\hat{\sigma}^2. \qquad (3.4.2)$$

Example 3.6. CD4+ cell counts

An estimate of the variogram for the CD4+ data is pictured in Fig. 3.15. The diagram shows both the basic quantities (u_{ijk}, v_{ijk}) and a smooth estimate of $\gamma(u)$ which has been produced using kernel smoother. Note that there are few data available for time differences of less than six months or beyond six years. Also, to accentuate the shape of the smooth estimate, we have truncated the vertical axis at 180 000. The variogram smoothly increases with lag corresponding to decreasing correlation as observations are separated in time. This is in contrast to the apparent, albeit spurious, *rise* in the estimated autocorrelation at lag 6, as shown in Fig. 3.14. The explanation is that the non-parametric smoothing of the variogram recognises the sparsity of data at lag $u = 6$, and incorporates information from the sample variogram at smaller values of u. This illustrates one advantage of the sample variogram for irregularly spaced data, by comparison with the sample autocorrelation function based on artificially rounded measurement times. The horizontal line on Fig. 3.15 is the variogram-based estimate of the process variance. This is substantially larger than the value of the sample variogram at lag $u = 6$, indicating that the autocorrelation has not decayed to zero within the range of the data.

Incidentally, the enormous random fluctuations in the basic quantities (u_{ijk}, v_{ijk}) are entirely typical. The marginal sampling distribution of each v_{ijk} is proportional to chi-squared on 1 degree of freedom.

Example 3.7. Milk protein measurements

We now construct an estimate of the variogram for the milk protein data. In a designed experiment such as this, we are able to use a *saturated* model for the mean response profiles. By this, we mean that we fit a separate parameter for the mean response at each of the 19 times in each of the 3 treatment groups. The ordinary least-squares estimates of these 57 mean response parameters are just the corresponding observed means. The sample variogram of the resulting set of ordinary least-squares residuals is shown in Fig. 3.16. In contrast to the CD4+ example, the pattern is clear without further smoothing. The variogram rises steadily with increasing u, but levels off within the range spanned by the data. This levelling off suggests that the correlation has effectively decayed to zero by a time separation of about 10 weeks. The increase in $\hat{V}(u)$ at $u = 18$ is apparently inconsistent with this interpretation. However, this estimate is an average of rather few v_{ijk}, and is therefore unreliable. For a more detailed interpretation of this variogram estimate, see Section 5.4.1.

3.5 Further reading

The modern rebirth of exploratory data analysis was led by John W. Tukey. The books by Tukey (1977) and by Mosteller and Tukey (1977) are the best original sources. A simpler introduction is provided by Velleman and Hoaglin (1981). Background information on graphical methods in statistics, as well as on lowess, can be found in Chambers *et al.* (1983). Two stimulating books on graphical displays of information are by Edward Tufte (1983, 1990). An excellent review of smoothing methods is available in the first three chapters of Hastie and Tibshirani (1990). Further reading, with

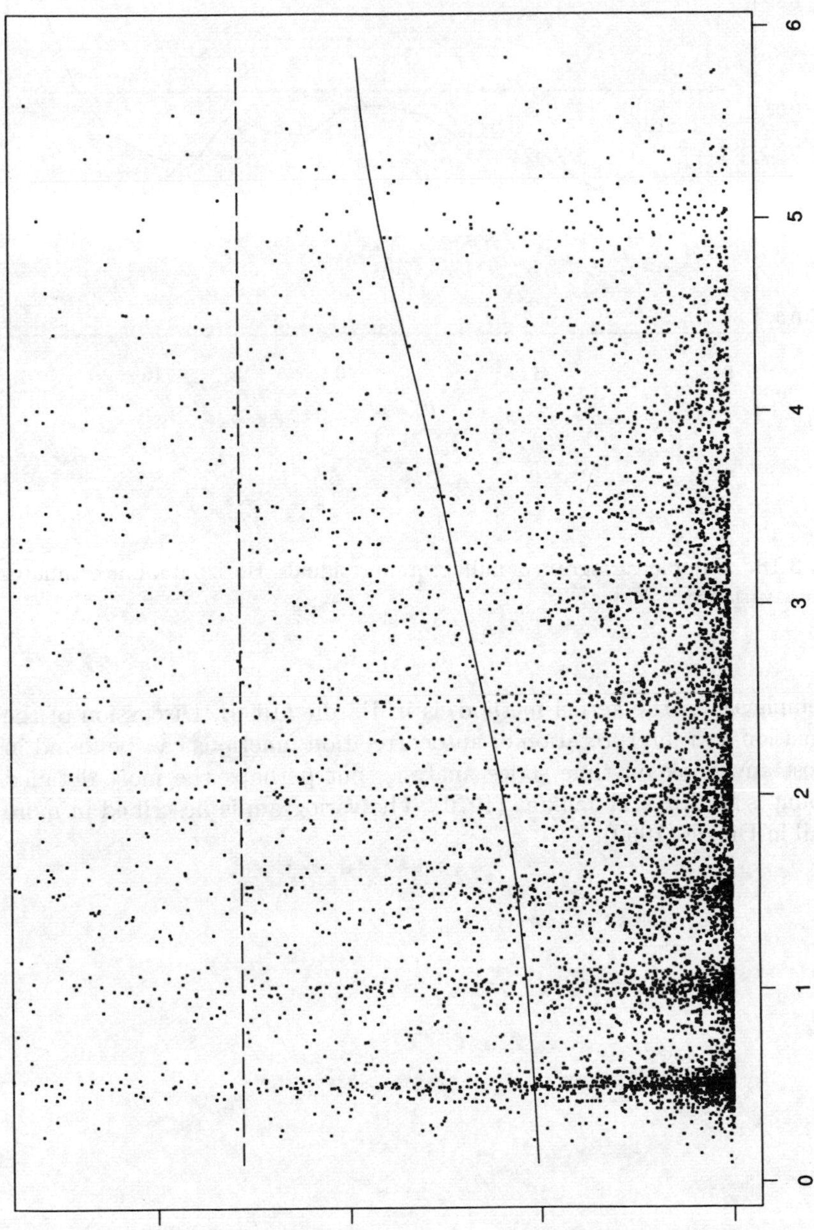

Fig. 3.15. Sample variogram of CD4+ residuals : individual points (u_{ijk}, v_{ijk}); —— : kernel smooth; - - - : residual variance.

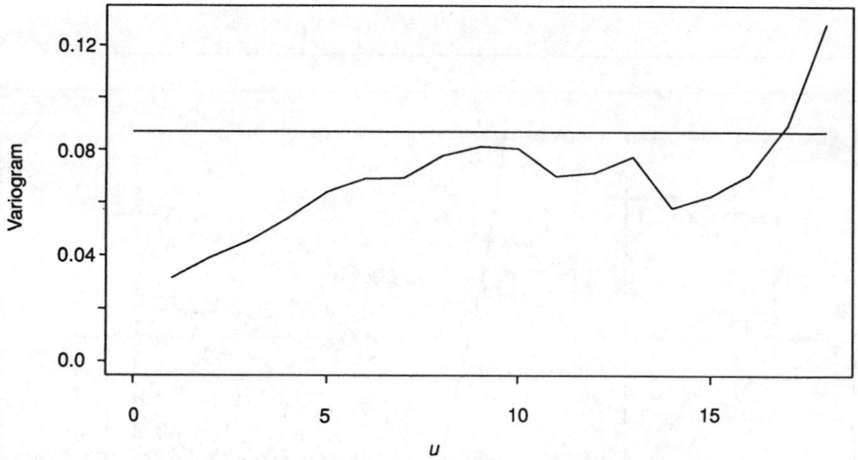

Fig. 3.16. Sample variogram of milk protein residuals. Horizontal line estimates process variance.

an emphasis on the kernel method, is in Härdle (1990). Discussion of the estimation and interpretation of autocorrelation functions can be found in almost any book on time series analysis, but perhaps the most detailed account is in Box and Jenkins (1970). The variogram is described in more detail in Diggle (1990).

4
General linear models for longitudinal data

4.1 Motivation

In this and the next chapter, our aim is to develop a general linear modelling framework for longitudinal data, in which the inferences we make about the regression parameters of primary interest recognize the likely correlation structure in the data. Two ways of achieving this are to build explicit parametric models of the covariance structure whose validity can be checked against the available data or, where possible, to use methods of inference which are robust to misspecification of the covariance structure. In this chapter, we formulate the general linear model for correlated data and discuss general approaches to parameter estimation. We use weighted least-squares estimation for parameters which describe the mean response, and maximum likelihood or restricted maximum likelihood for covariance parameters. In Chapter 5 we develop explicit parametric models for the covariance structure. These models are particularly useful when the data consist of relatively long, irregularly spaced or incomplete sequences of observations. For data consisting of short, complete sequences a non-parametric specification of the covariance structure will often suffice.

4.2 The general linear model with correlated errors

Let y_{ij}, $j = 1, \ldots, n$ be the sequence of observed measurements on the *ith* of m subjects, and t_j, $j = 1, \ldots, n$ be the corresponding times at which the measurements are taken on each unit. We relax the assumption of a common set of times, t_j, later. Associated with each y_{ij} are the values, x_{ijk}, $k = 1, \ldots, p$, of p explanatory variables. We assume that the y_{ij} are realizations of random variables Y_{ij} which follow the regression model

$$Y_{ij} = \beta_1 x_{ij1} + \ldots + \beta_p x_{ijp} + \epsilon_{ij},$$

where the ϵ_{ij} are random sequences of length n associated with each of the m subjects. In the classical linear model, the ϵ_{ij} would be mutually independent $N(0, \sigma^2)$ random variables. In our context, the longitudinal structure of the data means that we expect the ϵ_{ij} to be correlated within subjects.

For a formal analysis of the linear model, a matrix formulation is preferable. Let $\boldsymbol{y}_i = (y_{i1}, ..., y_{in})$ be the observed sequence of measurements on the ith subject and $\boldsymbol{y} = (\boldsymbol{y}_1, ..., \boldsymbol{y}_m)$ the complete set of $N = nm$ measurements from m units. Let X be an $N \times p$ matrix of explanatory variables, with $\{n(i-1)+j\}th$ row $(x_{ij1}, \ldots, x_{ijp})$. Let $\sigma^2 V$ be a block-diagonal matrix with non-zero $n \times n$ blocks $\sigma^2 V_0$, each representing the variance matrix for the vector of measurements on a single subject. Then, the general linear model for longitudinal data treats $\boldsymbol{y}$ as a realization of a multivariate Gaussian random vector, $\boldsymbol{Y}$, with

$$\boldsymbol{Y} \sim MVN(X\boldsymbol{\beta}, \sigma^2 V). \tag{4.2.1}$$

If we want to use the robust approach to analysing data generated by the model (4.2.1), the block-diagonal structure of $\sigma^2 V$ is crucial, because we will use the replication across units to estimate $\sigma^2 V$ without making any parametric assumptions about its form. Nevertheless, it is perhaps useful at this stage to consider what form the non-zero blocks $\sigma^2 V_0$ might take to anticipate the parametric modelling approach of Chapter 5.

4.2.1 *The uniform correlation model*

In this model, we assume that there is a positive correlation, ρ, between any two measurements on the same subject. In matrix terms, this corresponds to

$$V_0 = (1-\rho)I + \rho J, \tag{4.2.2}$$

where I denotes the $n \times n$ identity matrix and J the $n \times n$ matrix all of whose elements are 1.

A justification of this *uniform correlation* model is the following. Let the observed measurements, y_{ij}, be realizations of random variables, Y_{ij}, such that

$$Y_{ij} = \mu_{ij} + U_i + Z_{ij}, \quad i = 1, \ldots, m; \quad j = 1, \ldots, n, \tag{4.2.3}$$

where $\mu_{ij} = \mathrm{E}(Y_{ij})$, the U_i are mutually independent $N(0, \nu^2)$ random variables, the Z_{ij} are mutually independent $N(0, \tau^2)$ random variables and the U_i and Z_{ij} are independent of one another. Then, the covariance structure of the data corresponds to (4.2.2) with $\rho = \nu^2/(\nu^2 + \tau^2)$ and $\sigma^2 = \nu^2 + \tau^2$. Note that (4.2.3) gives an interpretation of the uniform correlation model as one in which a linear model for the mean response incorporates a random intercept term with variance ν^2 between subjects. It therefore provides the foundation for a wider class of random coefficient models which we consider in more detail in Chapter 9. Also, (4.2.3) provides a model-based justification for a so called *split-plot* analysis of variance approach to longitudinal data which we describe in Chapter 6.

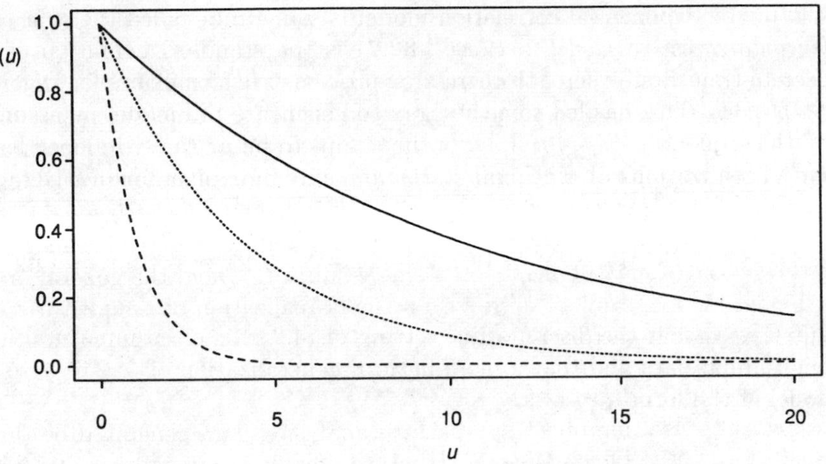

Fig. 4.1. The autocorrelation function for the exponential model. —— : $\phi = 0.1$; .... : $\phi = 0.25$; – – – : $\phi = 1.0$.

4.2.2 *The exponential correlation model*

In this model, V_0 has $jkth$ element, $v_{jk} = \text{Cov}(Y_{ij}, Y_{ik})$, of the form

$$v_{jk} = \sigma^2 \exp(-\phi \mid t_j - t_k \mid). \tag{4.2.4}$$

In contrast to the uniform correlation model, the correlation between a pair of measurements on the same unit decays towards zero as the time separation between the measurements increases. The rate of decay is faster for larger values of ϕ. Figure 4.1 illustrates this with $\phi = 0.1, 0.25$, and 1.0. Note that if the observation times, t_j, are equally spaced, say $t_{j+1} - t_j = d$ for all j, then (4.2.4) is expressible as

$$v_{jk} = \sigma^2 \rho^{|j-k|} \tag{4.2.5}$$

where $\rho = \exp(-\phi d)$ is the correlation between successive observations on the same subject.

A justification of (4.2.5) is to represent the random variables Y_{ij} as

$$Y_{ij} = \mu_{ij} + W_{ij}, \quad i = 1, \ldots, m; \quad j = 1, \ldots, n \tag{4.2.6}$$

where

$$W_{ij} = \rho W_{ij-1} + Z_{ij} \tag{4.2.7}$$

and the Z_{ij} are mutually independent $N\{0, \sigma^2(1-\rho^2)\}$ random variables, to give $\text{Var}(Y_{ij}) = \text{Var}(W_{ij}) = \sigma^2$ as required. In view of (4.2.6) and

(4.2.7), the exponential correlation model is sometimes called the *first order autoregressive model*, because (4.2.7) is the standard definition of a discrete-time first-order autoregressive process (see, for example, Diggle, 1990, p.34). This has led some authors to generalize the model by assuming the sequences, $W_{ij}, j = 1, \ldots, n$, in (4.2.6) to be mutually independent partial realizations of a general stationary autoregressive moving average process,

$$W_{ij} = \sum_{r=1}^{p} \rho_r W_{ij-r} + Z_{ij} + \sum_{s=1}^{q} \beta_s Z_{ij-s}. \tag{4.2.8}$$

Our view is that the discrete-time setting of (4.2.8) is often unnatural for longitudinal data and that a more natural generalization of (4.2.6) is to a model of the form

$$Y_{ij} = \mu_{ij} + W_i(t_j), \quad i = 1, \ldots, m; \quad j = 1, \ldots, n \tag{4.2.9}$$

where the sequences $W_i(t_j), j = 1, ..., n$ are realizations of mutually independent, continuous-time, stationary Gaussian processes $\{W_i(t), t \in R\}$ with a common covariance structure, $\gamma(u) = \text{Cov}\{W_i(t), W_i(t-u)\}$. Note that (4.2.9) automatically accomodates irregularly spaced times of measurement, t_j, and, incidentally, different times of measurement for different units, whereas either of the discrete-time models, (4.2.7) or (4.2.8), would be quite unnatural for irregularly spaced t_j.

4.3 Weighted least-squares estimation

We now return to the general formulation of (4.2.1), and consider the problem of estimating the regression parameter $\boldsymbol{\beta}$. The *weighted least-squares* estimate of $\boldsymbol{\beta}$, using a symmetric *weight matrix*, W, is the value, $\tilde{\boldsymbol{\beta}}_W$, which minimizes the quadratic form

$$(\boldsymbol{y} - X\boldsymbol{\beta})'W(\boldsymbol{y} - X\boldsymbol{\beta}). \tag{4.3.1}$$

Standard matrix manipulations give the explicit result

$$\tilde{\boldsymbol{\beta}}_W = (X'WX)^{-1}X'W\boldsymbol{y}. \tag{4.3.2}$$

Because $\boldsymbol{y}$ is a realization of a random vector $\boldsymbol{Y}$ with $E(\boldsymbol{Y}) = X\boldsymbol{\beta}$, the weighted least-squares estimator, $\tilde{\boldsymbol{\beta}}_W$, is unbiased, whatever the choice of W. Furthermore, since $\text{Var}(\boldsymbol{Y}) = \sigma^2 V$, then

$$\text{Var}(\tilde{\boldsymbol{\beta}}_W) = \sigma^2\{(X'WX)^{-1}X'W\}V\{WX(X'WX)^{-1}\}. \tag{4.3.3}$$

If $W = I$, the identity matrix, (4.3.2) reduces to the *ordinary least-squares* estimator

$$\tilde{\boldsymbol{\beta}}_I = (X'X)^{-1}X'\boldsymbol{y} \tag{4.3.4}$$

with

$$\text{Var}(\tilde{\boldsymbol{\beta}}_I) = \sigma^2(X'X)^{-1}X'VX(X'X)^{-1}. \tag{4.3.5}$$

If $W = V^{-1}$, the estimator becomes

$$\hat{\boldsymbol{\beta}} = (X'V^{-1}X)^{-1}X'V^{-1}\boldsymbol{y} \tag{4.3.6}$$

with

$$\text{Var}(\hat{\boldsymbol{\beta}}) = \sigma^2(X'V^{-1}X)^{-1}. \tag{4.3.7}$$

The 'hat' notation anticipates that $\hat{\boldsymbol{\beta}}$ is the maximum likelihood estimator for $\boldsymbol{\beta}$ under the multivariate Gaussian assumption (4.2.1).

This last remark suggests, correctly, that the most efficient weighted least-squares estimator for $\boldsymbol{\beta}$ uses $W = V^{-1}$. However, to identify this optimal weighting matrix we need to know the complete correlaton structure of the data – we do not need to know σ^2, because $\tilde{\boldsymbol{\beta}}_W$ is unchanged by proportional changes in all the elements of W. Because the correlation structure may be difficult to identify in practice, it is of interest to ask how much loss of efficiency might result from using a different W. Note that the relative efficiency of $\tilde{\boldsymbol{\beta}}_W$ and $\hat{\boldsymbol{\beta}}$ can be calculated from their respective variance matrices (4.3.3) and (4.3.6).

The relative efficiency of ordinary least-squares depends on the precise interplay between the matrices X and V, as described by Bloomfield and Watson (1975). However, our experience has been that under the conditions encountered in a wide range of longitudinal data applications, the relative efficiency is often quite good. As a simple example, consider $m = 10$ units each observed at $n = 5$ time-points $t_j = -2, -1, 0, 1, 2$. Let the mean response at time t be $\mu(t) = \beta_0 + \beta_1 t$. We shall illustrate the relative efficiency of the ordinary least-squares estimators for $\boldsymbol{\beta} = (\beta_0, \beta_1)$ under each of the covariance structures described in Section 4.2.

First, note that

$$(X'X)^{-1} = \begin{bmatrix} 50 & 0 \\ 0 & 100 \end{bmatrix}^{-1} = \begin{bmatrix} 0.02 & 0 \\ 0 & 0.01 \end{bmatrix}.$$

Now, suppose that V corresponds to the uniform correlation model (4.2.2) with

$$V_0 = (1-\rho)I + \rho J$$

for some $0 \le \rho < 1$. Then, straightforward matrix manipulations give

$$X'VX = \begin{bmatrix} 50(1+4\rho) & 0 \\ 0 & 100(1-\rho) \end{bmatrix}$$

and

$$\mathrm{Var}(\tilde{\boldsymbol{\beta}}_I) = \sigma^2 \begin{bmatrix} 0.02(1+4\rho) & 0 \\ 0 & 0.01(1-\rho) \end{bmatrix} \qquad (4.3.8)$$

by substitution of $(X'X)^{-1}$ and $X'VX$ into (4.3.5). Similarly,

$$V_0^{-1} = (1-\rho)^{-1}I - \rho\{(1-\rho)(1+4\rho)\}^{-1}J,$$

$$X'V^{-1}X = \begin{bmatrix} 50(1+4\rho)^{-1} & 0 \\ 0 & 100(1-\rho)^{-1} \end{bmatrix}$$

and

$$\mathrm{Var}(\hat{\boldsymbol{\beta}}) = \sigma^2 \begin{bmatrix} 0.02(1+4\rho) & 0 \\ 0 & 0.01(1-\rho) \end{bmatrix} \qquad (4.3.9)$$

from (4.3.7). Comparing (4.3.8) and (4.3.9), we see that they are identical, which implies that ordinary least-squares is fully efficient in this case. An intuitive explanation is that with a common correlation between any two equally spaced measurements on the same unit, there is no reason to weight measurements differently. However, this is not always the case. For example, it would not be true if the number of measurements varied between units because, with $\rho > 0$, units with more measurements would then convey more information in total but *less* information per measurement, than units with fewer measurements.

Now suppose that V corresponds to the exponential correlation model (4.2.5), in which V_0 has $jkth$ element

$$v_{jk} = \sigma^2 \rho^{|j-k|}.$$

Straightforward matrix manipulations now give $X'VX$ as

$$\begin{bmatrix} 10(5+8\rho+6\rho^2+4\rho^3+2\rho^4) & 0 \\ 0 & 20(5+4\rho-\rho^2-4\rho^3-4\rho^4) \end{bmatrix}$$

and $\mathrm{Var}(\tilde{\boldsymbol{\beta}}_I)$ as

$$\sigma^2 \begin{bmatrix} 0.004(5+8\rho+6\rho^2+4\rho^3+2\rho^4) & 0 \\ 0 & 0.002(5+4\rho-\rho^2-4\rho^3-4\rho^4) \end{bmatrix}. \qquad (4.3.10)$$

The matrix V_0^{-1} is tri-diagonal, with $jkth$ element $(1-\rho^2)^{-1}w_{jk}$, and

$$w_{11} = w_{55} = 1,$$
$$w_{22} = w_{33} = w_{44} = 1 + \rho^2,$$

$$w_{jj+1} = w_{jj-1} = -\rho.$$

Then,

$$X'V^{-1}X = (1-\rho^2)^{-1}\begin{bmatrix} 10(5-8\rho+3\rho^2) & 0 \\ 0 & 20(5-4\rho+\rho^2) \end{bmatrix}$$

and

$$\text{Var}(\hat{\beta}) = \sigma^2(1-\rho^2)\begin{bmatrix} 0.1(5-8\rho+3\rho^2)^{-1} & 0 \\ 0 & 0.05(5-4\rho+\rho^2)^{-1} \end{bmatrix}. \quad (4.3.11)$$

The relative efficiencies, $e(\beta_k) = \text{Var}(\hat{\beta}_k)/\text{Var}(\tilde{\beta}_k)$, in this case are obtained from (4.3.10) and (4.3.11) as

$$e(\beta_0) = 25(1-\rho^2)\{(5-8\rho+3\rho^2)(5+8\rho+6\rho^2+4\rho^3+2\rho^4)\}^{-1}$$

and

$$e(\beta_1) = 25(1-\rho^2)\{(5-4\rho+\rho^2)(5+4\rho-\rho^2-4\rho^3-4\rho^4)\}^{-1}.$$

These are shown in Table 4.1 for a range of values of ρ. Note that all of the tabulated values are close to 1.

Comparisons of this kind suggest that in many circumstances where there is a balanced design, the ordinary least-squares estimator, $\tilde{\beta}$, is perfectly satisfactory for point estimation. However, this is not always the case. Consider, for example, a two-treatment crossover design in which $n = 3$ measurements are taken, at unit time intervals, on each of $m = 8$ subjects. The sequences of treatments given to the eight subjects are AAA, AAB, ABA, ABB, BAA, BAB, BBA, and BBB (Fitzmaurice *et al.*, 1993). The model for the data is

$$Y_{ij} = \beta_0 + \beta_1 x + \epsilon_{ij},$$

where x is a binary indicator for treatment B and the ϵ_{ij} follow the exponential correlation model, as in the previous example, with autocorrelation

Table 4.1. Relative efficiency of ordinary least-squares estimation in the linear regression example.

ρ	0.1	0.2	0.3	0.4	0.5
$e(\beta_0)$	0.998	0.992	0.983	0.973	0.963
$e(\beta_1)$	0.997	0.989	0.980	0.970	0.962

ρ	0.6	0.7	0.8	0.9	0.99
$e(\beta_0)$	0.955	0.952	0.956	0.970	0.996
$e(\beta_1)$	0.952	0.955	0.952	0.955	0.961

ρ between successive measurements on any subject. The relative efficiencies $e(\beta_0)$ and $e(\beta_1)$ are given in Table 4.2. We see that ordinary least-squares is tolerably efficient for β_0, but horribly inefficient for β_1 when ρ is large. In this example, efficient estimation of β_1 requires careful balancing of between-subject and within-subject comparisons of the two treatments, and the appropriate balance depends critically on the correlation structure.

Table 4.2. Relative efficiency of ordinary least-squares estimation in the crossover example.

ρ	0.1	0.2	0.3	0.4	0.5
$e(\beta_0)$	0.993	0.974	0.946	0.914	0.880
$e(\beta_1)$	0.987	0.947	0.883	0.797	0.692

ρ	0.6	0.7	0.8	0.9	0.99
$e(\beta_0)$	0.846	0.815	0.788	0.766	0.751
$e(\beta_1)$	0.571	0.438	0.297	0.150	0.015

Even when ordinary least-squares is reasonably efficient, it is clear from the form of $\text{Var}(\tilde{\boldsymbol{\beta}})$ given at (4.3.5) that interval estimation for $\boldsymbol{\beta}$ still requires information about $\sigma^2 V$, the variance matrix of the data. In particular, the usual formula for the variance of the least-squares estimator,

$$\text{Var}(\tilde{\boldsymbol{\beta}}) = \sigma^2 (X'X)^{-1} \tag{4.3.12}$$

assumes that $V = I$, the identity matrix, and can be seriously misleading when this is not so.

A naive use of ordinary least-squares would be to ignore the correlation structure in the data and to base interval estimation for $\boldsymbol{\beta}$ on the variance formula (4.3.12) with σ^2 replaced by its usual estimator, the residual mean square

$$\tilde{\sigma}^2 = (nm - p)^{-1}(\boldsymbol{y} - X\tilde{\boldsymbol{\beta}})'(\boldsymbol{y} - X\tilde{\boldsymbol{\beta}}). \tag{4.3.13}$$

There are two sources of error in this naive approach when $V \neq I$. Firstly, formula (4.3.12) is wrong for $\text{Var}(\tilde{\boldsymbol{\beta}})$. Secondly, $\tilde{\sigma}^2$ is no longer an unbiased estimator for σ^2. To assess the combined effect of these two sources of error, we can compare the diagonal elements of $\text{Var}(\tilde{\boldsymbol{\beta}})$ as given by (4.3.5) with the corresponding diagonal elements of the matrix $\text{E}(\tilde{\sigma}^2)(X'X)^{-1}$. Some numerical examples are given in Chapter 1 and in Diggle (1990, Chapter 3), where the conclusion is that in the presence of positive autocorrelation, naive use of ordinary least squares can seriously over- or underestimate the variance of $\tilde{\boldsymbol{\beta}}$, depending on the design matrix.

4.4 Maximum likelihood estimation under Gaussian assumptions

One strategy for parameter estimation in the general linear model is to consider simultaneous estimation of the parameters of interest, $\boldsymbol{\beta}$, and of the covariance parameters, σ^2 and V_0, using the likelihood function. Recall that V is a block-diagonal matrix with common non-zero blocks V_0. Under the Gaussian assumption (4.2.1), the log-likelihood for observed data $\boldsymbol{y}$ is

$$L(\boldsymbol{\beta}, \sigma^2, V_0) = -0.5\{nm \log(\sigma^2) + m \log(| V_0 |) + \sigma^{-2}(\boldsymbol{y} - X\boldsymbol{\beta})'V^{-1}(\boldsymbol{y} - X\boldsymbol{\beta})\}. \tag{4.4.1}$$

For given V_0, the maximum likelihood estimator for $\boldsymbol{\beta}$ is the weighted least-squares estimator (4.3.6), namely

$$\hat{\boldsymbol{\beta}}(V_0) = (X'V^{-1}X)^{-1}X'V^{-1}\boldsymbol{y}. \tag{4.4.2}$$

Substitution into (4.4.1) gives

$$L\{\hat{\boldsymbol{\beta}}(V_0), \sigma^2, V_0\} = -0.5\{nm \log \sigma^2 + m \log(| V_0 |) + \sigma^{-2}\text{RSS}(V_0)\}, \tag{4.4.3}$$

where

$$\text{RSS}(V_0) = \{\boldsymbol{y} - X\hat{\boldsymbol{\beta}}(V_0)\}'V^{-1}\{\boldsymbol{y} - X\hat{\boldsymbol{\beta}}(V_0)\}.$$

Now, differentiation of (4.4.3) with respect to σ^2 gives the maximum likelihood estimator for σ^2, again for fixed V_0, as

$$\hat{\sigma}^2(V_0) = \text{RSS}(V_0)/(nm). \tag{4.4.4}$$

Substitution of (4.4.3) and (4.4.4) into (4.4.1) now gives a reduced log-likelihood for V_0 which, apart from a constant term, is

$$L_r(V_0) = L\{\hat{\boldsymbol{\beta}}(V_0), \hat{\sigma}^2(V_0), V_0\} = -0.5m\{n \log \text{RSS}(V_0) + \log(| V_0 |)\}. \tag{4.4.5}$$

Finally, maximization of $L_r(V_0)$ yields $\hat{V}_0$ and, by substitution into (4.4.2) and (4.4.4), the maximum likelihood estimators, $\hat{\boldsymbol{\beta}} \equiv \hat{\boldsymbol{\beta}}(\hat{V}_0)$ and $\hat{\sigma}^2 \equiv \hat{\sigma}^2(\hat{V}_0)$.

In general, maximization of (4.4.5) with respect to the distinct elements of V_0 requires numerical optimization techniques. The dimensionality of the optimization problem for V_0 is $\frac{1}{2}n(n-1)$ if we assume a common variance σ^2 at each of the n time-points, and $\frac{1}{2}n(n+1) - 1$ otherwise. Also, the major computational work in evaluating $L(V_0)$ consists of calculating the determinant and inverse of a symmetric, positive-definite, $n \times n$ matrix.

Note that in using maximum likelihood for simultaneous estimation of $\boldsymbol{\beta}, \sigma^2$ and V_0, the form of the design matrix X is involved explicity in the estimation of σ^2 and V_0. One consequence of this is that if we assume the wrong form for X, we may not even get consistent estimators for σ^2 and V_0. To combat this problem, a sensible strategy is to use an over-elaborate model for the mean response profiles in estimating the covariance structure of the data. When the data are from a designed experiment with no continuously varying explanatory variables, we recommend incorporating a separate parameter for the mean response at each time within each treatment, so defining a *saturated* model for the mean response profiles. This guarantees consistent estimates of the covariance structure which we can then use in assessing whether a more economical parametrization of the mean structure can be justified.

In many observational studies this strategy is not feasible. In particular, whenever the data include one or more continuously varying covariates which are thought to affect the mean response, we need to decide whether to incorporate the covariate as a linear effect, using a single additional column in the X matrix, or as a quadratic or more general non-linear effect. In these circumstances, the concept of a saturated model breaks down.

Even when it is feasible, the saturated model strategy runs into another problem. For g treatments and n times of observation it requires $p = ng$ parameters to define the mean structure, and if this number is relatively large, the maximum likelihood estimators for σ^2 and V_0 may be seriously biased. For example, it is well known that when $V_0 = I$, an unbiased estimator for σ^2 requires a divisor $(nm - p)$ rather than nm in (4.4.4), and the problem is exacerbated by the autocorrelation structure in the data.

Maximum likelihood estimation therefore presents us with a conflict. We may need to use a design matrix with a large number of columns so as to obtain consistent estimates of the covariance structure, whereas approximately unbiased estimation requires a small number of columns. If we can be confident that a small number of columns will define an adequate model, the conflict is not a serious one. Otherwise, we need to consider other methods of estimation. One such is the method of restricted maximum likelihood.

4.5 Restricted maximum likelihood estimation

The method of *restricted maximum likelihood*, or REML estimation, was introduced by Patterson and Thompson (1971) as a way of estimating variance components in a general linear model. The objection to the standard maximum likelihood procedure is that it produces biased estimators. This is well known in the case of the general linear model with independent errors,

$$\boldsymbol{Y} \sim MVN(X\boldsymbol{\beta}, \sigma^2 I). \tag{4.5.1}$$

In this case, the maximum likelihood estimator for σ^2 is $\hat{\sigma}^2 = \text{RSS}/(nm)$, where RSS denotes the residual sum of squares, whereas the usual unbiased estimator is $\tilde{\sigma}^2 = \text{RSS}/(nm - p)$, where p is the number of elements of $\boldsymbol{\beta}$. In fact, $\tilde{\sigma}^2$ is the REML estimator for σ^2 in the model (4.5.1).

In the case of the general linear model with dependent errors,

$$\boldsymbol{Y} \sim MVN(X\boldsymbol{\beta}, \sigma^2 V), \tag{4.5.2}$$

the REML estimator is defined as a maximum likelihood estimator based on a linearly transformed set of data $\boldsymbol{Y}^* = A\boldsymbol{Y}$ such that the distribution of $\boldsymbol{Y}^*$ does not depend on $\boldsymbol{\beta}$. One way to achieve this is by taking A to be the matrix which converts $\boldsymbol{Y}$ to ordinary least-squares residuals,

$$A = I - X(X'X)^{-1}X'. \tag{4.5.3}$$

Then, $\boldsymbol{Y}^*$ has a singular multivariate Gaussian distribution with mean zero, whatever the value of $\boldsymbol{\beta}$. To obtain a non-singular distribution, we could use only $nm - p$ rows of the matrix, A, defined at (4.5.3). It turns out that the resulting estimators for σ^2 and V do not depend on *which* rows we use, nor indeed on the particular choice of A: any full-rank matrix with the property that $\text{E}(\boldsymbol{Y}^*) = \boldsymbol{0}$ for all $\boldsymbol{\beta}$ will give the same answer. Furthermore, from an operational viewpoint we do not need to make the transformation from $\boldsymbol{Y}$ to $\boldsymbol{Y}^*$ explicit. The algebraic details are as follows. Less mathematically inclined readers may wish to skim through the next few pages.

For the theoretical development it is convenient to re-absorb σ^2 into V, and write our model for the response vector $\boldsymbol{Y}$ as

$$\boldsymbol{Y} \sim MVN(X\boldsymbol{\beta}, H),$$

where $H \equiv H(\boldsymbol{\alpha})$, the variance matrix of $\boldsymbol{Y}$, is characterized by a vector of parameters $\boldsymbol{\alpha}$. Now, let A be the matrix defined in (4.5.3), and B the $nm \times (nm - p)$ matrix defined by the requirements that $BB' = A$ and $B'B = I$, where I denotes the $(nm-p) \times (nm-p)$ identity matrix. Finally, let $\boldsymbol{Z} = B'\boldsymbol{Y}$.

Now, for fixed $\boldsymbol{\alpha}$ the maximum likelihood estimator for $\boldsymbol{\beta}$ is the generalized least-squares estimator,

$$\hat{\boldsymbol{\beta}} = (X'H^{-1}X)^{-1}X'H^{-1}\boldsymbol{Y} = G\boldsymbol{Y},$$

say. Also, the respective pdf's of $\boldsymbol{Y}$ and $\hat{\boldsymbol{\beta}}$ are

$$f(\boldsymbol{y}) = (2\pi)^{-\frac{1}{2}nm} \mid H \mid^{-\frac{1}{2}} \exp\{-\tfrac{1}{2}(\boldsymbol{y} - X\boldsymbol{\beta})'H^{-1}(\boldsymbol{y} - X\boldsymbol{\beta})\}$$

and

$$g(\hat{\boldsymbol{\beta}}) = (2\pi)^{-\frac{1}{2}p} \mid X'H^{-1}X \mid^{\frac{1}{2}} \exp\{-\tfrac{1}{2}(\hat{\boldsymbol{\beta}} - \boldsymbol{\beta})'(X'H^{-1}X)(\hat{\boldsymbol{\beta}} - \boldsymbol{\beta})\}.$$

Furthermore, $\text{E}(\boldsymbol{Z}) = \boldsymbol{0}$ and $\boldsymbol{Z}$ and $\hat{\boldsymbol{\beta}}$ are independent, whatever the value of $\boldsymbol{\beta}$, as we now show.

Firstly,

$$\mathrm{E}(\boldsymbol{Z}) = B'\mathrm{E}(\boldsymbol{Y}) = B'X\boldsymbol{\beta} = B'BB'X\boldsymbol{\beta},$$

since $B'B = I$. But since $BB' = A$, this gives

$$\mathrm{E}(\boldsymbol{Z}) = B'AX\boldsymbol{\beta},$$

and

$$AX = \{I - X(X'X)^{-1}X'\}X = X - X = 0,$$

hence $\mathrm{E}(\boldsymbol{Z}) = \mathbf{0}$ as required.

Secondly,

$$\begin{aligned}\mathrm{Cov}(\boldsymbol{Z}, \hat{\boldsymbol{\beta}}) &= \mathrm{E}\{\boldsymbol{Z}(\hat{\boldsymbol{\beta}} - \boldsymbol{\beta})'\} \\ &= \mathrm{E}\{B'\boldsymbol{Y}(\boldsymbol{Y}'G' - \boldsymbol{\beta}')\} \\ &= B'\mathrm{E}(\boldsymbol{Y}\boldsymbol{Y}')G' - B'\mathrm{E}(\boldsymbol{Y})\boldsymbol{\beta}' \\ &= B'\{\mathrm{Var}(\boldsymbol{Y}) + \mathrm{E}(\boldsymbol{Y})\mathrm{E}(\boldsymbol{Y})'\}G' - B'\mathrm{E}(\boldsymbol{Y})\boldsymbol{\beta}'.\end{aligned}$$

Now, substituting $\mathrm{E}(\boldsymbol{Y}) = X\boldsymbol{\beta}$ into this last expression gives

$$\mathrm{Cov}(\boldsymbol{Z}, \hat{\boldsymbol{\beta}}) = B'(H + X\boldsymbol{\beta}\boldsymbol{\beta}'X')G' - B'X\boldsymbol{\beta}\boldsymbol{\beta}'.$$

Also,

$$\begin{aligned}X'G' &= X'H^{-1}X(X'H^{-1}X)^{-1} \\ &= I\end{aligned}$$

and

$$\begin{aligned}B'HG' &= B'HH^{-1}X(X'H^{-1}X)^{-1} \\ &= B'X(X'H^{-1}X)^{-1} = 0,\end{aligned}$$

because $B'X = B'AX = 0$ as in the proof that $\mathrm{E}(\boldsymbol{Z}) = 0$. It follows that $\mathrm{Cov}(\boldsymbol{Z}, \hat{\boldsymbol{\beta}}) = B'X\boldsymbol{\beta}\boldsymbol{\beta}' - B'X\boldsymbol{\beta}\boldsymbol{\beta}' = 0$. Finally, in the multivariate Gaussian setting, zero covariance is equivalent to independence. It follows that the algebraic form of the (singular) multivariate Gaussian pdf of $\boldsymbol{Z}$, expressed in terms of $\boldsymbol{Y}$, is proportional to the ratio $f(\boldsymbol{y})/g(\hat{\boldsymbol{\beta}})$. To obtain the explicit form of this ratio, we use the following standard result for the general linear model, that

$$(\boldsymbol{y} - X\boldsymbol{\beta})'H^{-1}(\boldsymbol{y} - X\boldsymbol{\beta}) = (\boldsymbol{y} - X\hat{\boldsymbol{\beta}})'H^{-1}(\boldsymbol{y} - X\hat{\boldsymbol{\beta}}) + (\hat{\boldsymbol{\beta}} - \boldsymbol{\beta})'(X'H^{-1}X)(\hat{\boldsymbol{\beta}} - \boldsymbol{\beta}).$$

Then, the pdf of $\boldsymbol{Z} = B'\boldsymbol{Y}$ is proportional to

$$\frac{f(\boldsymbol{y})}{g(\hat{\boldsymbol{\beta}})} = (2\pi)^{-\frac{1}{2}(nm-p)} \mid H \mid^{-\frac{1}{2}} \mid X'H^{-1}X \mid^{-\frac{1}{2}} \times \exp\{-\tfrac{1}{2}(\boldsymbol{y} - X\hat{\boldsymbol{\beta}})'H^{-1}(\boldsymbol{y} - X\hat{\boldsymbol{\beta}})\}, \qquad (4.5.4)$$

where the omitted constant of proportionality is the Jacobian of the transformation from $\boldsymbol{Y}$ to $(\boldsymbol{Z}, \hat{\boldsymbol{\beta}})$. Harville (1974) shows that the Jacobian reduces to $\mid (X'X) \mid^{-\frac{1}{2}}$, which does not depend on any of the parameters in the model and can therefore be ignored for making inferences about $\boldsymbol{\alpha}$ or $\boldsymbol{\beta}$. Note that the right hand side of (4.5.4) is independent of A, and the same result would therefore hold for any $\boldsymbol{Z}$ such that $E(\boldsymbol{Z}) = 0$ and $\text{Cov}(\boldsymbol{Z}, \hat{\boldsymbol{\beta}}) = 0$.

The practical implication of (4.5.4) is that the REML estimator, $\tilde{\boldsymbol{\alpha}}$, maximizes the log-likelihood

$$L^*(\boldsymbol{\alpha}) = -\tfrac{1}{2}\log \mid H \mid -\tfrac{1}{2}\log \mid X'H^{-1}X \mid -\tfrac{1}{2}(\boldsymbol{y} - X\hat{\boldsymbol{\beta}})'H^{-1}(\boldsymbol{y} - X\hat{\boldsymbol{\beta}}),$$

whereas the maximum likelihood estimator $\hat{\boldsymbol{\alpha}}$ maximizes

$$L(\boldsymbol{\alpha}) = -\tfrac{1}{2}\log \mid H \mid -\tfrac{1}{2}(\boldsymbol{y} - X\hat{\boldsymbol{\beta}})'H^{-1}(\boldsymbol{y} - X\hat{\boldsymbol{\beta}}).$$

It follows from this last result that the algorithm to implement REML estimation for the model (4.5.2) incorporates only a simple modification to the maximum likelihood algorithm derived in Section 4.4. Recall that we consider m units with n measurements per unit and that $\sigma^2 V$ is a block-diagonal matrix with non-zero $n \times n$ blocks $\sigma^2 V_0$ representing the variance matrix of the measurements on any one unit. Also, for given V_0 we write

$$\hat{\boldsymbol{\beta}}(V_0) = (X'V^{-1}X)^{-1}X'V^{-1}\boldsymbol{y} \qquad (4.5.5)$$

and

$$\text{RSS}(V_0) = \{\boldsymbol{y} - X\hat{\boldsymbol{\beta}}(V_0)\}'V^{-1}\{\boldsymbol{y} - X\hat{\boldsymbol{\beta}}(V_0)\}.$$

Then, the REML estimator for σ^2 is

$$\tilde{\sigma}^2(V_0) = \text{RSS}(V_0)/(nm - p), \qquad (4.5.6)$$

where p is the number of elements of $\boldsymbol{\beta}$. The REML estimator for V_0 maximizes the reduced log-likelihood

$$L^*(V_0) = -\tfrac{1}{2}m\{n \log \text{RSS}(V_0) + \log(\mid V_0 \mid)\} - \tfrac{1}{2}\log(\mid X'V^{-1}X \mid). \qquad (4.5.7)$$

Finally, substitution of the resulting estimator $\tilde{V}_0$ into (4.5.5) and (4.5.6) gives the REML estimators $\tilde{\boldsymbol{\beta}} = \hat{\boldsymbol{\beta}}(\tilde{V}_0)$ and $\tilde{\sigma}^2 = \tilde{\sigma}^2(\tilde{V}_0)$.

It is instructive to compare the functions $L(V_0)$ and $L^*(V_0)$ defined at (4.4.5) and (4.5.7), respectively. The difference is the addition of the term $\frac{1}{2}\log(| X'V^{-1}X |)$ in (4.5.7). Note that $X'V^{-1}X$ is a $p \times p$ matrix, so this term is typically of order p, whereas $L(V_0)$ is of order nm suggesting, correctly, that the distinction between maximum likelihood and REML estimation is important only when p is relatively large. This is, of course, precisely the situation which can arise when we use a saturated or nearly saturated model for the mean response so as to obtain consistent estimation of the covariance structure, as we recommended in Section 4.4. Also, the order-of-magnitude argument breaks down when V is near-singular, that is, when there are strong correlations amongst the responses on an individual experimental unit, as is not uncommon in longitudinal studies.

Many authors have discussed the relative merits of maximum likelihood and REML estimators for covariance parameters. Early work, summarized by Patterson and Thompson (1971), concentrated on the estimation of variance components in designed experiments. Harville (1974) gives a Bayesian interpretation. More recently, Cullis and McGilchrist (1990) and Verbyla and Cullis (1990) apply REML in the longitudinal data setting, whilst Tunnicliffe-Wilson (1989) uses it for time series estimation under the name of marginal likelihood. One of Tunnicliffe-Wilson's examples shows very clearly how REML copes much more effectively with a near-singular variance matrix than does maximum likelihood estimation. General theoretical results are harder to find, because the two methods are asymptotically equivalent as either or both of m and n tend to infinity for fixed p. When p tends to infinity, comparisons unequivocally favour REML; for example, the usual procedure for a paired-sample t-test can be interpreted as a REML procedure with $n = 2$ observations on each pair and $p = m + 1$ parameters to define the m pair means and the single treatment effect.

In summary, maximum likelihood and REML estimators will often give very similar results. However, when they do differ substantially, our view is that REML estimators are preferable. Henceforth, we use the 'hat' notation to refer to maximum likelihood *or* REML estimators, except when the context does not make it clear which is intended.

4.6 Robust estimation of standard errors

The essential idea of the robust approach to inference for $\boldsymbol{\beta}$ is to use the generalized least-squares estimator $\tilde{\boldsymbol{\beta}}_W$ defined by (4.3.2),

$$\tilde{\beta}_W = (X'WX)^{-1}X'W\boldsymbol{y}, \tag{4.6.1}$$

in conjunction with an estimated variance matrix

$$\hat{R}_W = \{(X'WX)^{-1}X'W\}\hat{V}\{WX(X'WX)^{-1}\}, \tag{4.6.2}$$

where $\hat{V}$ is consistent for V whatever the true covariance structure. Note that in (4.6.2) we have re-absorbed the scale parameter σ^2 into V. For inference, we proceed as if

$$\tilde{\boldsymbol{\beta}}_W \sim MVN(\boldsymbol{\beta}, \hat{R}_W). \tag{4.6.3}$$

In this approach, we call W^{-1} the *working covariance matrix*, to distinguish it from the true covariance matrix, V. Typically, we use a relatively simple form for W^{-1} which we hope captures the qualitative structure of V. However, the crucial difference between this and a parametric modelling approach is that a poor choice of W will affect only the efficiency of our inferences for $\boldsymbol{\beta}$, not their validity. In particular, confidence intervals and tests of hypotheses derived from (4.6.3) will be asymptotically correct whatever the true form of V. This idea has a long history, dating back at least to Huber (1967). More recent references are White (1982), Liang and Zeger (1986) and Royall (1986).

The simplest possible implementation is to use the ordinary least-squares estimator $\tilde{\boldsymbol{\beta}}$. This is equivalent to assuming that the measurements within a subject are uncorrelated, since we are then using $W^{-1} = I$ for the working covariance matrix. Note, incidentally, that equations (4.3.2) and (4.3.3) do not change if the elements of W are multiplied by any constant, so that it would strictly be more correct to say that W^{-1} is *proportional* to the working covariance matrix. For data with a smoothly decaying autocorrelation structure, a more efficient implementation might be obtained from a block-diagonal W^{-1}, with non-zero elements of the form $\exp\{-c \mid t_j - t_k \mid\}$, where c is a positive constant chosen roughly to match the rate of decay anticipated for the actual autocovariances of the data. In our experience, more elaborate choices for W^{-1} are often unnecessary.

For designed experiments in which we choose to fit the saturated model for the mean response, the explicit form of the robust estimator $\hat{V}$ required in (4.6.2) is obtained using the REML principle as follows. Suppose that measurements are made at each of n time-points t_j on m_h experimental units in the hth of g experimental treatment groups. Write the complete set of measurements as

$$y_{hij}, \quad h = 1, \ldots, g; \quad i = 1, \ldots m_h; \quad j = 1, \ldots, n.$$

The saturated model for the mean response is

$$\mathrm{E}(Y_{hij}) = \mu_{hj}, \; h = 1, \ldots, g; \; j = 1, \ldots, n,$$

and a saturated model for the covariance structure is that $V = \mathrm{Var}(\boldsymbol{Y})$ is block-diagonal, with all non-zero blocks equal to V_0, a positive definite but otherwise arbitrary $n \times n$ matrix.

This model falls within the general framework of Section 4.5, with a rather special form for the design matrix X. For example, with $g = 2$ treatments and replications $m_1 = 2$ and $m_2 = 3$ we have

$$X = \begin{bmatrix} I & O \\ I & O \\ O & I \\ O & I \\ O & I \end{bmatrix},$$

where I and O are, respectively, the $n \times n$ identity matrix and the $n \times n$ matrix of zeros. The extension to general g and $m_1, \ldots, m_g$ is obvious. It is then a straightforward exercise to deduce that the estimators for the μ_{hj} are the corresponding sample means,

$$\hat{\mu}_{hj} = m_h^{-1} \sum_{i=1}^{m_h} y_{hij},$$

and that the REML estimator for V_0 is

$$\hat{V}_0 = \left(\sum_{i=1}^{g} m_i - g \right)^{-1} \sum_{h=1}^{g} \sum_{i=1}^{m_h} (\boldsymbol{y}_{hi} - \hat{\boldsymbol{\mu}}_h)(\boldsymbol{y}_{hi} - \hat{\boldsymbol{\mu}}_h)', \qquad (4.6.4)$$

where $\boldsymbol{y}_{hi} = (y_{hi1}, \ldots, y_{hin})'$ and $\hat{\boldsymbol{\mu}}_h = (\hat{\mu}_{h1}, \ldots, \hat{\mu}_{hn})'$. Then, the required estimate $\hat{V}$ is the block-diagonal matrix with non-zero blocks $\hat{V}_0$. Note that at this stage, the *interpretation* of the ordinary least-squares fit would be highly problematical, with abundant scope for partial confounding between time-varying covariate effects and time effects *per se*. The sole purpose of the ordinary least-squares fit is to provide a consistent estimate of V_0.

When the saturated model strategy is not feasible, typically when the data are from observational studies with continuously varying covariates, it is usually no longer possible to obtain an explicit expression for the REML estimate of V_0. However, the same basic idea applies. We make no assumption about the form of V_0, use an X matrix corresponding to the most elaborate model we are prepared to entertain for the mean response, and obtain the REML estimate, $\hat{V}_0$, by numerical maximization of (4.5.7).

To make robust inferences for $\boldsymbol{\beta}$, we now substitute $\hat{V}$ into (4.6.2), and use (4.6.3). Note that for these inferences about $\boldsymbol{\beta}$ we will typically use an X matrix with many fewer than the ng columns of the X matrix corresponding to the saturated model. Furthermore, if we want to test linear hypotheses about $\boldsymbol{\beta}$ within this model we can use the standard approach for the general linear model. Suppose, for example, we wish to test the hypothesis $Q\boldsymbol{\beta} = \boldsymbol{0}$, where Q is a full-rank $q \times p$ matrix for some $q < p$. Then we deduce from (4.6.3) that

$$Q\hat{\boldsymbol{\beta}}_W \sim MVN(Q\boldsymbol{\beta}, Q\hat{R}_W Q').$$

An appropriate test statistic for the hypothesis that $Q\boldsymbol{\beta} = \mathbf{0}$ is

$$T = \hat{\boldsymbol{\beta}}'_W Q'(Q\hat{R}_W Q')^{-1} Q\hat{\boldsymbol{\beta}}_W, \tag{4.6.5}$$

and the approximate null sampling distribution of T is chi-squared on q degrees of freedom.

When measurement times are not common to all units, the robust approach can still be useful in the following, modified form. The required variance matrix V is still block-diagonal, but the non-zero blocks corresponding to the sets of measurements within units are no longer constant between units. Write V_{0i} for the $n_i \times n_i$ variance matrix of the set of measurements on the ith unit and $\boldsymbol{\mu}_i$ for the mean vector of these measurements. Estimate $\boldsymbol{\mu}_i$ by $\hat{\boldsymbol{\mu}}_i$, the ordinary least-squares estimates from the most complicated model we are prepared to entertain for the mean response. Then, an estimate of V_{0i} is

$$\hat{V}_{0i} = (\boldsymbol{y}_i - \hat{\boldsymbol{\mu}}_i)(\boldsymbol{y}_i - \hat{\boldsymbol{\mu}}_i)'. \tag{4.6.6}$$

Our robust estimate of V is then $\hat{V}$, the block-diagonal matrix with non-zero blocks $\hat{V}_{0i}$ defined by (4.6.6).

A few comments on this procedure are in order. Firstly, the choice of a 'most complicated model' for the mean response is less clear-cut than in the case of a common set of measurement times for all units. At one extreme, if sets of measurement times are essentially unique for units as in the CD4+ data of example 1 in Section 1.2, a saturated treatments by times model would result in $\hat{\boldsymbol{\mu}}_i = \boldsymbol{y}_i$ and $\hat{V}_{0i} = O$, a matrix of zeros. In this situation it is necessary, and sensible, to incorporate some smoothness into the assumed form for the variation in mean response over time. But it is not entirely obvious precisely how to do this in practice. Fortunately, in many applications the replication of measurement times between units is sufficient to allow a sensible fit to a saturated treatments by times model provided, as always, that there are no important time-varying covariates.

A second comment is that $\hat{V}_{0i}$ defined by (4.6.6) is a *terrible* estimator for V_{0i}! For example, it is singular of rank 1, so the implied $\hat{\boldsymbol{V}}$ will be singular of rank m, the number of units. This is less serious than it sounds, because we only use $\hat{V}$ in (4.6.2) to evaluate the $p \times p$ matrix $\hat{R}_W$. The matrix multiplication in (4.6.2) achieves, in effect, the same ends as the explicit averaging over replicates in (4.6.4), to produce a useful estimate of R_W. Note that this argument breaks down unless p, the number of parameters defining the mean response for the most complicated model, is much less than m, the number of experimental units. It also breaks down when, in effect, any one regression parameter is estimated from only a small part of the data. For example, if (4.6.6) is used in conjunction with a saturated model for the mean response in a designed esperiment, it corresponds to using the within-treatment sample variance matrices to estimate the corresponding blocks of V_C. This is in contrast to (4.6.4), which pools the sample variance matrices from all g treatment groups to achieve a more precise estimate of V_0 under the declared assumption that V_0 is independent of treatment. For a further discussion, see White (1982), Zeger, Liang and Self (1985), Liang and Zeger (1986) and Zeger and Liang (1986).

Example 4.1. Growth of Sitka spruce with and without ozone
We now use the approach described above to analyse the data from Example 1.3. We recall that the data consist of measurements on 79 sitka spruce trees over two growing seasons. The trees were grown in four controlled environment chambers, of which the first two, containing 27 trees each, were treated with introduced ozone at 70 ppb whilst the remaining two, containing 12 and 13 trees, were controls. As the response variable we use a log-size measurement, $y = \log(hd^2)$ where h denotes height and d denotes stem diameter. The question of scientific interest concerns the effect of ozone on the growth pattern, rather than the growth pattern itself. Chamber effects were thought to be negligible and we shall see that this is borne out in the results.

Figure 4.2 shows the observed mean response in each of the four chambers. In the first year, the four curves are initially close but diverge progressively, with the control chambers showing a higher mean response than the treated chambers. In the second year, the curves are roughly parallel.

Turning to the covariance structure, we computed the REML estimates, $\hat{V}_0$, separately for each of the two years using a saturated model for the means in each case. That is, we used (4.6.4) with $g = 4$ in each case, and $n = 5$ and 8 for the 1988 and 1989 data, respectively. The resulting estimates are, for 1988,

$$\hat{V}_0 = \begin{matrix} 0.445 & 0.404 & 0.373 & 0.370 & 0.370 \\ & 0.397 & 0.372 & 0.375 & 0.376 \\ & & 0.370 & 0.372 & 0.371 \\ & & & 0.401 & 0.401 \\ & & & & 0.410 \end{matrix}$$

and, for 1989,

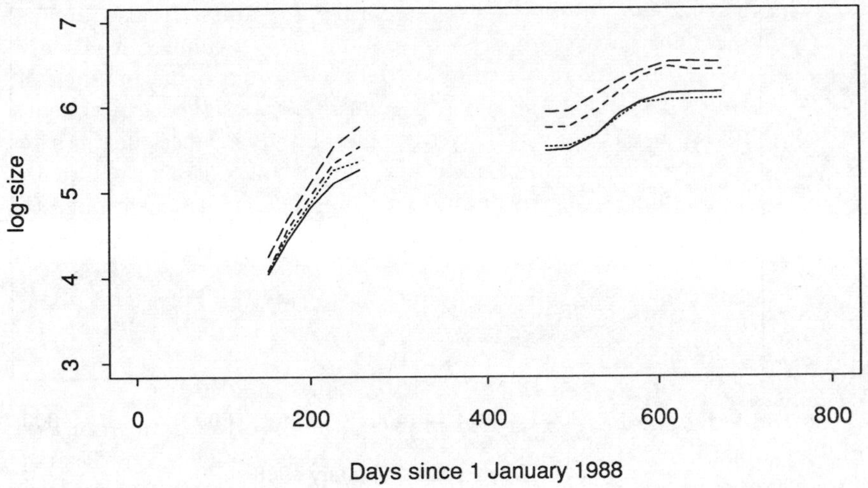

Fig. 4.2. Sitka spruce data: mean response profiles in each of the four growth chambers.

$$\hat{V}_0 = \begin{matrix} 0.457 & 0.454 & 0.427 & 0.417 & 0.433 & 0.422 & 0.408 & 0.418 \\ & 0.451 & 0.425 & 0.415 & 0.431 & 0.420 & 0.406 & 0.416 \\ & & 0.409 & 0.398 & 0.412 & 0.404 & 0.390 & 0.401 \\ & & & 0.396 & 0.410 & 0.402 & 0.388 & 0.400 \\ & & & & 0.434 & 0.422 & 0.405 & 0.418 \\ & & & & & 0.415 & 0.399 & 0.412 \\ & & & & & & 0.394 & 0.402 \\ & & & & & & & 0.416 \end{matrix}$$

The most striking feature of each matrix $\hat{V}_0$ is its near-singularity. This arises because the dominant component of variation in the data is a random intercept between trees, as a result of which the variance structure approximates to that of the uniform correlation model 5.2.2 with ρ close to 1.

Using the estimated covariance matrices $\hat{V}_0$, we can construct pointwise standard errors for the observed mean responses in the four chambers. These standard errors vary between about 0.1 and 0.2, suggesting that differences between chambers within experimental groups are negligible. We therefore proceed to model the data ignoring chambers.

Figure 4.3 shows the observed mean response in each of the two treatment groups. The two curves diverge progressively during 1988, the second year of the experiment, and are approximately parallel during 1989. The overall growth pattern is clearly non-linear, nor could it be well approximated by a low-order polynomial in time. For this reason, and because

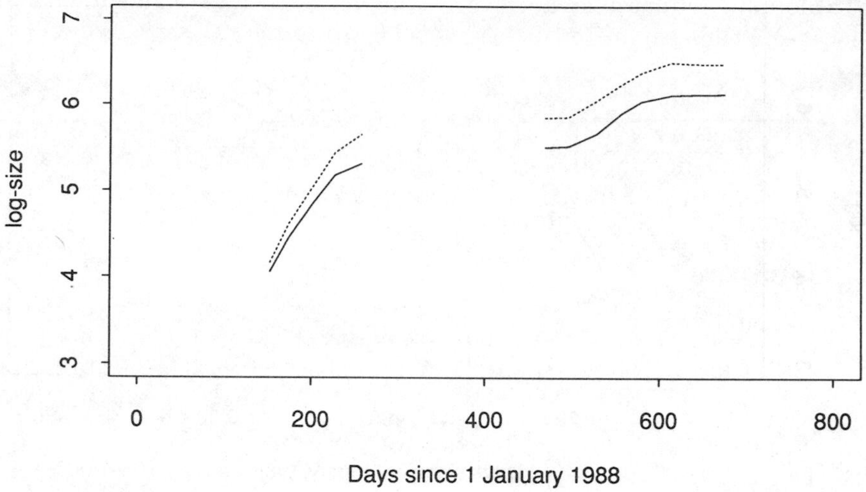

Fig. 4.3. Sitka spruce data : mean response profiles in control and ozone-treated groups. - - - - : control; —— : ozone-treated.

the primary inferential focus is on the ozone effect, we make no attempt to model the overall growth pattern parametrically. Instead, we simply use a separate parameter, β_j say, for the treatment mean response at the jth time-point and concentrate our modelling efforts on the control versus treatment contrast.

For the 1988 data, we assume that this contrast is linear in time. Thus, if $\mu_1(t)$ and $\mu_2(t)$ represent the mean response at time t for treated and control trees, respectively, then

$$\mu_1(t_j) = \beta_j, \quad j = 1, \ldots, 5$$

and

$$\mu_2(t_j) = \beta_j + \tau + \gamma t_j, \;\; j = 1, \ldots, 5. \tag{4.6.7}$$

To estimate the parameters β_j, τ and γ in (4.6.7) we use ordinary least-squares, that is, $W = I$ in (4.6.1). To estimate the variance matrix of the parameter estimates, we use the estimated $\hat{V}_0$ in (4.6.2). The resulting estimates and standard errors are given in Table 4.3(a). The hypothesis of no treatment effect is that $\tau = \gamma = 0$. The test statistic for this hypothesis, derived from (4.6.5), is $T = 9.79$ on two degrees of freedom, corresponding to $p = 0.007$; this constitutes strong evidence of a negative treatment effect, that is, ozone suppresses growth. Note that for the linear model-fitting, we used a scaled time variable, $x = t/100$, where t is measured in days since 1 January 1988.

Table 4.3. Ordinary least-squares estimates, and robust standard errors, for mean value parameters in the model fitted to the sitka spruce data

(a) 1988

Parameter	β_1	β_2	β_3	β_4	β_5	τ	γ
Estimate	4.060	4.470	4.483	5.179	5.316	-0.221	0.213
Std. Error	0.090	0.086	0.083	0.086	0.087	0.220	0.077

(b) 1989

β_1	β_2	β_3	β_4	β_5	β_6	β_7	β_8	τ
5.504	5.516	5.679	5.901	6.040	6.127	6.128	6.130	0.354
0.091	0.090	0.087	0.086	0.089	0.088	0.086	0.088	0.156

For the 1989 data, we assume that the control versus treatment contrast is constant in time. Thus

$$\mu_1(t_j) = \beta_j, \quad j = 6, \ldots, 13$$

and

$$\mu_2(t_j) = \beta_j + \tau, \quad j = 6, \ldots, 13. \tag{4.6.8}$$

To estimate the parameters in (4.6.8) and their standard errors, we again use $W = I$ and the estimated variance matrix $\hat{V}_0$ in (4.6.2). The resulting values are given in Table 4.3(b). The hypothesis of no treatment effect is that $\tau = 0$. The test statistic derived from (4.6.5) is $T = 5.15$ on one degree of freedom, corresponding to $p = 0.023$; alternatively, an approximate 95 per cent confidence interval for τ is 0.354 ± 0.306.

Figure 4.4 shows the estimated treatment mean response, $\hat{\mu}_1(t_j) = \hat{\beta}_j$, together with pointwise 95 per cent confidence limits calculated as plus and minus two estimated standard errors; note that under the assumed model, both control and treated data contain information on the β_j. Figure 4.5 shows the observed mean differences between control and treated trees, with the estimated control versus treatment contrast. The model appears to fit the data well, and the analysis provides strong evidence that ozone suppresses early growth.

Incidentally, we obtained virtually identical results for the Sitka spruce data using an exponential working covariance matrix, in which the $(j,k)th$ element of W^{-1} is equal to $0.9^{|j-k|}$. This agrees with our general experience that the precise choice among reasonable working covariance matrices is usually not critical.

The robust approach described in this section is extremely simple to implement. The REML estimates of the covariance structure are simple to compute provided that the experimental design allows the fitting of a

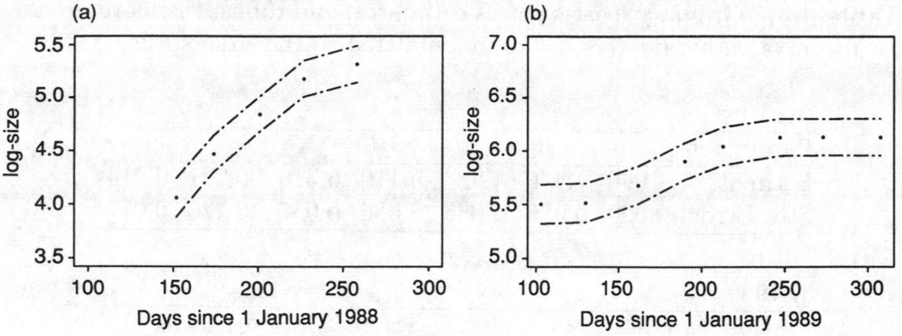

Fig. 4.4. Sitka spruce data: estimated response profiles, shown as solid dots, and 95 per cent pointwise confidence limits for the ozone-treated group: (a) 1988 growing season; (b) 1989 growing season

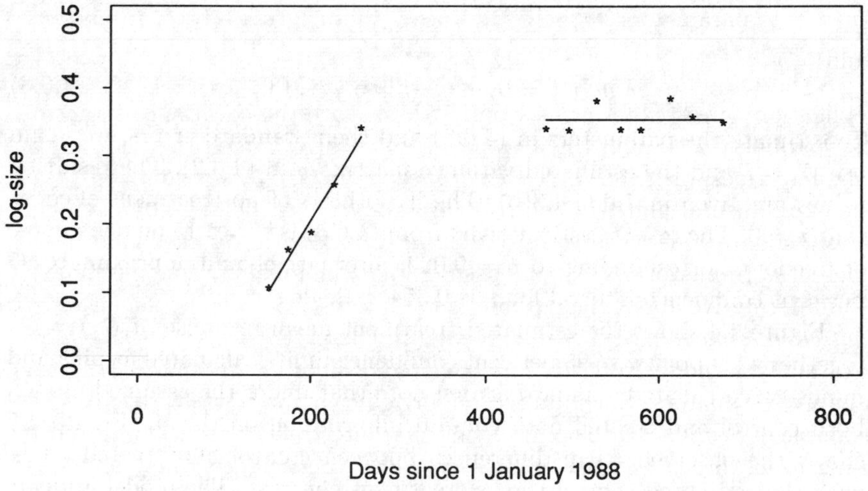

Fig. 4.5. Sitka spruce data: observed and fitted differences in mean response profiles between the control and ozone-treated groups. $\star$: observed; —— : fitted.

saturated model for the mean response, and the remaining calculations involve only standard matrix manipulations. By design, consistent inferences for the mean response parameters follow from a correct specification of the mean structure, whatever the true covariance structure.

However, there are good reasons why we should nevertheless consider careful, explicit modelling of the covariance structure.

One argument, that of efficiency of estimation, has two distinct strands. Firstly, the theoretically optimal weighted least-squares estimate uses a weight matrix whose inverse is proportional to the true covariance matrix, so it would seem reasonable to use the data to estimate this optimal weight matrix. Secondly, when there are n measurements per experimental unit, the robust approach uses $\frac{1}{2}n(n+1)$ parameters to describe the covariance matrix, all of which must be estimated from the data. In contrast, the true covariance structure may involve many fewer parameters, which can themselves be estimated more accurately than the unconstrained variance matrix. Taken together, these arguments constitute a case for likelihood-based inference within an explicit parametric model for both the mean and covariance structure. However, as we have already argued, the resulting gains in efficiency are often modest, and may well be outweighed by the potential loss of consistency when the covariance structure is wrongly specified.

A more basic practical argument concerns the number, n, of measurements per experimental unit. When n is large, the objection to estimating $\frac{1}{2}n(n+1)$ parameters in the covariance structure gains force. In extreme cases, n can exceed the available replication.

A third argument concerns missing values. The robust approach uses the replication across experimental units to estimate the covariance structure non-parametrically, and this becomes problematical when there are many missing values, or when the times of measurement are not common to all the experimental units. In contrast, a parametric modelling approach can accommodate general patterns of missing values within a likelihood-based inferential framework. This does assume that the missing value mechanism is uninformative about the parameters of interest; the whole issue of *informative missing values* will be addressed in Chapter 11.

Finally, although robustly implemented ordinary least-squares will usually give estimators which are reasonably efficient, we saw in the crossover example of Section 4.3 that this is not guaranteed. This argues in favour of using robust *weighted* least-squares, in which the weighting matrix is chosen to reflect qualitatively reasonable assumptions about the correlation structure.

To summarize, the robust approach is usually satisfactory when the data consist of short, essentially complete, sequences of measurements observed at a common set of times on many experimental units, and care is taken in the choice of a working correlation matrix. In other circumstances, it is worth considering a parametric modelling approach. The next chapter takes up this theme.

5
Parametric models for covariance structure

5.1 Introduction

In this chapter, we continue with the general linear model (4.2.1) for the data, but assume that the covariance structure of the sequence of measurements on each experimental unit is to be specified by the values of a few unknown parameters. Examples include the uniform correlation model (4.2.2) and the exponential correlation model (4.2.4), each of which uses two parameters to define the covariance structure.

As discussed in Section 4.6, a parametric modelling approach is particularly useful for data in which the measurements on different units are not made at a common set of times. For this reason, we use a slightly more general notation than in Chapter 4. Let $\boldsymbol{y}_i = (y_{i1}, \ldots, y_{in_i})$ be the vector of n_i measurements on the ith unit and $\boldsymbol{t}_i = (t_{i1}, \ldots, t_{in_i})$ the corresponding set of times at which these measurements are made. If there are m units altogether, write $\boldsymbol{y} = (\boldsymbol{y}_1, \ldots, \boldsymbol{y}_m)$, $\boldsymbol{t} = (\boldsymbol{t}_1, \ldots, \boldsymbol{t}_m)$ and $N = \Sigma_{i=1}^m n_i$.

We assume that the $\boldsymbol{y}_i$ are realizations of mutually independent Gaussian random vectors $\boldsymbol{Y}_i$, with

$$\boldsymbol{Y}_i \sim MVN\{X_i\boldsymbol{\beta}, V_i(\boldsymbol{t}_i, \boldsymbol{\alpha})\}. \tag{5.1.1}$$

In (5.1.1), X_i is an $n_i \times p$ matrix of explanatory variables. The unknown parameters are $\boldsymbol{\beta}$ of dimension p, and $\boldsymbol{\alpha}$ of dimension q. Note that the mean and covariance structures are parametrized separately. Also, when convenient we shall write the model for the entire set of data $\boldsymbol{y}$ as

$$\boldsymbol{Y} \sim MVN\{X\boldsymbol{\beta}, V(\boldsymbol{t}, \boldsymbol{\alpha})\}. \tag{5.1.2}$$

In (5.1.2), the $N \times p$ matrix X is obtained by stacking the unit-specific matrices X_i, and the $N \times N$ matrix $V(\cdot)$ is block-diagonal, with non-zero blocks $V_i(\cdot)$.

Our notation emphasizes that the natural setting for most longitudinal data is in continuous time. Because of this, we shall derive specific models by assuming that the sequences, Y_{ij}, $j = 1, \ldots, n_i$, are sampled from independent copies of an underlying continuous-time stochastic process, $\{Y(t),\ t \in R\}$. Thus, $Y_{ij} = Y_i(t_{ij}),\ j = 1, \ldots, n_i;\ i = 1, \ldots, m$. In the next section, we give examples of models which fall within the general framework of (5.1.1), and show how different kinds of stationary and non-stationary behaviour arise naturally. In later sections, we develop methods for fitting the models to data, describe several applications, and consider *semi-parametric* models in which the linear model for the mean response is replaced by a non-parametric specification.

The principal tools which we shall use to describe the properties of each model for the stochastic process $\{Y(t)\}$ are the covariance function and its close relation, the variogram. Recall from Section 3.4 that the *variogram* of a stochastic process $\{Y(t)\}$ is the function

$$\gamma(u) = \tfrac{1}{2}\mathrm{E}\left[\{Y(t) - Y(t-u)\}^2\right], \quad u \geq 0.$$

For a *stationary process* $Y(t)$, if $\rho(u)$ denotes the correlation between $Y(t)$ and $Y(t-u)$, and $\sigma^2 = \mathrm{Var}\{Y(t)\}$, then

$$\gamma(u) = \sigma^2\{1 - \rho(u)\}.$$

5.2 Models

In order to develop a useful set of models we need to understand, at least qualitatively, what are the likely sources of random variation in longitudinal data. In the earlier chapters, we have shown several examples of the kinds of random variation which occur in practice. Our experience with these and many other sets of data leads us to the view that we would wish to be able to include in our models at least three qualitatively different sources of random variation.

- **Random effects**: When units are sampled at random from a population, various aspects of their behaviour may show stochastic variation between units. Perhaps the simplest example of this is when the general level of the response profile varies between units, that is, some units are intrinsically high responders, others low responders. This was very evident, for example, in the data of Example 1.3 concerning the growth of Sitka spruce trees.
- **Serial correlation**: At least part of any unit's observed measurement profile may be a response to time-varying stochastic processes operating within that unit. For example, in the data of Example 1.4 concerning the protein content of milk samples, the measured sequences of protein contents must, to some extent, reflect the biochemical processes operating within each cow. This type of stochastic

variation results in a correlation between pairs of measurements on the same unit which depends on the time separation between the pair of measurements. Typically, the correlation becomes weaker as the time separation increases.

- **Measurement error**: Especially when the individual measurements involve some kind of sampling *within* units, the measurement process may itself add a component of variation to the data. The data on protein content of milk samples again illustrate the point. Here, two samples taken simultaneously from a cow would have different measured protein contents, because the measurement process involves an assay technique which itself introduces a component of random variation.

There are many different ways in which these qualitative features could be incorporated into specific models. The following additive formulation is tractable and, in our experience, useful.

Firstly, we make explicit the separation between mean and covariance structures by writing (5.1.2) as

$$\boldsymbol{Y} = X\boldsymbol{\beta} + \boldsymbol{\epsilon}. \tag{5.2.1}$$

It follows that

$$\boldsymbol{\epsilon} \sim MVN\{0, V(\boldsymbol{t}, \boldsymbol{\alpha})\}. \tag{5.2.2}$$

Now, using ϵ_{ij} to denote the element of $\boldsymbol{\epsilon}$ which corresponds to the jth measurement on the ith unit, we assume an additive decomposition of ϵ_{ij} into random effects, serially correlated variation and measurement error. This can be expressed formally as

$$\epsilon_{ij} = \boldsymbol{d}'_{ij}\boldsymbol{U}_i + W_i(t_{ij}) + Z_{ij}. \tag{5.2.3}$$

In this decomposition, the Z_{ij} are a set of N mutually independent Gaussian random variables, each with mean zero and variance τ^2. The $\boldsymbol{U}_i$ are a set of m mutually independent r-element Gaussian random vectors, each with mean vector zero and covariance matrix G, say. The $\boldsymbol{d}_{ij}$ are r-element vectors of explanatory variables attached to individual measurements. The $W_i(t_{ij})$ are sampled from m independent copies of a stationary Gaussian process with mean zero, variance σ^2 and correlation function $\rho(u)$. Note that the $\boldsymbol{U}_i$, the $\{W_i(t_{ij})\}$ and the Z_{ij} correspond to random effects, serial correlation and measurement error, respectively.

In applications, the assumed additive structure may be more reasonable after a transformation of the data. For example, a logarithmic transformation would convert an underlying multiplicative structure to an additive one.

To say anything more useful about (5.2.3) we need some more notation. Write $\boldsymbol{\epsilon}_i = (\epsilon_{i1}, \ldots, \epsilon_{in_i})$ for the vector of random variables, ϵ_{ij}, associated with the ith unit. Let D_i be the $n_i \times r$ matrix with jth row $\boldsymbol{d}_{ij}$. Let H_i be the $n_i \times n_i$ matrix with $(j,k)th$ element $h_{ijk} = \rho(\mid t_{ij} - t_{ik} \mid)$, that is, h_{ijk} is the correlation between $W_i(t_{ij})$ and $W_i(t_{ik})$. Finally, let I_i be the $n_i \times n_i$ identity matrix. Then, the covariance matrix of $\boldsymbol{\epsilon}_i$ is

$$\mathrm{Var}(\boldsymbol{\epsilon}_i) = D_i G D_i' + \sigma^2 H_i + \tau^2 I_i. \tag{5.2.4}$$

In the rest of this section, we look at particular examples of (5.2.4). Because all of our models have the property that measurements from different units are independent, we drop the subscript i, and write (5.2.4) as

$$\mathrm{Var}(\boldsymbol{\epsilon}) = \boldsymbol{DGD}' + \sigma^2 H + \tau^2 I. \tag{5.2.5}$$

In (5.2.5), $\boldsymbol{\epsilon} = (\epsilon_1, \ldots, \epsilon_n)$ denotes a generic sequence of n measurements from one unit. In what follows, we write $\boldsymbol{t} = (t_1, \ldots, t_n)$ for the corresponding set of times at which the measurements are made.

5.2.1 *Pure serial correlation*

For our first example we assume that neither random effects nor measurement errors are present, so that (5.2.3) reduces to

$$\epsilon_j = W(t_j)$$

and (5.2.5) correspondingly simplifies to

$$\mathrm{Var}(\boldsymbol{\epsilon}) = \sigma^2 H.$$

Now, σ^2 is the variance of each ϵ_j, and the correlations amongst the ϵ_j are determined by the autocorrelation function $\rho(u)$. Specifically,

$$\mathrm{Cov}(\epsilon_j, \epsilon_k) = \sigma^2 \rho(\mid t_j - t_k \mid).$$

The corresponding variogram is

$$\gamma(u) = \sigma^2\{1 - \rho(u)\}. \tag{5.2.6}$$

Thus, $\gamma(0) = 0$ and $\gamma(u) \to \sigma^2$ as $u \to \infty$. Typically, $\gamma(u)$ is an increasing function of u because the correlation, $\rho(u)$, decreases with increasing time-separation, u.

A popular choice for $\rho(u)$ is the exponential correlation model,

$$\rho(u) = \exp(-\phi u), \tag{5.2.7}$$

for some value of $\phi > 0$. Figure 5.1 shows the variogram (5.2.6) of this model for $\sigma^2 = 1$ and $\phi = 0.1, 0.25,$ and 1.0, together with a simulation of a sequence of $n = 25$ measurements at times $t_j = 1, 2, \ldots, 25$. As ϕ increases, the strength of the autocorrelation decreases; in the three examples shown, the autocorrelation between successive measurements is approximately 0.905, 0.779, and 0.368, respectively. The corresponding variograms and simulated realizations reflect this; as ϕ increases, the variograms approach their asymptote more quickly and the simulated realizations appear less smooth.

In many applications the empirical variogram differs from the exponential correlation model by showing an initially slow increase as u increases from zero, then a sharp rise and finally a slower increase again as it approaches its asymptote. A model which captures this behaviour is the so-called Gaussian correlation function,

$$\rho(u) = \exp(-\phi u^2), \tag{5.2.8}$$

for some $\phi > 0$. Figure 5.2 shows the variogram and a simulated realization for $\sigma^2 = 1$ and each of $\phi = 0.1, 0.25$ and 1.0, with times of observation $t_j = 1, 2, \ldots, 25$, as in Fig. 5.1.

A panel-by-panel comparison between Figs. 5.1 and 5.2 is interesting. For both models the parameter ϕ has a qualitatively similar effect: as ϕ increases the variogram rises more sharply and the simulated realizations are less smooth. Also, for each value of ϕ, the correlation between successive unit-spaced measurements is the same for the two models. When $\phi = 1.0$ the correlations at larger time-separations decay rapidly to zero and the corresponding simulations of the two models are almost identical (the same underlying random number stream is used for all the simulations). For smaller values of ϕ the simulated realisation of the Gaussian model has a smoother appearance than its exponential counterpart.

The most important qualitative difference between the Gaussian and exponential correlation functions is their behaviour near $u = 0$. Note that the domain of the correlation function is extended to all real u by the requirement that $\rho(-u) = \rho(u)$. Then, the exponential model (5.2.7) is continuous but not differentiable at $u = 0$, whereas the Gaussian model (5.2.8) is infinitely differentiable. The same remarks apply to the variogram, which is given by $\gamma(u) = \sigma^2\{1 - \rho(u)\}$. If we now recall the definition of the variogram as $\gamma(u) = \frac{1}{2}\mathrm{E}[\{\epsilon(t) - \epsilon(t-u)\}^2]$, we see why these considerations of mathematical smoothness in the correlation functions translate to physical smoothness in the simulated realizations. In particular, if we consider a Gaussian model with correlation parameter ϕ_1, and an exponential model with correlation parameter ϕ_2, then whatever the values of ϕ_1 and ϕ_2, there is some positive value, u_0 say, such that for time-separations less than u_0, the mean squared difference between $R(t)$ and $R(t - u_0)$ is smaller for the Gaussian than for the exponential. In other words, on a sufficiently small time-scale the Gaussian model will appear to be smoother than the exponential.

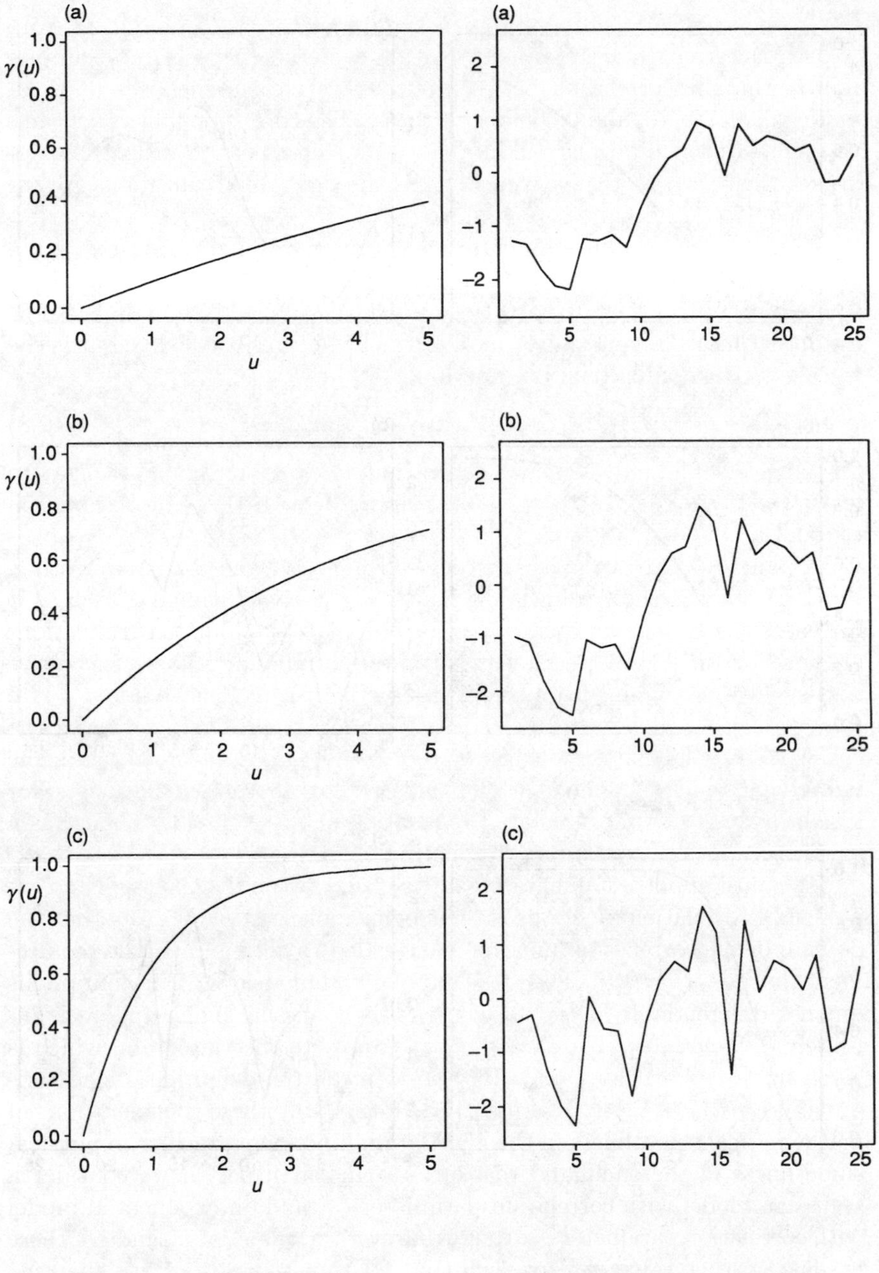

Fig. 5.1. Variograms (left-hand panels) and simulated realizations (right-hand panels) for the exponential correlation model: (a) $\phi = 0.1$; (b) $\phi = 0.25$; (c) $\phi = 1.0$.

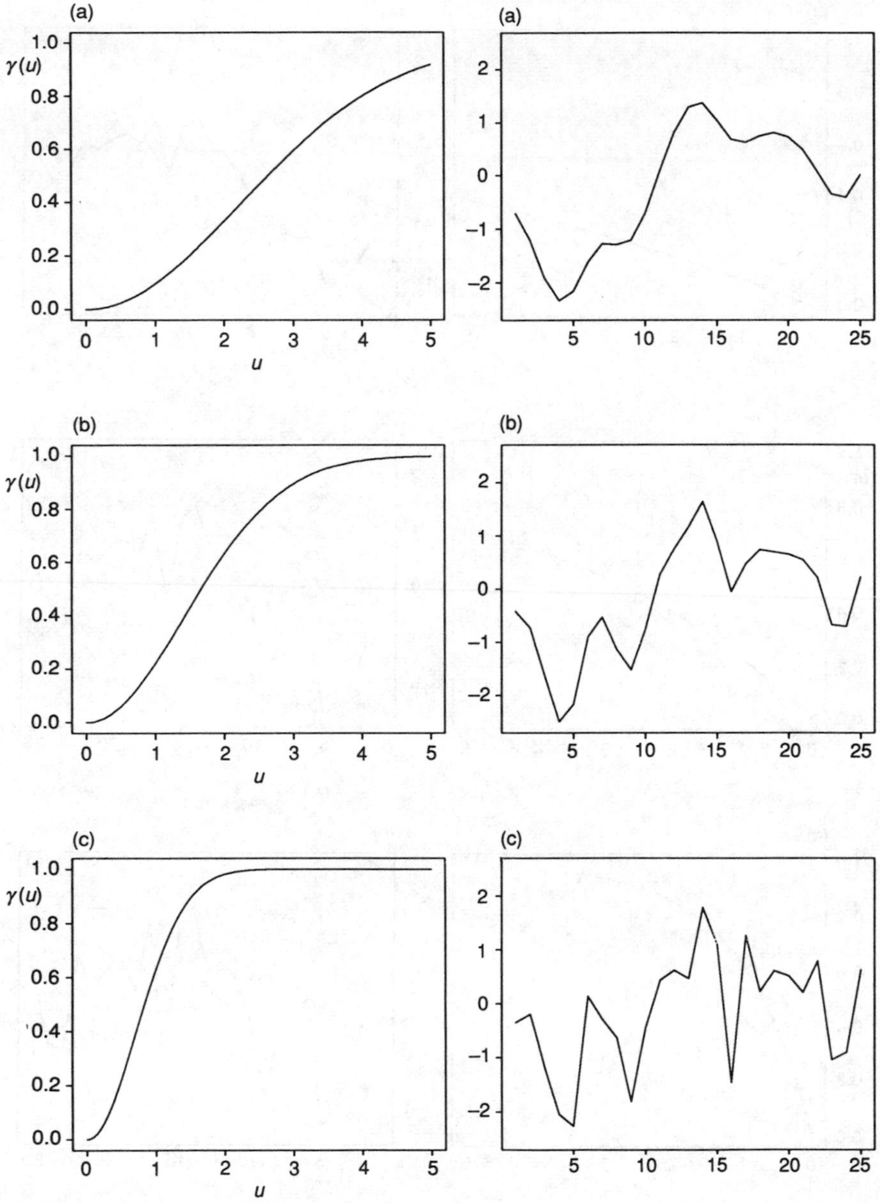

Fig. 5.2. Variograms (left-hand panels) and simulated realizations (right-hand panels) for the Gaussian correlation model: (a) $\phi = 0.1$; (b) $\phi = 0.25$; (c) $\phi = 1.0$.

Another way to build pure serial correlation models is to assume an explicit dependence of the current ϵ_j on a limited number of its predecessors, $\epsilon_{j-1}, \ldots, \epsilon_1$. This approach has a long history in time series analysis, going back at least to Yule (1927). For longitudinal data, the approach was proposed by Kenward (1987), who adopted Gabriel's (1972) terminology of *ante-dependence of order p* to describe a model in which the conditional distribution of ϵ_j given its predecessors $\epsilon_{j-1}, \ldots, \epsilon_1$ depends only on $\epsilon_{j-1}, \ldots, \epsilon_{j-p}$. A sequence of random variables ϵ_j with this property is more usually called a *pth order Markov model*, after the Russian mathematician who first studied stochastic processes of this kind. Ante-dependence modelling therefore has close links with the study of Markov processes (Cox and Miller, 1965), and with recent developments in the graphical modelling of multivariate data (Whittaker, 1990).

The simplest example of an ante-dependence model for a zero-mean sequence is the *first-order autoregressive process*. In this model, the ϵ_j can be generated from the relationship

$$\epsilon_j = \alpha \epsilon_{j-1} + Z_j, \tag{5.2.9}$$

where the Z_j are a sequence of mutually independent $N(0, \sigma^2)$ random variables, and the process is initiated by $\epsilon_0 \sim N\{0, \sigma^2/(1-\alpha^2)\}$. An equivalent formulation, with the same initial conditions, specifies the conditional distribution of ϵ_j given ϵ_{j-1} as

$$\epsilon_j \mid \epsilon_{j-1} \sim N(\alpha \epsilon_{j-1}, \sigma^2).$$

Care needs to be taken in defining the analogous ante-dependence models for the time-sequences of responses, Y_j. To take a very simple illustration, suppose that we wish to incorporate a linear relationship between the response, Y, and an explanatory variable, x, with serially correlated random variation about the mean response. One possibility is to assume that

$$Y_j = \beta_1 x_j + \epsilon_j,$$

where the ϵ_js follow the first-order autoregressive process defined by (5.2.9), with parameters α_1 and σ_1^2. Another is to assume that the conditional distribution of Y_j given Y_{j-1} is

$$Y_j \mid Y_{j-1} \sim N(\beta_2 x_j + \alpha_2 Y_{j-1}, \sigma_2^2).$$

Both are valid models, but they are not equivalent and cannot be made so by any choices for the two sets of parameter values. We take up this point in Chapter 7.

One very convenient feature of ante-dependence models is that the joint probability density function of ϵ follows easily from the specified form of the conditional distributions. Let $f_c(\epsilon_j \mid \epsilon_{j-1}, \ldots, \epsilon_{j-p}; \boldsymbol{\alpha})$ be the conditional probability density function of ϵ_j given $\epsilon_{j-k}, k = 1, \ldots, p$, and $f_o(\epsilon_1, \ldots, \epsilon_p; \boldsymbol{\alpha})$ the implied joint probability density function of $(\epsilon_1, \ldots, \epsilon_p)$. Then the joint probability density function of $\boldsymbol{\epsilon}$ is

$$f(\epsilon_1, \ldots, \epsilon_n; \boldsymbol{\alpha}) = f_o(\epsilon_1, \ldots, \epsilon_p; \boldsymbol{\alpha}) \prod_{j=p+1}^{n} f_c(\epsilon_j \mid \epsilon_{j-1}, \ldots, \epsilon_{j-p}; \boldsymbol{\alpha}).$$

Typically, $f_c(\cdot)$ is specified explicitly, from which it may or may not be easy to deduce $f_o(\cdot)$. However, if p is small and n large, there is usually only a small loss of information in conditioning on $\epsilon_1, \ldots, \epsilon_p$, in which case the contribution to the overall likelihood function is simply the product of the $n - p$ conditional densities, $f_c(\cdot)$. This makes estimation of $\boldsymbol{\alpha}$ very straightforward. For example, if $f_c(\cdot)$ is specified as a Gaussian probability density function with mean depending linearly on the conditioning variables, so that the conditional distribution of ϵ_j, given $\epsilon_{j-k}, k = 1, \ldots, p$, is $N\left(\Sigma_{k=1}^{p} \alpha_k \epsilon_{j-k}, \alpha_0^2\right)$ then the conditional maximum likelihood estimates of the α_k are obtained simply by ordinary least-squares regression, treating the observed values of $\epsilon_{j-1}, \ldots, \epsilon_{j-p}$ as a set of p explanatory variables attached to the response, ϵ_j.

Ante-dependence models are appealing for equally spaced data, less so for unequally spaced data. For example, it would be hard to give a natural interpretation to the α_k parameters in the above model if the measurements were not equally spaced in time. Similarly, ante-dependence models do not cope easily with data for which the times of measurement are not common to all units.

One exception to the above remark is the exponential correlation model, which can also be interpreted as an ante-dependence model of order 1. Specifically, if $\boldsymbol{\epsilon} = (\epsilon_1, \ldots, \epsilon_n)$ has a multivariate Gaussian distribution with covariance structure

$$\text{Cov}(\epsilon_j, \epsilon_k) = \sigma^2 \exp(-\phi \mid t_j - t_k \mid),$$

then the conditional distribution of ϵ_j given all its predecessors depends only on the value of ϵ_{j-1}. Furthermore

$$\epsilon_j \mid \epsilon_{j-1} \sim N\left(\alpha_j \epsilon_{j-1}, \sigma^2(1 - \alpha_j^2)\right), \quad j = 2, \ldots, n$$

where

$$\alpha_j = \exp\left(-\phi \mid t_j - t_{j-1} \mid\right),$$

and

$$\epsilon_1 \sim N(0, \sigma^2).$$

As a final comment on ante-dependence models, we note that their Markovian structure essentially rules out any straightforward extension to accommodate measurement error or random effects.

5.2.2 *Serial correlation plus measurement error*

These are models for which there are no random effects in (5.2.3), so that $Y_j = W(t_j) + Z_j$, and (5.2.5) reduces to

$$\text{Var}(\epsilon) = \sigma^2 H + \tau^2 I.$$

Now, the variance of each ϵ_j is $\sigma^2+\tau^2$, and if the elements of H are specified by a correlation function $\rho(u)$, so that $h_{ij} = \rho(| t_i - t_j |)$, then the variogram becomes

$$\gamma(u) = \tau^2 + \sigma^2\{1 - \rho(u)\}. \tag{5.2.10}$$

A characteristic property of models with measurement error is that $\gamma(u)$ does not tend to zero as u tends to zero. If the data include duplicate measurements at the same time, we can estimate $\gamma(0) = \tau^2$ directly as one-half the average squared difference between such duplicates. Otherwise, estimation of τ^2 involves explicit or implicit extrapolation based on an assumed parametric model, and the estimate $\gamma(0)$ may be strongly model-dependent. Figure 5.3 illustrates this point. It shows an empirical variogram calculated from measurements at unit time-spacing, and two theoretical variograms which fit the data equally well but show very different extrapolations to $u = 0$.

5.2.3 *Random intercept plus serial correlation plus measurement error*

The simplest example of our general model (5.2.3) in which all three components of variation are present takes U to be a univariate, zero-mean Gaussian random variable with variance ν^2, and $d_j = 1$. Then, the realized value of U represents a random intercept, that is, an amount by which *all* measurements on the unit in question are raised or lowered relative to the population average. The variance matrix (5.2.5) becomes

$$\text{Var}(\epsilon) = \nu^2 J + \sigma^2 H + \tau^2 I,$$

where J is an $n \times n$ matrix with all of its elements equal to 1. The variogram has the same form (5.2.10) as for the serial correlation plus measurement error model,

$$\gamma(u) = \tau^2 + \sigma^2\{1 - \rho(u)\},$$

except that now the variance of each ϵ_j is $\text{Var}(\epsilon_j) = \nu^2 + \sigma^2 + \tau^2$ and the limit of $\gamma(u)$ as $u \to \infty$ is *less* than $\text{Var}(\epsilon_j)$. Figure 5.4 shows this behaviour, using $\rho(u) = \exp(-u^2)$ as the correlation function of the serially correlated component.

The first explicit formulation of this model, with all three components of variation in a continuous-time setting, appears to be in Diggle (1988). Note that in fitting the model to data, the information on ν^2 derives from the replication of units within treatment groups.

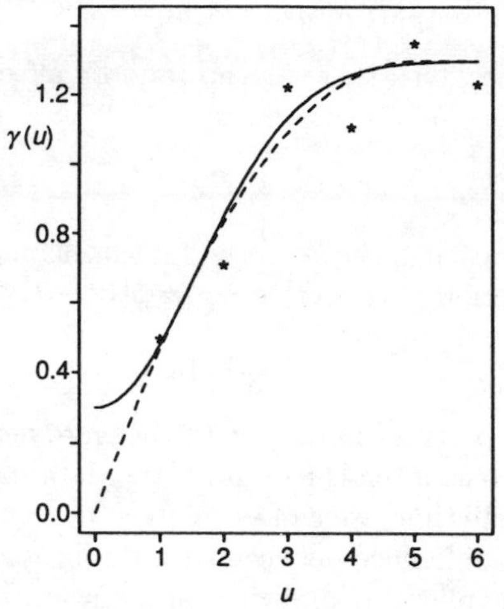

Fig. 5.3. A sample variogram and two different fitted models. $\star$: sample variogram; —— : fitted model with a non-zero intercept; - - - : fitted model with a zero intercept.

5.2.4 *Random effects plus measurement error*

It is interesting to speculate on reasons for the relatively late infiltration of time series models into longitudinal data-analysis. Perhaps the explanation is the following. Whilst serial correlation would appear to be a natural feature of any longitudinal data model, in specific applications its effects may be dominated by the combination of random effects and measurement error. In terms of Fig. 5.4, if σ^2 is much smaller than either τ^2 or ν^2, the increasing curve of the variogram is squeezed between the two horizontal dotted lines, and becomes an unnecessary refinement of the model.

If we eliminate the serially correlated component altogether, (5.2.3) reduces to

$$\epsilon_j = \mathbf{d}_j' \boldsymbol{U} + Z_j.$$

The simplest model of this kind incorporates a scalar random intercept, U, with $d_j = 1$ for all j, to give

$$\operatorname{Var}(\epsilon) = \nu^2 J + \tau^2 I.$$

The variance of each ϵ_j is $\nu^2 + \tau^2$, and the correlation between any two measurements on the same unit is

$$\rho = \nu^2/(\nu^2 + \tau^2).$$

As noted earlier, this uniform correlation structure is sometimes called a 'split-plot model' because of its formal equivalence with the correlation

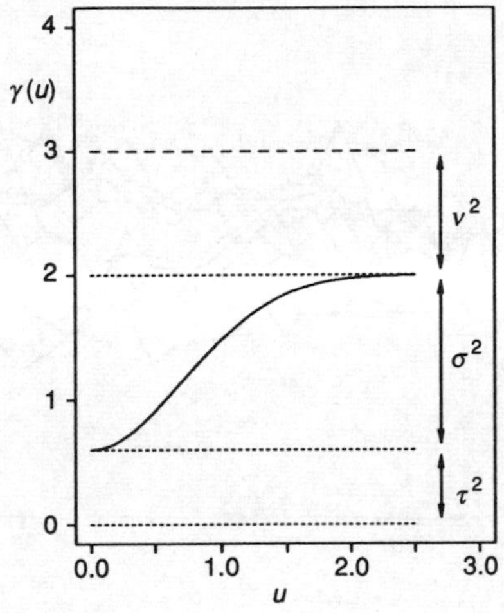

Fig. 5.4. The variogram for a model with a random intercept, serial correlation, and measurement error.

structure induced by the randomization for a classical split-plot experiment. However, the randomization argument gives no justification for its use with longitudinal data.

More general specifications of the random effect component lead to models with non-stationary structure. For example, if $\boldsymbol{U}$ is a pair of independent Gaussian random variables with variances ν_1^2 and ν_2^2, and $\boldsymbol{d}_j = (1, t_j)$, specifying a model with a linear time-trend for each unit but random intercept and slope between units, then $\text{Var}(\epsilon_j) = \nu_1^2 + t_j^2\nu_2^2 + \tau^2$, and for $j \neq k$,

$$\text{Cov}(\epsilon_j, \epsilon_k) = \nu_1^2 + t_j t_k \nu_2^2.$$

Figure 5.5 shows a simulated realization of this model with $\tau^2 = 0.25$, $\nu_1^2 = 0.5$, and $\nu_2^2 = 0.01$. The fanning out of the collection of response profiles over time is a common characteristic of growth data, for which non-stationary random effects models of this kind are often appropriate. Indeed, this kind of random effects model is often referred to as *the* growth curve model, following its systematic development in a seminal paper by Rao (1965) and many subsequent papers including Fearn (1977), Verbyla (1986) and Verbyla and Venables (1988).

Sandland and McGilchrist (1979) advocate a different kind of non-stationary model for growth data. If $S(t)$ denotes the 'size' of an animal or plant at time t, measured on a logarithmic scale, then the relative growth rate (RGR) is defined to be the first derivative of $S(t)$. Sandland and

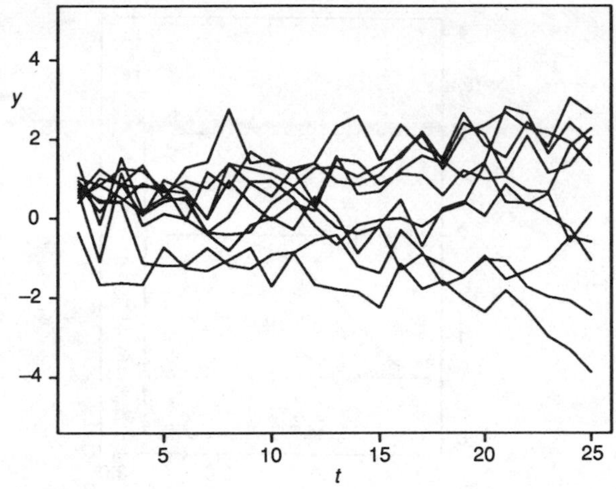

Fig. 5.5. A simulated realization of a model with random intercepts and random slopes.

McGilchrist argue that the RGR is often well modelled by a stationary random process. For models of this kind, the variance of $S(t)$ increases linearly over time, in contrast to the quadratic increase of the random slope and intercept model. Also, the source of the increasing variability is within, rather than between, units. In practice, quite long sequences of measurements may be needed to distinguish these two kinds of behaviour. Cullis and McGilchrist (1990) advocate using the Sandland and McGilchrist model routinely in the longitudinal data setting. Note that for equally spaced data the model can be implemented by analysing successive differences, $R_t = S(t) - S(t-1)$, using a stationary model for the sequence $\{R_t\}$.

5.3 Model-fitting

In fitting a model to a set of data, our objective is typically to answer questions about the process which generated the data. Note that 'model' and 'process' are not synonymous. We expect only that the former will be a useful approximation to the latter in the sense that it will contain a small number of parameters whose values can be interpreted as answers to the scientific questions posed by the data. Typically, the model will contain further parameters which are not of interest in themselves, but whose values affect the inferences which we make about the parameters of direct interest.

In many applications, inference about the model parameters is an end in itself. In others, we may wish to use the fitted model for prediction of the underlying continuous-time response profile associated with an individual unit. We take up this issue in Section 5.6. Here, we consider only the model-fitting process, which we divide into four stages:

(1) *Formulation* — choosing the general form of the model;

(2) *Estimation* — attaching numerical values to parameters;

(3) *Inference* — calculating confidence intervals or testing hypotheses about parameters of direct interest;

(4) *Diagnostics* — checking that the model fits the data.

5.3.1 *Formulation*

Formulation of a model is essentially a continuation of the exploratory data analysis considered in Chapter 3, but directed towards the specific aspects of the data which our model aims to describe. In particular, we assume here that any essentially model-independent issues such as the identification and treatment of outliers have already been resolved. The focus of attention is then the mean and covariance structure of the data.

With regard to the mean, time-plots of observed averages within treatment groups are simple and effective aids to model formulation for well-replicated data. When there are few measurements at any one time, non-parametric smoothing of the data is helpful.

For the covariance structure, we use residuals obtained by subtracting from each measurement the ordinary least-squares estimate of the corresponding mean response, based on the most elaborate model we are prepared to contemplate for the mean response. As discussed in Section 4.4, for data from designed experiments in which the only relevant explanatory variables are the treatment labels we can use a saturated model, fitting a separate mean response for each combination of treatment and time. For data in which the times of measurement are not common to all units, or when there are continuously varying covariates which may affect the mean response non-linearly, the choice of a 'most elaborate model' is less obvious.

Time-plots, scatterplot matrices and empirical variogram plots of these residuals can then be used to give an idea of the underlying structure. Non-stationary variation in the time-plots of the residuals suggests the need for a transformation or an inherently non-stationary model such as a random effects model. Once an apparently stationary pattern of variation has been achieved, the empirical variogram can be used to estimate the underlying covariance structure. In particular, if all units are observed at a common set of times the empirical variogram of the residuals is essentially unbiased for the underlying theoretical variogram (Diggle, 1990, Chapter 5). Note that if an incorrect parametric model is used to define the residuals, the empirical variogram will be biased to an unknown extent.

5.3.2 *Estimation*

Our objective in this second stage is to attach numerical values to the parameters in the model, whose general form is

$$\boldsymbol{Y} \sim MVN\{X\boldsymbol{\beta}, \sigma^2 V(\boldsymbol{\alpha})\}. \tag{5.3.1}$$

We write $\hat{\boldsymbol{\beta}}, \hat{\sigma}^2$, and $\hat{\boldsymbol{\alpha}}$ for the estimates of the model parameters, $\boldsymbol{\beta}, \sigma^2$ and $\boldsymbol{\alpha}$.

Particular methods of estimation have been derived for special cases of (5.3.1). However, it is straightforward to derive a general likelihood method. The resulting algorithm is computationally feasible provided that the sequences of measurements on individual units are not too long. If computing efficiency is paramount, more efficient algorithms can be derived to exploit the properties of special covariance structures. For example, estimation in ante-dependence models can exploit their Markovian structure.

The general method follows the same lines as the development in Section 4.4, but making use of the parametric structure of the variance matrix, $V(\boldsymbol{\alpha})$. Also as in Section 4.4, we favour REML estimates over classical maximum likelihood estimates to reduce the bias in the estimation of $\boldsymbol{\alpha}$, although this is less important if the dimensionality of $\boldsymbol{\beta}$ is small.

For given $\boldsymbol{\alpha}$, equation (4.5.5) holds in a modified form as

$$\hat{\boldsymbol{\beta}}(\boldsymbol{\alpha}) = \left(X'V(\boldsymbol{\alpha})^{-1}X\right)^{-1} X'V(\boldsymbol{\alpha})^{-1}\boldsymbol{y}, \tag{5.3.2}$$

and (4.5.6) likewise holds in the form

$$\hat{\sigma}^2(\boldsymbol{\alpha}) = \text{RSS}(\boldsymbol{\alpha})/(N-p), \tag{5.3.3}$$

where

$$\text{RSS}(\boldsymbol{\alpha}) = \{\boldsymbol{y} - X\hat{\boldsymbol{\beta}}(\boldsymbol{\alpha})\}'V(\boldsymbol{\alpha})^{-1}\{\boldsymbol{y} - X\hat{\boldsymbol{\beta}}(\boldsymbol{\alpha})\} \tag{5.3.4}$$

and $N = \Sigma_{i=1}^{m} n_i$ is the total number of measurements on all m units. Finally, the REML estimate $\hat{\boldsymbol{\alpha}}$ maximizes

$$\begin{aligned} L^*(\boldsymbol{\alpha}) = -\frac{1}{2}\,[N\log\{\text{RSS}(\boldsymbol{\alpha})\} + \log \mid V(\boldsymbol{\alpha}) \mid \\ + \log \mid X'V(\boldsymbol{\alpha})^{-1}X \mid], \end{aligned} \tag{5.3.5}$$

and the resulting REML estimate of $\boldsymbol{\beta}$ is $\hat{\boldsymbol{\beta}} = \hat{\boldsymbol{\beta}}(\hat{\boldsymbol{\alpha}})$.

The nature of the computations involved in maximizing $L^*(\boldsymbol{\alpha})$ becomes clear if we recognize explicitly the block-diagonal structure of $V(\boldsymbol{\alpha})$. Write $\boldsymbol{y}_i$ for the vector of n_i measurements on the ith unit and $\boldsymbol{t}_i$ for the corresponding vector of times at which these measurements are taken. Write

$V_i(\boldsymbol{t}_i; \boldsymbol{\alpha})$ for the non-zero $n_i \times n_i$ block of $V(\boldsymbol{\alpha})$ corresponding to the covariance matrix of $\boldsymbol{Y}_i$. The notation emphasizes that the source of the differences amongst the variance matrices of the different units is variation in the corresponding times of measurements. Finally, write X_i for the $n_i \times p$ matrix consisting of those rows of X which correspond to the explanatory variables measured on the *ith* unit. Then (5.3.4) and (5.3.5) can be written as

$$\text{RSS}(\boldsymbol{\alpha}) = \sum_{i=1}^{m} \{\boldsymbol{y}_i - X_i \hat{\boldsymbol{\beta}}(\boldsymbol{\alpha})\}' V_i(\boldsymbol{t}_i; \boldsymbol{\alpha})^{-1} \{\boldsymbol{y}_i - X_i \hat{\boldsymbol{\beta}}(\boldsymbol{\alpha})\} \tag{5.3.6}$$

and

$$L^*(\boldsymbol{\alpha}) = -\frac{1}{2}\left[N \log\{\text{RSS}(\boldsymbol{\alpha})\} + \sum_{i=1}^{m} \log \mid V_i(\boldsymbol{t}_i; \boldsymbol{\alpha}) \mid + \log\left(\sum_{i=1}^{m} \mid X_i' V_i(\boldsymbol{t}_i; \boldsymbol{\alpha})^{-1} X_i \mid \right)\right]. \tag{5.3.7}$$

Each evaluation of $L^*(\boldsymbol{\alpha})$ therefore involves m determinants of order $p \times p$ for the second term in (5.3.7), and at most m determinants and inverses of the $V_i(\boldsymbol{t}_i; \boldsymbol{\alpha})$. In most applications, the number of distinct $V_i(\boldsymbol{t}_i; \boldsymbol{\alpha})$ is much fewer than m. Furthermore, the dimensionality of $\boldsymbol{\alpha}$ is typically very small, say three or four at most, and in our experience it is seldom necessary to use sophisticated optimization algorithms. Either the simplex algorithm of Nelder and Mead (1965) or a quasi-Newton algorithm, neither of which requires the user to provide information on the derivatives of $L^*(\boldsymbol{\alpha})$, usually works well. The only exceptions in our experience have been when an over-elaborate model is fitted to sparse data and the covariance parameters, $\boldsymbol{\alpha}$, are fundamentally ill-determined. Finally in this subsection, note that whilst the REML estimates maximize $L^*(\boldsymbol{\alpha})$ given by (5.3.7), maximum likelihood estimates maximize

$$L(\boldsymbol{\alpha}) = -\frac{1}{2}\left[N \log\{\text{RSS}(\alpha)\} + \sum_{i=1}^{m} \log \mid V_i(\boldsymbol{t}_i; \boldsymbol{\alpha}) \mid \right]. \tag{5.3.8}$$

5.3.3 *Inference*

Inference about $\boldsymbol{\beta}$ can be based on the result (5.3.2) which, in conjunction with (5.3.1), implies that

$$\hat{\boldsymbol{\beta}}(\boldsymbol{\alpha}) \sim MVN\{\boldsymbol{\beta}, \sigma^2 (X' V(\boldsymbol{\alpha})^{-1} X)^{-1}\}. \tag{5.3.9}$$

We assume that (5.3.9) continues to hold, to a good approximation, if we substitute the REML estimates $\hat{\sigma}^2$ and $\hat{\boldsymbol{\alpha}}$ for the unknown values of σ^2 and $\boldsymbol{\alpha}$ in (5.3.9). This gives

$$\hat{\boldsymbol{\beta}} \sim MVN(\boldsymbol{\beta}, \hat{V}), \tag{5.3.10}$$

where

$$\hat{V} = \hat{\sigma}^2(X'V(\hat{\boldsymbol{\alpha}})^{-1}X)^{-1}.$$

The immediate application of (5.3.10) is to set standard errors on individual elements of $\boldsymbol{\beta}$. Almost as immediate is the calculation of confidence regions for general linear transformations of the form

$$\boldsymbol{\psi} = D\boldsymbol{\beta}, \tag{5.3.11}$$

where D is a full-rank, $r \times p$ matrix with $r \leq p$. Confidence regions for $\boldsymbol{\psi}$ follow from the result that if $\boldsymbol{\psi} = D\hat{\boldsymbol{\beta}}$, then

$$\hat{\boldsymbol{\psi}} \sim MVN(\boldsymbol{\psi}, D\hat{V}D'),$$

from which it follows in turn that if

$$T(\boldsymbol{\psi}) = (\hat{\boldsymbol{\psi}} - \boldsymbol{\psi})'(D\hat{V}D')^{-1}(\hat{\boldsymbol{\psi}} - \boldsymbol{\psi}) \tag{5.3.12}$$

then $T(\boldsymbol{\psi}) \sim \chi_r^2$. Let $c_r(q)$ denote the q-critical value of χ_r^2, so that $P\{\chi_r^2 \geq c_r(q)\} = q$. Then, a $100(1-q)$ per cent confidence region for $\boldsymbol{\psi}$ is

$$\{\boldsymbol{\psi} : T(\boldsymbol{\psi}) \leq c_r(q)\}. \tag{5.3.13}$$

Using the well known duality between hypothesis tests and confidence regions, a test of a hypothesized value for $\boldsymbol{\psi}$, say $H_0 : \boldsymbol{\psi} = \boldsymbol{\psi}_0$, consists of rejecting H_0 at the $100q$ per cent level if $T(\boldsymbol{\psi}_0) > c_r(q)$. Note in particular that a statistic to test $H_0 : \boldsymbol{\psi} = \mathbf{0}$ is

$$T_0 = \hat{\boldsymbol{\psi}}'(D\hat{V}D')^{-1}\hat{\boldsymbol{\psi}}, \tag{5.3.14}$$

whose null sampling distribution is χ_r^2.

In implementing the above method of inference, the nuisance parameters, σ^2 and $\boldsymbol{\alpha}$, are estimated once only, from the maximal model for $\boldsymbol{\beta}$. The choice of the maximal model is sometimes difficult, especially when the times of measurement are highly variable between units or when there are continuously varying covariates. An alternative, 'step-up' approach may then be useful, beginning with a very simple model for $\boldsymbol{\beta}$ and considering the maximized log-likelihoods from a sequence of progressively more elaborate models. At each stage, the maximized log-likelihood is $L(\hat{\boldsymbol{\alpha}})$, where

$L(\boldsymbol{\alpha})$ is defined by (5.3.8) with the current design matrix X, and $\hat{\boldsymbol{\alpha}}$ maximizes $L(\boldsymbol{\alpha})$. Let L_k be the maximized log-likelihood associated with the kth model in this nested sequence, and p_k the number of columns in the corresponding design matrix. The log-likelihood ratio statistic to test the adequacy of the kth model within the $(k+1)st$ is

$$W_k = 2(L_{k+1} - L_k)$$

If the kth model is correct, W_k is distributed as chi-squared on $p_{k+1} - p_k$ degrees of freedom.

This likelihood ratio testing approach to inference is convenient when the data are unbalanced and there is no obvious maximal model. However, the earlier approach based on the multivariate Gaussian distribution for $\hat{\boldsymbol{\beta}}$ is in our view preferable when applicable because it more clearly separates the inferences about $\boldsymbol{\beta}$ and $\boldsymbol{\alpha}$.

5.3.4 *Diagnostics*

Diagnostic checking of the model against the data completes the model-fitting process. The aim of diagnostic checking is to compare the data with the fitted model in such a way as to highlight any systematic discrepancies. Since our models are essentially models for the mean and covariance structure of the data, simple and highly effective checks are to superimpose the fitted mean response profiles on a time-plot of the average observed response within each combination of treatment and time, and to superimpose the fitted variogram on a plot of the empirical variogram.

Simple plots of this kind can be very effective in revealing inconsistencies between data and model which were missed at earlier stages. If so, these can be incorporated into a revised model, and the fitting process repeated.

More formal diagnostic criteria for regression models are discussed by Cook and Weisberg (1982) and Atkinson (1985) in the context of linear models for cross-sectional data and, in a non-parametric setting, by Hastie and Tibshirani (1990).

5.4 Examples

In this section we give two applications of the model-based approach. These involve:

(1) the data set of Example 1.4 on the protein content of milk samples from cows on each of three different diets,

(2) a set of data on the body-weights of cows in a 2 by 2 factorial experiment.

Example 5.1. Protein contents of milk samples

These data were shown in Fig. 1.4. They consist of measurements of protein content in up to 19 weekly samples taken from each of 79 cows allocated to one of three different diets: 25 cows received a barley diet, 27 cows a mixed diet of barley and lupins, and 27 cows a diet of lupins only. The initial inspection of the data reported in Chapter 1 suggested that the mean response profiles are approximately parallel, showing an initial sharp drop associated with a settling-in period, followed by an approximately constant mean response over most of the experimental period and a possible gentle rise towards the end. The empirical variogram, shown in Fig. 3.16, exhibits a smooth rise with increasing lag, levelling out within the range spanned by the data. Both the mean and covariance structure therefore seem amenable to parametric modelling.

Our provisional model for the mean response profiles takes the form

$$\mu_g(t) = \begin{cases} \beta_{0g} + \beta_1 t, & t \leq 3 \\ \beta_{0g} + 3\beta_1 + \beta_2(t-3) + \beta_3(t-3)^2, & t > 3 \end{cases} \tag{5.4.1}$$

where $g = 1, 2, 3$ denotes treatment group, and time, t, is measured in weeks. Our provisional model for the covariance structure takes the form of (5.2.10) with an exponential correlation function. For the model-fitting, it is convenient to extract σ^2 as a scale parameter and reparametrize to $\alpha_1 = \tau^2/\sigma^2$ and $\alpha_2 = \nu^2/\sigma^2$. With this parametrization the theoretical variance of each measurement is $\sigma^2(1 + \alpha_1 + \alpha_2)$ and the theoretical variogram is

$$\gamma(u) = \sigma^2\{\alpha_1 + 1 - \exp(-\alpha_3 u)\}.$$

The REML estimates of the model parameters, with estimated standard errors and correlation matrix of the mean parameters, are given in Table 5.1. This information can now be used to make inferences about the mean response profiles as discussed in Section 5.3.3. We give two examples.

Because our primary interest is in whether the diets affect the mean response profiles, we first test the hypothesis that $\beta_{01} = \beta_{02} = \beta_{03}$. To do this, we need only consider the REML estimates and variance submatrix of $\boldsymbol{\beta}_0 = (\beta_{01}, \beta_{02}, \beta_{03})$. This submatrix is

$$\hat{V}_1 = \begin{bmatrix} 0.0029 & 0.0015 & 0.0015 \\ 0.0015 & 0.0028 & 0.0014 \\ 0.0015 & 0.0014 & 0.0028 \end{bmatrix}.$$

Now, using (5.3.11) and (5.3.12) on this three-parameter system, we proceed as follows. The hypothesis of interest is $\boldsymbol{\psi} = D\boldsymbol{\beta}_0 = \mathbf{0}$, where

$$D = \begin{bmatrix} 1 & -1 & 0 \\ 0 & 1 & -1 \end{bmatrix}$$

Table 5.1. REML estimates for the model fitted to data on protein content of milk samples

(a)Mean response

$$\mu_i(t) = \begin{cases} \beta_{0g} + \beta_1 t, & t \leq 3 \\ \beta_{0g} + 3\beta_1 + \beta_2(t-3) + \beta_3(t-3)^2, & t > 3 \end{cases}$$

Parameter	Est.	SE	Corr. matrix					
β_{01}	4.15	0.054	1.00					
β_{02}	4.05	0.053	0.52	1.00				
β_{03}	3.94	0.053	0.52	0.53	1.00			
β_1	-0.229	0.016	-0.61	-0.62	-0.62	1.00		
β_2	0.0078	0.0079	-0.60	-0.06	-0.06	-0.33	1.00	
β_3	-0.00056	0.00050	0.01	0.01	0.02	0.24	-0.93	

(b) Covariance structure

$$\gamma(u) = \sigma^2\{\alpha_1 + 1 - \exp(-\alpha_3 u)\}$$
$$\text{Var}(Y) = \sigma^2(1 + \alpha_1 + \alpha_2)$$

Parameter	Estimate
σ^2	0.0635
α_1	0.3901
α_2	0.1007
α_3	0.1674

The REML estimate of $\boldsymbol{\psi}$ is

$$\hat{\boldsymbol{\psi}} = D\hat{\boldsymbol{\beta}}_0 = (0.10, 0.11).$$

From (5.3.12), the test statistic is

$$T_0 = \hat{\boldsymbol{\psi}}'(D\hat{V}_1 D')^{-1}\hat{\boldsymbol{\psi}} = 15.98,$$

which we refer to critical values of χ_2^2. Since $P\{\chi_2^2 > 15.98\} = 0.0003$, we clearly reject $\boldsymbol{\psi} = \mathbf{0}$ and conclude that diet affects the mean response profile. Furthermore, the ordering of the three estimates, $\hat{\beta}_{og}$, is sensible, with the parameter estimate for the mixed diet lying between those for the two pure diets.

A question of secondary interest is whether there is a rise in the mean response towards the end of the experiment. The hypothesis to test this is $\beta_2 = \beta_3 = 0$. The variance submatrix of $\hat{\boldsymbol{\beta}} = (\hat{\beta}_2, \hat{\beta}_3)$ is $\hat{V}_2$, where

$$10^6 \hat{V}_2 = \begin{bmatrix} 62.094 & -3.631 \\ -3.631 & 0.246 \end{bmatrix}$$

The test statistic is

$$T_0 = \hat{\boldsymbol{\beta}} \hat{V}_2^{-1} \hat{\boldsymbol{\beta}} = 1.29$$

which we again refer to critical values of χ_2^2. Since $P\{\chi_2^2 > 1.29\} = 0.525$, we do not reject the null hypothesis in favour of a late rise in the mean response. Refitting the simpler model with $\beta_2 = \beta_3 = 0$ leaves the other parameter estimates essentially unchanged, and does not alter the conclusions about the treatment effects. Results for the model with $\beta_2 = \beta_3 = 0$, but using maximum likelihood estimation, are given in Diggle (1990, section 5.6).

Verbyla and Cullis (1990) suggest that the mean response curves in the three treatment groups may not be parallel, and fit a model in which the difference, $\mu_1(t) - \mu_3(t)$, is linear in t. A model which allows the difference in mean response between any two treatments to be linear in t replaces β_{02} and β_{03} in (5.4.1) by $\beta_{021} + \beta_{022}t$ and $\beta_{031} + \beta_{032}t$, respectively. Using this eight-parameter specification of the mean response profiles as our working model, a hypothesis of possible interest in that $\beta_{022} = \beta_{032} = 0$, i.e. that (5.4.1) gives an adequate description of the data. Fitting this enlarged model, we obtain

$$(\hat{\beta}_{022}, \hat{\beta}_{032}) = (-0.0042, -0.0095)$$

with estimated variance matrix, $\hat{V}_3$, where

$$10^6 \hat{V}_3 = \begin{bmatrix} 33.41 & 18.92 \\ 18.92 & 41.22 \end{bmatrix}.$$

The statistic to test $\beta_{022} = \beta_{032} = 0$ is $T_0 = 2.210$, which is to be compared with critical values of χ_2^2. Since $P\{\chi_2^2 > 2.210\} = 0.33$, there is no real evidence against $\beta_{022} = \beta_{032} = 0$. Even if we consider $\hat{\beta}_{032}$ alone, the statistic to test $\beta_{032} = 0$ is $T_0 = 2.208$, on one degree of freedom, which is still not significant. This analysis is not directly comparable with the one reported in Verbyla and Cullis (1990), as they consider only the last 16 time-points and use a non-parametric description of the mean response profile, $\mu_1(t)$, for cows on the barley diet.

Our provisional model for the data is summarized by the parameter estimates in Table 5.1 but with the simplification that $\beta_2 = \beta_3 = 0$. Figure 5.6 compares the empirical and fitted mean response profiles and variograms. The fit to the variogram appears to be satisfactory. The rise in the empirical variogram at $u = 18$ can be discounted as it is based on only 41

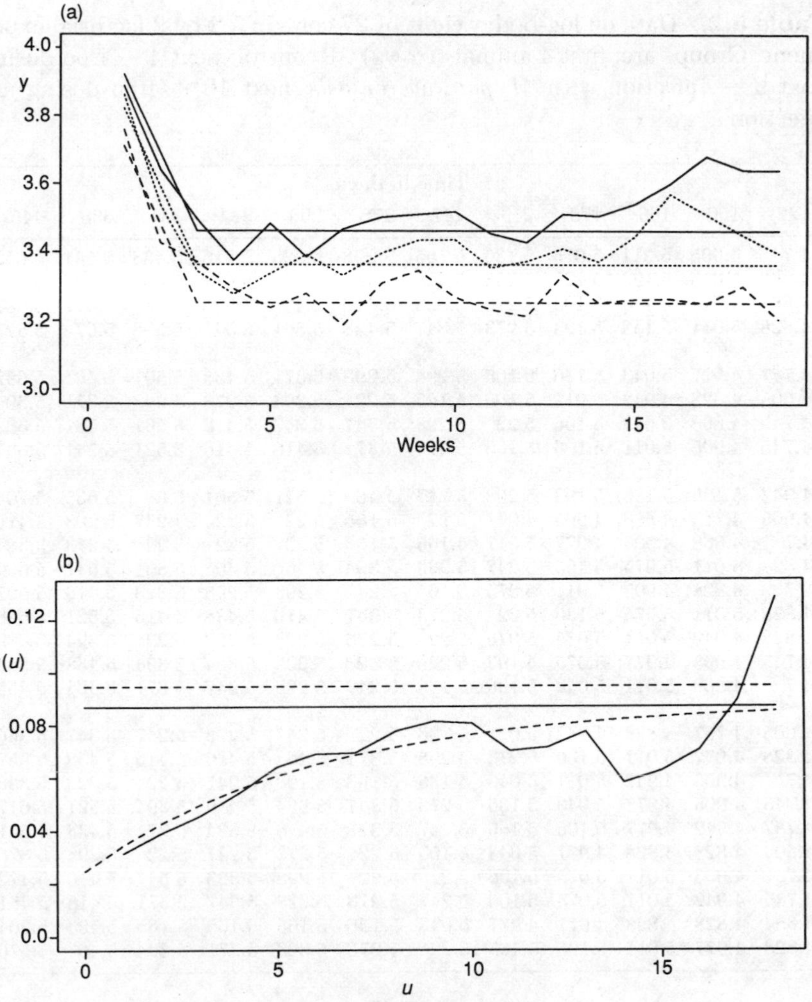

Fig. 5.6. Milk protein data: comparison between observed and fitted mean response profiles and variograms: (a) mean response profiles —— : barley diet; .... : mixed diet ; – – – : lupins diet; (b) variograms —— : sample variogram; – – – : fitted variogram.

observed pairwise differences. Since these pairwise differences derive from 41 different animals and are therefore approximately independent, elementary calculations give a 95 per cent confidence interval for the mean of $\hat{\gamma}(18)$ as 0.127 ± 0.052, which includes the fitted value, $\gamma(18) = 0.085$.

At first sight, the fit to the mean response profiles appears less satisfactory. However, the apparent lack of fit is towards the end of the experiment, by which time the responses from almost half of the animals

Table 5.2. Data on log-bodyweight of 27 cows in a 2 by 2 factorial experiment. Groups are: first 4 animals (rows) — control; next 4 — iron dosing; next 9 — infection with *M. paratuberculosis*; next 10 — iron dosing and infection

Time in days											
122	150	166	179	219	247	276	296	324	354	380	445
4.7	4.905	5.011	5.075	5.136	5.165	5.298	5.323	5.416	5.438	5.541	5.652
4.868	5.075	5.193	5.22	5.298	5.416	5.481	5.521	5.617	5.635	5.687	5.768
4.868	5.011	5.136	5.193	5.273	5.323	5.416	5.46	5.521	5.58	5.617	5.687
4.828	5.011	5.136	5.193	5.273	5.347	5.438	5.561	5.541	5.598	5.67	5.521
4.787	4.977	5.043	5.136	5.106	5.298	5.298	5.371	5.438	5.501	5.561	5.652
4.605	4.828	4.942	5.011	5.106	5.165	5.22	5.298	5.273	5.298	5.371	5.394
4.745	4.868	5.043	5.106	5.22	5.298	5.347	5.347	5.416	5.501	5.561	5.58
4.745	4.905	5.011	5.106	5.165	5.273	5.371	5.416	5.416	5.521	5.541	5.635
4.942	5.106	5.136	5.193	5.298	5.347	5.46	5.521	5.561	5.58	5.635	5.704
4.605	4.745	4.868	4.905	4.977	5.22	5.165	5.22	5.22	5.247	5.298	5.416
4.7	4.868	4.905	4.977	5.011	5.106	5.165	5.22	5.22	5.22	5.273	5.384
4.828	5.011	5.075	5.165	5.247	5.323	5.394	5.46	5.46	5.501	5.541	5.609
4.7	4.828	4.905	5.011	5.075	5.165	5.247	5.298	5.298	5.323	5.416	5.505
4.828	5.011	5.075	5.136	5.22	5.273	5.347	5.416	5.438	5.416	5.521	5.628
4.828	4.942	5.011	5.075	5.075	5.22	5.273	5.298	5.323	5.298	5.394	5.489
4.745	4.905	4.977	5.075	5.193	5.22	5.298	5.323	5.394	5.394	5.438	5.583
4.7	4.868	5.011	5.043	5.106	5.165	5.247	5.298	5.347	5.371	5.438	5.455
4.605	4.787	4.828	4.942	5.011	5.136	5.22	5.247	5.273	5.247	5.347	5.366
4.828	4.977	5.011	5.136	5.273	5.298	5.371	5.46	5.416	5.416	5.438	5.557
4.7	4.905	4.942	5.011	5.043	5.136	5.193	5.193	5.247	5.22	5.323	5.338
4.745	4.905	4.977	5.043	5.136	5.273	5.347	5.394	5.416	5.394	5.521	5.617
4.787	4.942	4.977	5.106	5.165	5.247	5.323	5.416	5.394	5.371	5.438	5.521
4.605	4.828	4.828	4.977	5.043	5.165	5.22	5.273	5.247	5.22	5.298	5.375
4.7	4.905	5.011	5.075	5.106	5.22	5.22	5.298	5.323	5.347	5.416	5.472
4.745	4.942	5.011	5.075	5.106	5.247	5.273	5.323	5.347	5.371	5.416	5.481
4.654	4.828	4.828	4.977	4.977	5.043	5.136	5.165	5.165	5.165	5.193	5.204
4.828	4.977	5.011	5.106	5.165	5.22	5.273	5.323	5.371	5.394	5.46	5.576

are missing. Clearly, this increases the variability in the observed mean responses. Also, the strong positive correlations between successive measurements mean that a succession of positive residuals is less significant than it would be for uncorrelated data. Finally, and most interestingly, we might question whether calving date is independent of the measurement process. If later-calving cows are also more likely to produce milk with a lower protein content, we would expect the observed mean responses to rise towards the end of the study. The implications for the interpretation of the fitted model are subtle, and will be pursued in Chapter 11.

Example 5.2. Body-weights of cows

These data, provided by Dr Andrew Lepper (CSIRO Division of Animal

Table 5.2. — *continued*

Time in days										
478	508	536	569	599	627	655	668	723	751	781
5.687	5.737	5.814	5.799	5.784	5.844	5.886	5.914	5.979	5.927	5.94
5.799	5.872	5.886	5.872	5.914	5.966	5.991	6.016	6.087	6.098	6.153
5.72	5.753	5.784	5.784	5.784	5.814	5.829	5.872	5.927	5.9	5.991
5.737	5.844	5.858	5.872	5.886	5.927	5.94	5.979	6.052	6.028	6.12
5.67	5.737	5.784	5.768	5.784	5.784	5.829	5.858	5.914	5.9	5.94
5.438	5.493	5.521	5.497	5.541	5.561	5.58	5.58	5.635	5.561	5.652
5.687	5.72	5.737	5.72	5.737	5.753	5.768	5.784	5.844	5.844	5.9
5.687	5.704	5.784	5.768	5.768	5.814	5.829	5.858	5.94	5.94	6.004
5.784	5.823	5.858	5.9	5.94	5.991	6.016	6.064	6.052	6.016	5.979
5.501	5.521	5.58	5.58	5.635	5.67	5.72	5.753	5.799	5.829	5.858
5.438	5.438	5.501	5.501	5.541	5.598	5.58	5.635	5.687	5.72	5.704
5.687	5.704	5.72	5.704	5.704	5.72	5.737	5.768	5.858	5.9	5.94
5.561	5.58	5.561	5.635	5.687	5.72	5.72	5.737	5.784	5.814	5.799
5.67	5.687	5.72	5.72	5.799	5.858	5.872	5.914	5.94	5.991	6.016
5.541	5.58	5.617	5.67	5.704	5.753	5.768	5.814	5.872	5.927	5.927
5.617	5.652	5.687	5.72	5.753	5.768	5.814	5.844	5.886	5.886	5.886
5.617	5.635	5.704	5.737	5.784	5.768	5.814	5.844	5.886	5.94	5.927
5.416	5.46	5.541	5.481	5.501	5.635	5.652	5.598	5.635	5.635	5.598
5.617	5.67	5.72	5.72	5.799	5.858	5.886	5.914	5.979	6.004	6.028
5.371	5.394	5.438	5.416	5.501	5.561	5.541	5.58	5.652	5.67	5.704
5.617	5.617	5.67	5.635	5.652	5.687	5.652	5.617	5.687	5.768	5.814
5.521	5.561	5.635	5.617	5.687	5.72	5.737	5.737	5.768	5.768	5.704
5.371	5.416	5.501	5.501	5.521	5.561	5.617	5.635	5.72	5.737	5.768
5.501	5.541	5.598	5.598	5.598	5.652	5.67	5.704	5.737	5.768	5.784
5.501	5.541	5.598	5.598	5.635	5.687	5.704	5.72	5.829	5.844	5.9
5.22	5.273	5.371	5.347	5.46	5.58	5.635	5.67	5.753	5.799	5.844
5.652	5.617	5.687	5.67	5.72	5.784	5.784	5.784	5.829	5.814	5.844

Health, Melbourne) consist of body-weights of 27 cows, measured at 23 unequally spaced times over a period of about 22 months. They are listed in Table 5.2. For the analysis, we use a time-scale which runs from 0 to 66, each unit representing 10 days. One animal showed an abnormally low weight gain throughout the experiment and we have removed it from the model-fitting analysis. The remaining 26 animals were allocated amongst treatments in a 2 by 2 factorial design. The two factors were presence/absence of iron dosing and of infection by the organism *M. paratuberculosis*. The replications in the four groups were: 4 × control, 3 × iron only, 9 × infection only, 10 × iron and infection. For further details of the biological background to these data, see Lepper *et al.* (1989).

We use a log-transformation of the bodyweights as the response variable, to stabilize the variance over time. The resulting data are shown in Fig. 5.7. The empirical variogram of ordinary least-squares residuals from a saturated model for the mean response, that is, fitting a separate parameter for each combination of time and treatment, is shown in Fig. 5.8. As in the previous section, the model described in Section 5.2.3 seems reasonable, although the detailed behaviour is rather different than that of the milk protein data. For these data, the measurement error variance, τ^2, seems to be relatively small, as we would expect, whereas the between-animal variance, ν^2, is very large. Also, the behaviour of the empirical variogram near the origin suggests that a Gaussian correlation function might be more appropriate than an exponential. With a reparametrization to $\alpha_1 = \tau^2/\sigma^2$ and $\alpha_2 = \nu^2/\sigma^2$, this suggests a model in which the theoretical variance of log-weight is $\sigma^2(1 + \alpha_1 + \alpha_2)$ and the theoretical variogram takes the form

$$\gamma(u) = \sigma^2\{\alpha_1 + 1 - \exp(-\alpha_3 u^2)\}.$$

Fitting the mean response profiles into the framework of the general linear model, $\boldsymbol{\mu} = X\boldsymbol{\beta}$, proves to be difficult. This is not untypical of growth studies in which the mean response shows an initially sharp increase before gradually levelling off as it approaches an asymptote. Low-degree polynomials provide poor approximations to this behaviour. A way forward is to use 23 parameters to describe the control mean response at each of the 23 time-points, and to model the factorial effects, or differences between groups, parametrically. This would not be sensible if the mean response profile were of intrinsic interest, but this is not so here. Interest focuses on the factorial effects. This situation is not uncommon, and the approach of modelling only contrasts between treatments, rather than treatment means themselves, has been advocated by a number of authors including Evans and Roberts (1979), Cullis and McGilchrist (1990) and Verbyla and Cullis (1990).

Figure 5.9 shows the observed mean response in the control group, and the differences in mean response for the other three groups relative to the control. These differences do seem amenable to a linear modelling approach. The fitted curves are based on a model in which each treatment contrast is a quadratic function of time, whilst the control mean is described by a separate parameter at each of the 23 time-points. Note that the fit to the observed control mean is nevertheless not exact because it includes indirect information from the other three treatment groups, as a consequence of our using a parametric model for the treatment contrasts. The standard errors of the fitted control means range from 0.044 to 0.046. Figure 5.10 shows the fit to the empirical variogram. This seems to be satisfactory, and confirms that the dominant component of variation is between animals; the estimated parameters in the covariance structure are $\hat{\sigma}^2 = 0.0016, \hat{\alpha_1} = 0.353, \hat{\alpha_2} = 4.099$, and $\hat{\alpha_3} = 0.0045$.

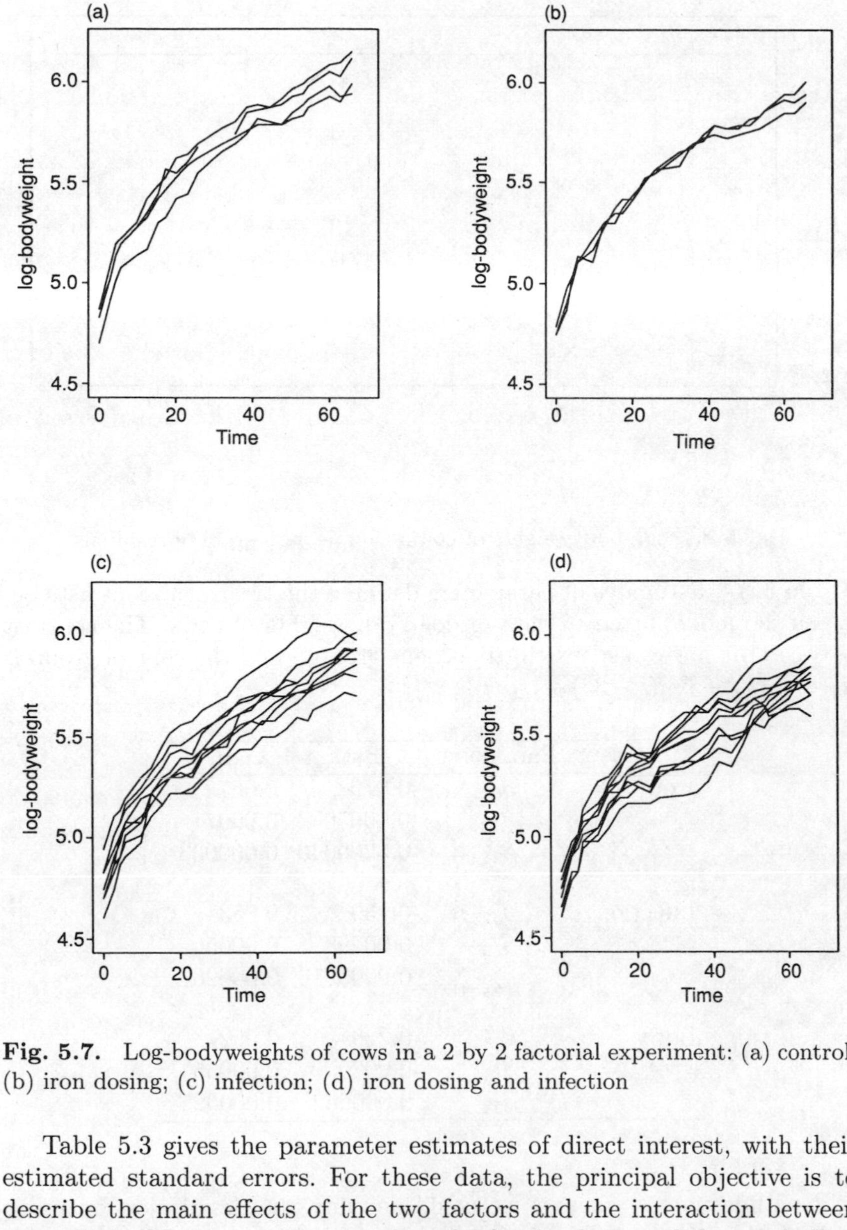

Fig. 5.7. Log-bodyweights of cows in a 2 by 2 factorial experiment: (a) control; (b) iron dosing; (c) infection; (d) iron dosing and infection

Table 5.3 gives the parameter estimates of direct interest, with their estimated standard errors. For these data, the principal objective is to describe the main effects of the two factors and the interaction between them. Note that in our provisional model for the data, each factorial effect is a quadratic function of time, representing the difference in mean log-body-weight between animals treated at the higher and lower levels of the effect in question.

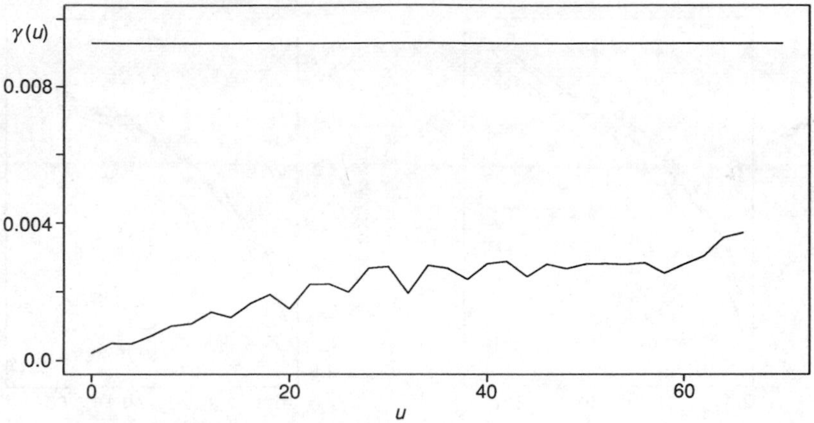

Fig. 5.8. Log-bodyweights of cows: sample variogram of residuals.

Table 5.3. Estimates of parameters defining the treatment contrasts between the model fitted to data on log-bodyweights of cows. The contrast between the mean response in treatment group g and the control group is $\mu_g(t) = \beta_{0g} + \beta_{1g}(t - 33) + \beta_{2g}(t - 33)^2$

Treatment	Parameter	Est.	SE
Iron	β_{01}	-0.079	0.067
	β_{11}	-0.00050	0.00071
	β_{21}	-0.0000040	0.000033
Infection	β_{02}	-0.159	0.053
	β_{12}	-0.00098	0.00056
	β_{22}	0.0000516	0.00026
Both	β_{03}	-0.239	0.052
	β_{13}	-0.0020	0.00055
	β_{23}	0.000062	0.000025

One question is whether we could simplify the description by using linear rather than quadratic effects. Using the notation of Table 5.3, this amounts to a test of the hypothesis that $\beta_{21} = \beta_{22} = \beta_{23} = 0$. The chi-squared test statistic to test this hypothesis is $T_0 = 10.785$ on three degrees of freedom, corresponding to a p-value of 0.020; we therefore retain the quadratic description of the factorial effects. Attention now focuses on the significance of the two main effects and of the interaction between them.

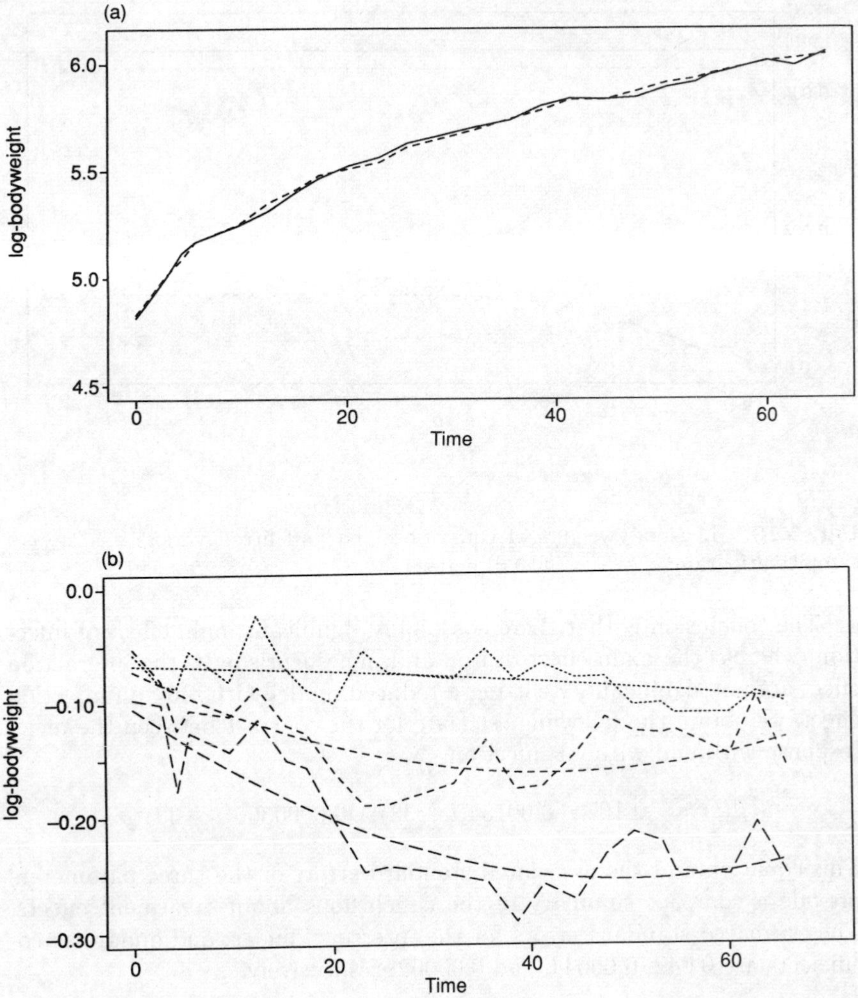

Fig. 5.9. Log-bodyweights of cows: observed and fitted mean response profiles: (a) control —— : observed; - - - : fitted; (b) difference between control and treated, with fitted contrasts shown as smooth curves .... : iron; - - - : infection; — — : both.

Again using the notation of Table 5.2, the hypothesis of no main effect for iron is $\beta_{01} = \beta_{11} = \beta_{21} = 0$, that of no main effect for infection is $\beta_{02} = \beta_{12} = \beta_{22} = 0$, and that of no interaction is $\beta_{03} = \beta_{01} + \beta_{02}$; $\beta_{13} = \beta_{11} + \beta_{12}$; $\beta_{23} = \beta_{21} + \beta_{22}$. The chi-squared statistics, T_0, each on three degrees of freedom, and their associated p-values, are as follows: Iron - $T_0 = 1.936$, $p = 0.586$; Infection - $T_0 = 14.32$, $p = 0.003$; Interaction - $T_0 = 0.521$, $p = 0.914$.

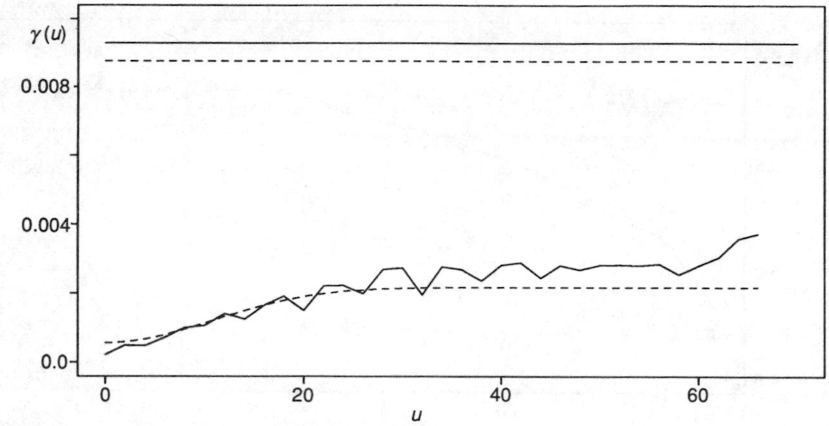

Fig. 5.10. Log-bodyweights of cows: observed and fitted variograms. —— : sample variogram; – – – : fitted model.

The conclusion is that there is a highly significant main effect of infection, whereas the main effect of iron and, not surprisingly, the interaction effect are not significant. Refitting a reduced model with only an infection effect, we obtain the following estimate for the contrast between the mean response with and without infection:

$$\hat{\mu}(t) = -0.167 - 0.00134(t - 33) + 0.0000566(t - 33)^2.$$

This equation and the associated standard errors of the three parameters provide a compact summary of the conclusions about treatment effects. The estimated standard errors for the intercept, linear, and quadratic coefficients are 0.042, 0.00044, and 0.00002, respectively.

5.5 Non-parametric modelling of the mean response

The second case-study in Section 5.4 introduced the idea of describing the mean response profile non-parametrically. There, the approach was to use a separate value for the mean response at each time-point, with no attempt to link the mean responses at different times. This assumes firstly that the complete mean response curve is not of direct interest, and secondly that the times of measurement are common to all (or at least a reasonable number) of the units, so as to give the required replication at each time. An instance in which neither of these assumptions holds is provided by the CD4+ data of Example 1.1. For these data, the CD4+ count as a function of time since seroconversion is of direct interest, and the sets of times of measurement are essentially unique to each person in the study. To handle this situation, we now consider how to fit *smooth* non-parametric models to the mean response profile as a function of time, whilst continuing to recognize the covariance structure within units.

In Section 3.3 we discussed the use of smooth, non-parametric curves as exploratory tools, drawing mainly on well-established methods for cross-sectional data. If we want to use these methods for confirmatory analyses, we need to consider more carefully how the correlation structure of longitudinal data impinges on considerations of how much to smooth the data, and how to make inferences.

To develop ideas, we assume that there are no experimental treatments or other explanatory variables. The data can be represented as $\{(y_{ij}, t_{ij}),\ j = 1, \ldots, n_i;\ i = 1, \ldots, m\}$, where n_i is the number of measurements on the *ith* of m units. We write $N = \Sigma_{i=1}^m n_i$ for the total number of measurements. Our model for the data takes the form

$$Y_{ij} = \mu(t_{ij}) + \epsilon_i(t_{ij}),$$

where $\{\epsilon_i(t),\ t \in R\}$ for $i = 1, \ldots, m$ are independent copies of a stationary random process, $\{\epsilon(t)\}$, with variance σ^2 and correlation function $\rho(u)$, and the mean response, $\mu(t)$, is a smooth function of t.

A simple, and intuitively appealing, non-parametric estimate of $\mu(t)$ is a weighted average of the data, with large weights given to measurements at times t_{ij} which are close to t. To implement this idea, we define a *kernel function*, $K(u)$, to be a symmetric, non-negative valued function taking large values close to $u = 0$ and small values when $\mid u \mid$ is large. For example, in what follows we use the Gaussian kernel,

$$K(u) = \exp(-u^2/2), \tag{5.5.1}$$

as in Section 3.3. Now, choose a positive number, h, and define weights

$$w_{ij}^*(t) = h^{-1} K\{(t_{ij} - t)/h\}. \tag{5.5.2}$$

Note that $w_{ij}^*(t)$ is large when t_{ij} is close to t and *vice-versa*, and that the rate at which the $w_{ij}^*(t)$ decrease as $\mid t_{ij} - t \mid$ increases is governed by the value of h, a small value giving a rapid decrease. Define standardized weights,

$$w_{ij}(t) = w_{ij}^*(t)\{\sum_{i=1}^{m}\sum_{j=1}^{n_i} w_{ij}^*(t)\}^{-1},$$

so that $\Sigma_{i=1}^m \Sigma_{j=1}^{n_i} w_{ij}(t) = 1$ for any value of t. Then, a non-parametric estimate of $\mu(t)$ is

$$\hat{\mu}(t) = \sum_{i=1}^{m}\sum_{j=1}^{n_i} w_{ij}(t) y_{ij}. \tag{5.5.3}$$

A useful refinement of (5.5.3) is an adaptive kernel estimator, in which we replace the constant, h, by a function of t such that $h(t)$ is small when there is a high density of data close to t, and *vice-versa*. This is consistent with the view that for repeated data the observed average at time t is a reasonable non-parametric estimate of $\mu(t)$ without any smoothing over a range of neighbouring times, that is, in effect setting $h = 0$. Also, using a function $h(t)$ with this qualitative behaviour can be shown more generally to improve the estimates of $\mu(t)$. See, for example, Silverman (1984) who demonstrates that a variable kernel estimate is approximately equivalent to a smoothing spline. In the remainder of this section, we consider in detail the case of a constant value of h. The essential ideas apply also to the case of adaptive kernel estimation where, typically, the function $h(t)$ is indexed by a scalar quantity, b say, which controls the overall degree of smoothing and takes on the role played by the constant, h, in non-adaptive kernel estimation.

Whilst the choice of h has a direct and major impact on the resulting estimate $\hat{\mu}(t)$, the choice of kernel function is generally held to be of secondary importance. The Gaussian kernel (5.5.1) is intuitively sensible but not uniquely compelling.

Estimators of this kind were introduced by Priestley and Chao (1972) for independent data, and have been studied in the longitudinal data setting by Hart and Wehrly (1986), Müller (1988), Altman (1990), Rice and Silverman (1991), and Hart (1991). From our point of view, the most important question concerns the choice of h. In some applications, it will be sensible to choose h to have a particular substantive interpretation; for example, smoothing over a 'natural' time-window such as a day if we wish to eliminate ciradian variation. If we want to choose h automatically from the data, the main message from the above-cited work is that a good choice of h depends on the correlation structure of the data, and in particular that methods for choosing h based on a false assumption of independence between measurements can give very misleading results.

Rice and Silverman (1991) give an elegant, cross-validatory prescription for choosing h which makes no assumptions about the underlying correlation structure. Their detailed results are for smoothing spline estimates of $\mu(t)$, but the method is easily adaptable to kernel estimates. For a given h, let $\hat{\mu}^{(k)}(t)$ be the estimate of $\mu(t)$ obtained from (5.5.3) but omitting the kth subject, thus

$$\hat{\mu}^{(k)} = \Sigma_{i \neq k} \Sigma_{j=1}^{n_i} w_{ij}^{(k)}(t) y_{ij},$$

where $w_{ij}^{(k)}(t) = w_{ij}^*(t) / \left\{\Sigma_{i \neq k} \Sigma_{j=1}^{n_i} w_{ij}^*(t)\right\}$. Then, the Rice and Silverman prescription chooses h to minimize the quantity

$$S(h) = \sum_{i=1}^{m} \sum_{j=1}^{n_i} \left\{ y_{ij} - \hat{\mu}^{(i)}(t_{ij}) \right\}^2 . \tag{5.5.4}$$

The rationale for minimizing (5.5.4) is that it is estimating the mean-square error of $\hat{\mu}(t)$ for $\mu(t)$ averaged across the design points, t_{ij}, up to an additive constant which does not depend on h. To see this, write

$$E\left[\{y_{ij} - \hat{\mu}^{(i)}(t_{ij})\}^2\right] = E\left[\left(\{y_{ij} - \mu(t_{ij})\} + \{\mu(t_{ij}) - \hat{\mu}^{(i)}(t_{ij})\}\right)^2\right]$$
$$= E\left[\{y_{ij} - \mu(t_{ij})\}^2\right] + 2E\left[\{y_{ij} - \mu(t_{ij})\}\{\mu(t_{ij}) - \hat{\mu}^{(i)}(t_{ij})\}\right] + E\left[\{\mu(t_{ij}) - \hat{\mu}^{(i)}(t_{ij})\}^2\right].$$

Now, the first term on the right hand side of this expression is equal to $\text{Var}(y_{ij})$ and does not depend on h, whilst the second is zero because $\text{E}(y_{ij}) = \mu(t_{ij})$ and y_{ij} is independent of $\hat{\mu}^{(i)}(t_{ij})$, by construction. Thus

$$\text{E}[\{y_{ij} - \hat{\mu}^{(i)}(t_{ij})\}^2] = \text{Var}(y_{ij}) + \text{MSE}^{(i)}(t_{ij}, h), \tag{5.5.5}$$

where $\text{MSE}^{(i)}(t,h)$ is the mean square error of $\hat{\mu}^{(i)}(t)$ for $\mu(t)$. Substitution of (5.5.5) into (5.5.4) gives the result.

One interesting thing about the Rice and Silverman prescription is that it does not explicitly involve the covariance structure of the data. However, this structure is accommodated implicitly by the device of leaving out all observations from a single subject in defining the criterion $S(h)$, whereas the standard method of cross-validation would leave out single observations to define a criterion

$$S_0(h) = \sum_{i=1}^{m}\sum_{j=1}^{n_i}\{y_{ij} - \hat{\mu}^{(ij)}(t_{ij})\}^2.$$

Direct computation of (5.5.4) would be very time-consuming with large numbers of subjects. An easier computation, again adapted from Rice and Silverman (1991), is the following. To simplify the notation, we temporarily suppress the dependence of $w_{ij}(t)$ on t, and write $w_i = \Sigma_{j=1}^{n_i} w_{ij}$, so that $\Sigma_{i=1}^{m} w_i = 1$. Note that

$$y_{ij} - \hat{\mu}^{(i)}(t_{ij}) = y_{ij} - \{\hat{\mu}(t_{ij}) - \sum_{j=1}^{n_i} w_{ij}y_{ij}\}/(1 - w_i)$$
$$= y_{ij} - \{\hat{\mu}(t_{ij}) - \sum_{j=1}^{n_i} w_{ij}y_{ij}\}\{1 + w_i/(1 - w_i)\}$$
$$= \{y_{ij} - \hat{\mu}(t_{ij})\}$$
$$+ \{w_i/(1 - w_i)\}\{\sum_{j=1}^{n_i} w_{ij}y_{ij}/w_i - \hat{\mu}(t_{ij})\}. \tag{5.5.6}$$

Using (5.5.6) to compute $S(h)$ avoids the need for explicit computation of the m leave-one-out estimates, $\hat{\mu}^{(i)}(t)$. Further computational savings can be made by collecting the time-points, t_{ij}, into a reduced set, say t_r, $r = 1, \ldots, p$, and computing only the p estimates $\hat{\mu}(t_r)$ for each value of h.

To make inferences about $\hat{\mu}(t)$, we need to know the covariance structure of the data. Let V be the $N \times N$ block-diagonal covariance matrix of the data, where $N = \Sigma_{i=1}^{m} n_i$. For fixed h, the estimate $\hat{\mu}(t)$ defined at (5.5.3) is a linear combination of the data-vector, $\boldsymbol{y}$, say $\hat{\mu}(t) = \boldsymbol{w}(t)'\boldsymbol{y}$. Now, for p values $t_r, r = 1, \ldots, p$, let $\hat{\boldsymbol{\mu}}$ be the vector with rth element $\hat{\mu}(t_r)$, and W the $p \times N$ matrix with r^{th} row $\boldsymbol{w}(t_r)$. Then,

$$\hat{\boldsymbol{\mu}} = W\boldsymbol{y}, \tag{5.5.7}$$

and $\hat{\boldsymbol{\mu}}$ has an approximate multivariate Gaussian sampling distribution with

$$\mathrm{E}(\hat{\boldsymbol{\mu}}) = W\mathrm{E}(\boldsymbol{y})$$

and

$$\mathrm{Var}(\hat{\boldsymbol{\mu}}) = WVW'. \tag{5.5.8}$$

Note that in general, $\hat{\mu}(t)$ is a biased estimator for $\mu(t)$, but that the bias will be small if h is small so that the weights, w_{ij}, decay quickly to zero with increasing $\mid t_{ij} - t \mid$. Conversely, the variance of $\hat{\mu}(t)$ typically increases as h decreases. The need to balance bias against variance is characteristic of non-parametric smoothing problems, and explains the use of a mean-squared-error criterion for choosing h automatically from the data. In practice, a sensible choice of h is one which tends to zero as the number of subjects increases, in which case the bias becomes negligible and we can use (5.5.8) to attach standard errors to the estimates $\hat{\mu}(t)$.

The covariance structure of the data can be estimated from the time-sequences of residuals, $r_{ij} = y_{ij} - \hat{\mu}(t_{ij})$. In particular, the empirical variogram of the r_{ij} can be used to formulate an appropriate model. Likelihood-based estimation of parameters in the assumed covariance structure is somewhat problematical. From a practical point of view the computations will be extensive if the number of subjects is large and there is little replication of the observation times, t_{ij}, which is precisely the situation in which non-parametric smoothing is most likely required. Also, the absence of a parametric model for the mean puts the inference on a rather shaky foundation. Simple curve-fitting (moment) estimates of covariance parameters may be preferable.

A final point to note is that expressions (5.5.7) and (5.5.8) for the mean vector and variance matrix of $\hat{\boldsymbol{\mu}}$ assume that both V and the value of h are specified without reference to the data. The effect of estimating parameters in V will be small when the number of units is large, whilst the effect of calculating h from the data should also be small when $\mu(t)$ is a smooth function although the theoretical issues are less well understood here. The references cited above give detailed discussion of these and other issues.

We now consider how the analysis can be extended to incorporate experimental treatments. For a non-parametric analysis we can simply apply the method separately to subjects from each group. From an interpretive point of view it is preferable to use a common value of the smoothing constant, h, in which case the mean squared error criterion (5.5.4) for choosing h from the data consists of a sum of contributions from within each treatment group. Covariance structure is estimated from the empirical variogram of the residuals, pooled across treatment groups if the data justify a common covariance structure in all groups.

Another possibility is to model treatment contrasts parametrically using a linear model. This gives the following semi-parametric specification of a model for the complete set of data,

$$Y_{ij} = \boldsymbol{x}'_{ij}\boldsymbol{\beta} + \mu(t_{ij}) + \epsilon_i(t_{ij}), \tag{5.5.9}$$

where $\boldsymbol{x}_{ij}$ is a p-element vector of covariates. This formulation covers situations in which explanatory variables other than indicators of treatment group are relevant, as will be the case in many observational studies. For the extended model (5.5.9), the kernel method for estimating $\mu(t)$ can be combined iteratively with a generalized least squares calculation for $\hat{\boldsymbol{\beta}}$, as follows:

(1) Given the current estimate, $\hat{\boldsymbol{\beta}}$, calculate residuals, $r_{ij} = y_{ij} - \boldsymbol{x}'_{ij}\hat{\boldsymbol{\beta}}$, and use these in place of y_{ij} to calculate a kernel estimate, $\hat{\mu}(t)$.

(2) Given $\hat{\mu}$, calculate residuals, $r_{ij} = y_{ij} - \hat{\mu}(t)$, and update the estimate $\hat{\boldsymbol{\beta}}$ using generalized least squares,

$$\hat{\boldsymbol{\beta}} = (X'V^{-1}X)^{-1}X'V^{-1}\boldsymbol{r},$$

where X is the $N \times p$ matrix with rows $\boldsymbol{x}'_{ij}$, V is the assumed block-diagonal covariance matrix of the data and $\boldsymbol{r}$ is the vector of residuals, r_{ij}.

(3) Repeat steps (1) and (2) for convergence.

This algorithm is an example of the back-fitting algorithm described by Hastie and Tibshirani (1990). Typically, few iterations are required unless there is a near-confounding between the linear and non-parametric parts of the model. This might well happen if, for example, we included a polynomial time-trend in the linear part. Further details of the algorithm, and a discussion of the asymptotic properties of the resulting estimators, are given in Zeger and Diggle (1994) and Moyeed and Diggle (1994).

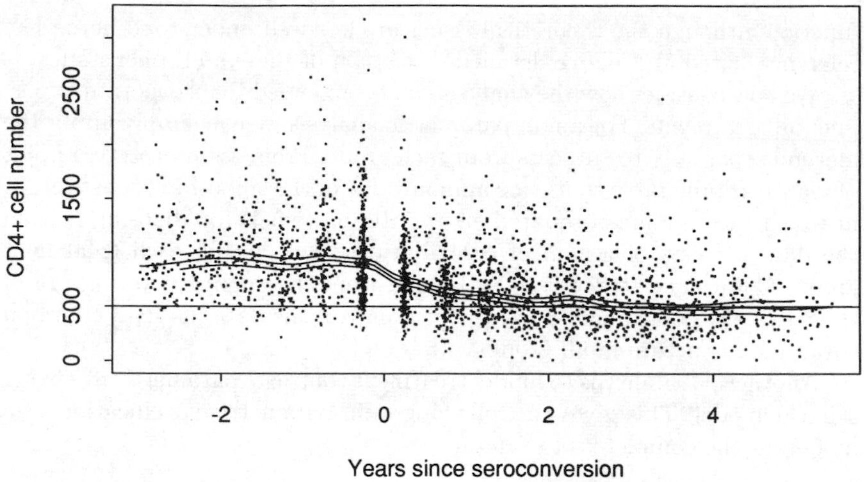

Fig. 5.11. CD4+ cell counts with kernel estimate and pointwise confidence limits for the mean response profile.

Example 5.3. Estimation of the population mean CD4+ curve
For the CD4+ data there are $N = 2376$ observations of CD4+ cell numbers on $m = 369$ men infected with the HIV virus. Time is measured in years with the origin at the date of seroconversion, which is known approximately for each individual. Explanatory variables which we include in the linear part of the model are smoking (packs per day); recreational drug use (yes/no); numbers of sexual partners; and depressive symptoms as measured by the CESD scale. For the non-parametric part of the model, we use a mildly adaptive kernel estimator, with $h(t) = b\{f(t)\}^{-0.25}$ where $f(t)$ is a crude estimate of the local density of observations times, t_{ij}, and b is chosen by cross-validation, as described above.

Figure 5.11 shows the data with the estimate $\hat{\mu}(t)$ and pointwise confidence limits calculated as plus and minus two standard errors. The standard errors were calculated from a parametric model for the covariance structure of the kind introduced in Section 5.2.3. We assume that the variance of each measurement is $\tau^2 + \sigma^2 + \nu^2$ and that the variogram within each unit is $\gamma(u) = \tau^2 + \sigma^2\{1 - \exp(-\alpha u)\}$. Figure 5.12 shows the empirical variogram and a parametric fit obtained by an *ad hoc* procedure, which gave estimates $\hat{\tau}^2 = 14.1, \hat{\sigma}^2 = 16.1, \hat{\nu}^2 = 6.9$, and $\hat{\alpha} = 0.22$.

Figure 5.11 suggests that the mean number of CD4+ cells is approximately constant at close to 1000 cells prior to seroconversion. Within the first six months after seroconversion, the mean drops to around 700. Subsequently, the rate of loss is much slower. For example, it takes nearly three

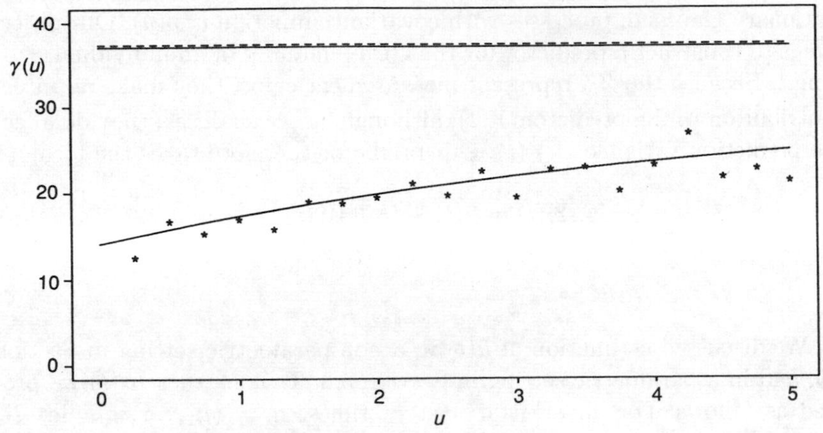

Fig. 5.12. CD4+ cell counts: observed and fitted variograms $\star$: sample variogram – – – : sample variance —— : fitted model

years before the mean number reaches 500, the level at which it is currently recommended that prophylactic AZT therapy should begin (Volberding *et al.*, 1990). As with Example 5.1 on the protein contents of milk samples, the interpretation of the fitted mean response is complicated by the possibility that subjects who become very ill may drop out of the study, and these subjects may also have unusually low CD4+ counts.

5.6 Estimation of individual trajectories

The CD4+ data provide an example in which the mean response is of interest, but not of sole interest. For counselling individual patients, it is that individual's history of CD4+ levels which is relevant. However, because observed CD4+ levels are highly variable over time, in part because of a substantial component of variation due to measurement error, the individual's observed trajectory is not reliable. This problem could arise in either a non-parametric or a parametric setting. Because our motivation is the CD4+ data, we will use the same modelling framework as in the previous section.

The data again consist of measurements, $y_{ij}, j = 1, \ldots, n_i;\ i = 1, \ldots, m$, and associated times, t_{ij}. The model for the data is that

$$Y_{ij} = \mu(t_{ij}) + U_i + W_i(t_{ij}) + Z_{ij},$$

where the U_i are mutually independent $N(0, \nu^2)$, the Z_{ij} are mutually independent $N(0, \tau^2)$ and the $\{W_i(t)\}$ are mutually independent, zero-mean stationary Gaussian processes with covariance function $\sigma^2\rho(u)$. Our objective is to construct a predictor for the CD4+ number of an individual, i, at time t. Because the Z_{ij} represent measurement error, they make no direct contribution to the predictor, $\hat{Y}(t)$, although, as we shall see, they do affect the prediction variance of $\hat{Y}(t)$. Our predictor therefore takes the form

$$\begin{aligned}\hat{Y}_i(t) &= \hat{\mu}(t) + \hat{U} + \hat{W}_i(t)\\ &= \hat{\mu}(t) + \hat{R}_i(t),\end{aligned}$$

say.

We discussed estimation of $\hat{\mu}(t)$ in a non-parametric setting in Section 5.5, and in a parametric setting in Section 5.3. To construct $\hat{R}_i(t)$ we proceed as follows. For an arbitrary set of times, $\boldsymbol{u} = (u_1, \ldots, u_p)$, let $\boldsymbol{R}_i$ be the p-element random vector with rth element $R_i(u_r)$. Let $\boldsymbol{Y}_i$ be the n_i-element vector with jth element $Y_{ij} = \mu(t) + R_i(t_{ij}) + Z_{ij}$, and write $\boldsymbol{t}_i = (t_{i1}, \ldots, t_{in_i})$. Then, as our predictor for $\boldsymbol{R}_i$ we use the conditional expectation

$$\hat{\boldsymbol{R}}_i = \mathrm{E}(\boldsymbol{R}_i \mid \boldsymbol{Y}_i).$$

Under our assumed model, $\boldsymbol{R}_i$ and $\boldsymbol{Y}_i$ have a joint multivariate Gaussian distribution. Furthermore, for any two vectors, $\boldsymbol{u}$ and $\boldsymbol{t}$ say, we can define a matrix, $G(\boldsymbol{u}, \boldsymbol{t})$, whose $(r, j)th$ element is the covariance between $R_i(u_r)$ and $R_i(t_j)$, namely $\nu^2 + \sigma^2\rho(\mid u_r - t_j \mid)$. Then, the covariance matrix of $\boldsymbol{Y}_i$ is $\tau^2 I + G(\boldsymbol{t}_i, \boldsymbol{t}_i)$ and the complete specification of the joint distribution of $\boldsymbol{R}_i$ and $\boldsymbol{Y}_i$ is

$$\begin{bmatrix} \boldsymbol{R}_i \\ \boldsymbol{Y}_i \end{bmatrix} \sim MVN\left\{\begin{bmatrix} 0 \\ \boldsymbol{\mu}_i \end{bmatrix}, \begin{bmatrix} G(\boldsymbol{u},\boldsymbol{u}) & G(\boldsymbol{u},\boldsymbol{t}_i) \\ G(\boldsymbol{t}_i,\boldsymbol{u}) & \tau^2 I + G(\boldsymbol{t}_i,\boldsymbol{t}_i) \end{bmatrix}\right\}$$

where $\boldsymbol{\mu}_i$ has j^{th} element $\mu(t_{ij})$.

Using standard properties of the multivariate Gaussian distribution, as given in Appendix A, it now follows that

$$\hat{\boldsymbol{R}}_i = G(\boldsymbol{u}, \boldsymbol{t}_i)\{\tau^2 I + G(\boldsymbol{t}_i, \boldsymbol{t}_{ij})\}^{-1}(\boldsymbol{Y}_i - \boldsymbol{\mu}_i), \qquad (5.6.1)$$

with variance matrix $\mathrm{Var}(\hat{\boldsymbol{R}}_i)$ defined as

$$\mathrm{Var}(\boldsymbol{R}_i \mid \boldsymbol{Y}_i) = G(\boldsymbol{u},\boldsymbol{u}) - G(\boldsymbol{u},\boldsymbol{t}_i)\{\tau^2 I + G(\boldsymbol{t}_i,\boldsymbol{t}_i)\}^{-1} G(\boldsymbol{t}_i,\boldsymbol{u}). \qquad (5.6.2)$$

Note that when $\tau^2 = 0$ and $\boldsymbol{u} = \boldsymbol{t}_i$, the predictor, $\hat{\boldsymbol{R}}_i$, reduces to $\boldsymbol{Y}_i - \boldsymbol{\mu}_i$ with zero variance, as it should – when there is no measurement error the data are a perfect prediction of the true CD4+ numbers at the observation times. More interestingly, when $\tau^2 > 0$, $\hat{\boldsymbol{R}}_i$ reflects a compromise between

$\boldsymbol{Y}_i - \boldsymbol{\mu}_i$ and zero, tending towards the latter as τ^2 increases. Again, this makes intuitive sense – if measurement error variance is large, an individual's observed CD4+ counts are unreliable and the prediction for that individual should be moved away from the individual's data and towards the population mean trajectory.

Formulae (5.6.1) and (5.6.2) hold also in the semi-parametric case, replacing $\boldsymbol{\mu}$ by $\boldsymbol{\mu} + X\boldsymbol{\beta}$, and in the linear model with $\boldsymbol{\mu} = X\boldsymbol{\beta}$. They do assume that $\mu(t), \boldsymbol{\beta}$ and the covariance matrix, V, are known exactly. They should hold approximately in large samples when $\mu(t), \boldsymbol{\beta}$ and V are estimated, although the precise effects of this are not well understood.

Example 5.4. Prediction of individual CD4+ trajectories

We now apply these ideas to the CD4+ data, using the same estimates of $\mu(t), \boldsymbol{\beta}$ and V as in Example 5.3. Figure 5.13 shows the data for two men along with their predicted trajectories, calculated from (5.6.1), the estimated population mean trajectory $\hat{\mu}(t)$ and covariance matrix $\hat{V}$. These predicted curves are referred to as *empirical Bayes estimates.* Note that the predicted trajectories smooth out the big fluctuations in observed CD4+ numbers, reflecting the substantial measurement error variance, τ^2. In particular, for individual A the predicted trajectory stays above the 500 threshold for AZT therapy throughout the study, whereas the observed CD4+ count reached this threshold two years earlier. For individual B, the fluctuations in observed CD4+ numbers are much smaller and the prediction tracks the data closely. Note that in both cases the general level of the observed data is preserved in the predicted trajectories. This reflects the large component of variance, ν^2, between individuals; the model recognizes that some men are intrinsically high responders and some low responders, and does not force the predictions towards the population mean level.

5.7 Further reading

The subject of parametric modelling for longitudinal data continues to generate an extensive literature. Within the scope of the linear model with correlated errors, most of the ideas are now well established, and the specific models proposed by different authors often amount to sensible variations on a familiar theme. Recent contributions not cited earlier include Pantula and Pollock (1985), Ware (1985), Jones and Ackerson (1990), and Munoz *et al.* (1992). Jones (1993) describes a state-space approach to linear modelling of longitudinal data with serial correlation, drawing on the ideas of Kalman filtering (Kalman, 1960). Non-linear regression models with correlated error structure have been developed by Lindstrom and Bates (1990) and by Glasbey (1980, 1986, 1988), the 1988 paper including a review of the relevant literature.

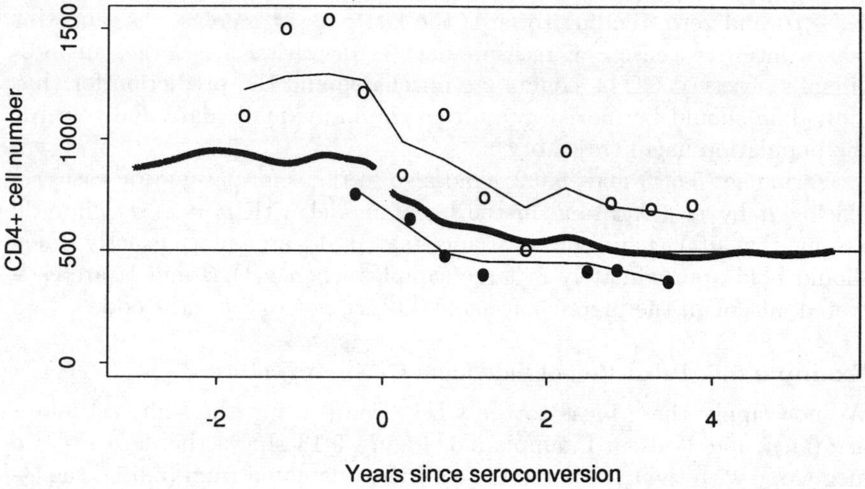

Fig. 5.13. CD4+ cell counts and empirical Bayes predictions for two subjects, with kernel estimate of population mean response profile.

6
Analysis of variance methods

6.1 Preliminaries

In this chapter we describe how simple methods for the analysis of data from designed experiments, in particular the analysis of variance (ANOVA) can be adapted to longitudinal studies. ANOVA has limitations which prevent its recommendation as a general approach for longitudinal data. The first is that it fails to exploit the potential gains in efficiency from modelling the covariance among repeated observations. A second is that, to a greater or lesser extent, ANOVA is a simple method only with a complete, balanced array of data. This second limitation has been substantially diminished by the development of general methodolodgy for incomplete data problems (Little and Rubin, 1987), but the first is fundamental. Whether it is crucial in practice depends on the details of the particular application, for example the size and shape of the data array. As noted in Section 4.7, modelling the covariance structure is most valuable for data consisting of a small number of long sequences of measurements.

Throughout this chapter, we denote the data by a triply subscripted array,

$$y_{hij}, \; h = 1, \ldots, g; \; i = 1, \ldots, m_h; \; j = 1 \ldots, n, \tag{6.1.1}$$

in which each y_{hij} denotes the jth observation from the ith unit within the hth treatment group. Furthermore, we assume a common set of times of observation, t_j, $j = 1, \ldots, n$, for all of the $m = \sum_{h=1}^{g} m_h$ sequences of observations. An emphasis on designed experiments is implicit in this choice of notation, although in principle we could augment the data-array by covariate vectors, $\boldsymbol{x}_{hij}$, attached to each y_{hij}, and so embrace many observational studies.

We write $\mu_{hj} = \mathrm{E}(Y_{hij})$ for the mean response at time t_j in treatment group h, and define the *mean response profile* in group h to be the vector $\boldsymbol{\mu}_h = (\mu_{h1}, \ldots, \mu_{hn})$. The primary objective of the analysis is to make inferences about the μ_{hj} with particular reference to differences amongst the g mean response profiles, $\boldsymbol{\mu}_h$.

6.2 Time-by-time ANOVA

A time-by-time ANOVA consists of n separate analyses, one for each subset of data corresponding to each time of observation, t_j. Each analysis is a conventional ANOVA, based on the appropriate underlying experimental design and incorporating relevant covariate information. Details can be found in any standard text, for example Snedecor and Cochran (1989), Winer (1977), and Mead and Curnow (1983).

The simplest illustration is the case of a one-way ANOVA for a completely randomized design. For the ANOVA table, we use the dot notation to indicate averaging over the relevant subscripts, so that

$$y_{h \cdot j} = m_h^{-1} \sum_{i=1}^{m_h} y_{hij}$$

and

$$\begin{aligned} y_{\cdot \cdot j} &= m^{-1} \sum_{h=1}^{g} \sum_{i=1}^{m_h} y_{hij} \\ &= m^{-1} \sum_{h=1}^{g} m_h y_{h \cdot j} \, . \end{aligned}$$

Then, the ANOVA table for the jth time-point is:

Source of variation	Sum of squares	df
Between treatments	$BTSS = \sum_{h=1}^{g} m_h (y_{h \cdot j} - y_{\cdot \cdot j})^2$	$g-1$
Residual	$RSS = TSS - BTSS$	$m-g$
Total	$TSS = \sum_{h=1}^{g} \sum_{i=1}^{m_h} (y_{hij} - y_{\cdot \cdot j})^2$	$m-1$

The F-statistic to test the hypothesis of no treatment effects is $F = \{BTSS/(g-1)\}/\{RSS/(m-g)\}$. Note that j is fixed throughout the analysis and should *not* be treated as a second indexing variable in a two-way ANOVA.

Whilst a time-by-time ANOVA has the virtue of simplicity, it suffers from two major weaknesses. Firstly, it cannot address questions concerning treatment effects which relate to the longitudinal development of the mean response profiles; for example, growth rates between successive t_j. This can be overcome in part by using observations, y_{hik}, at earlier times as covariates for the response y_{hij}. Kenward's (1987) ante-dependence models, which we discussed in Section 5.2.1, are a logical development of this idea.

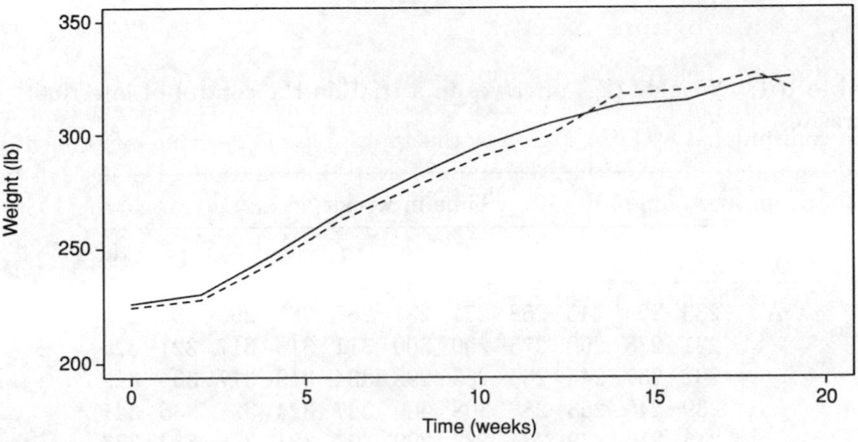

Fig. 6.1. Observed mean response profiles for data on weights of calves. —— : treatment A; – – – : treatment B.

Secondly, the inferences made within each of the n separate analyses are not independent, nor is it clear how they should be combined. For example, a succession of marginally significant group-mean differences may be collectively compelling with weakly correlated data, but much less so if there are strong correlations between successive observations on each unit.

Example 6.1. Weights of calves in a trial on intestinal parasites

Kenward (1987) describes an experiment to compare two different treatments, A and B say, for controlling intestinal parasites of calves. The data consist of the weights of 60 calves, 30 in each group, at each of 11 times of measurement, as listed in Table 6.1.

Figure 6.1 shows the two observed mean response profiles. The observed mean response for treatment B is initially below that for treatment A, but the order is reversed between the seventh and eighth measurement times.

A time-by-time analysis of these data consists of a two-sample t-test at each of the 11 times of measurement. Table 6.2 summarizes the results of these t-tests. None of the 11 tests attains conventional levels of significance and the analysis would appear to suggest that the two mean response profiles are identical.

An alternative analysis is a time-by-time analysis of successive weight gains, $d_{hij} = y_{hij} - y_{hij-1}$. The results of this second analysis are also given in Table 6.2. The striking features of the second analysis are the highly significant negative difference between the mean responses on treatments A and B at time 8, and the positive difference at time 11. This simple analysis leads to essentially the same conclusions as the more sophisticated

Table 6.1. Weights (kg) of calves in a trial on the control of intestinal parasites.

	Time in weeks										
Group	0	2	4	6	8	10	12	14	16	18	19
A	233	224	245	258	271	287	287	287	290	293	297
	231	238	260	273	290	300	311	313	317	321	326
	232	237	245	265	285	298	304	319	317	334	329
	239	246	268	288	308	309	327	324	327	336	341
	215	216	239	264	282	299	307	321	328	332	337
	236	226	242	255	263	277	290	299	300	308	310
	219	229	246	265	279	292	299	299	298	300	290
	231	245	270	292	302	321	322	334	323	337	337
	230	228	243	255	272	276	277	289	289	300	303
	232	240	247	263	275	286	294	302	308	319	326
	234	237	259	289	311	324	342	347	355	368	368
	237	235	258	263	282	304	318	327	336	349	353
	229	234	254	276	294	315	323	341	346	352	357
	220	227	248	273	290	308	322	326	330	342	343
	232	241	255	276	293	309	310	330	326	329	330
	210	225	242	260	272	277	273	295	292	305	306
	229	241	252	265	274	285	303	308	315	328	328
	204	198	217	233	251	258	272	283	279	295	298
	220	221	236	260	274	295	300	301	310	318	316
	233	234	250	268	280	298	308	319	318	336	333
	234	234	254	274	294	306	318	334	343	349	350
	200	207	217	238	252	267	284	282	282	284	288
	220	213	229	252	254	273	293	289	294	292	298
	225	239	254	269	289	308	313	324	327	347	344
	236	245	257	271	294	307	317	327	328	328	325
	231	231	237	261	274	285	291	301	307	315	320
	208	211	238	254	267	287	306	312	320	337	338
	232	248	261	285	292	307	312	323	318	328	329
	233	241	252	273	301	316	332	336	339	348	345
	221	219	231	251	270	272	287	294	292	292	299

Table 6.1. — *continued*

	Time in weeks										
Group	0	2	4	6	8	10	12	14	16	18	19
B	210	215	230	244	259	266	277	292	292	290	264
	230	240	258	277	277	293	300	323	327	340	343
	226	233	248	277	297	313	322	340	354	365	362
	233	239	253	277	292	310	318	333	336	353	338
	238	241	262	282	300	314	319	331	338	348	338
	225	228	237	261	271	288	300	316	319	333	330
	224	225	239	257	268	290	304	313	310	318	318
	237	241	255	276	293	307	312	336	336	344	328
	237	224	234	239	256	266	276	300	302	293	269
	233	239	259	283	294	313	320	347	348	362	352
	217	222	235	256	267	285	295	317	315	308	301
	228	223	246	266	277	287	300	312	308	328	333
	241	247	268	290	309	323	336	348	359	372	370
	221	221	240	253	273	282	292	307	306	317	318
	217	220	235	259	262	276	284	305	303	315	317
	214	221	237	256	271	283	287	314	316	320	298
	224	231	241	256	265	283	295	314	313	328	334
	200	203	221	236	248	262	276	294	291	311	310
	238	232	252	268	285	298	303	320	324	320	327
	230	222	243	253	268	284	290	316	314	330	330
	217	224	242	265	284	302	309	324	328	338	334
	209	209	221	238	256	267	281	295	301	309	289
	224	227	245	267	279	294	312	328	329	297	297
	230	231	244	261	272	283	294	318	320	333	338
	216	218	223	243	259	270	270	290	301	314	297
	231	239	254	276	294	304	317	335	333	319	307
	207	216	228	255	275	285	296	314	319	330	330
	227	236	251	264	276	287	297	315	309	313	294
	221	232	251	274	284	295	300	323	319	333	322
	233	238	254	266	282	294	295	310	320	327	326

Table 6.2. Time-by-time analysis of weights of calves. The analysis at each time is a two-sample t-test, using either the weights, y_{hij}, or the weight-gains, d_{hij}, as the response. The 5 per cent, 1 per cent, and 0.1 per cent critical values of t_{58} are 2.00, 2.16 and 3.47

	Test statistic at time										
	1	2	3	4	5	6	7	8	9	10	11
y	0.60	0.82	1.03	0.88	1.21	1.12	1.28	-1.10	-0.95	-0.53	0.85
d		0.51	0.72	-0.15	1.21	0.80	0.55	-6.85	0.21	0.72	4.13

ante-dependence analysis reported in Kenward (1987).

The overall message of Example 6.1 is that a simple analysis of longitudinal data can be highly effective *if* it focuses on exactly the right feature of the data, which in this case was the weight-gain between successive two-week periods rather than the weight itself. This leads us to a discussion of *derived variables* for longitudinal data analysis.

6.3 Derived variables

Given a vector of observations, $\boldsymbol{y}_{hi} = (y_{hi1}, \ldots, y_{hin})$, on a particular unit, a *derived variable* is a scalar-valued function, $u_{hi} = u(\boldsymbol{y}_{hi})$. The motivation for analysing a derived variable is two-fold. From a pragmatic point of view it reduces a multivariate problem to a univariate one; in particular applications, a single derived variable may convey the essence of the substantive issues raised by the data. For example, in growth studies the detailed growth process for each unit may be complex, yet for some purposes the interest may focus on something as simple as the average growth rate during the course of the experiment. By reducing each $\boldsymbol{y}_{hi}$ to a scalar quantity, u_{hi}, we avoid the issue of correlation within each sequence, and can again use standard ANOVA or regression methods for the data analysis. When there are two or more derived variables of interest, the problem of combined inference which we encountered with the time-by-time ANOVA again needs to be addressed. However, it is arguable that the practical importance of the combined inference question is reduced if the separate derived variables address substantially different questions and each one has a natural interpretation in its own right.

The derived variable approach goes back at least to Wishart (1938), and was systematized by Rowell and Walters (1976). Rowell and Walters set out the details of an analysis for data recorded at equally spaced time-points, t_j, in which the n-dimensional response, $\boldsymbol{y}_{hi}$, is transformed to a set of orthogonal polynomial coefficients of degree, $0, 1, \ldots, n-1$. The first two of these are immediately interpretable in terms of average response and average rate of change, respectively, during the course of the experiment. It is less clear what tangible interpretation can be placed on higher degree coefficients. Also, from an inferential point of view, it is important to recognize that the 'orthogonal' sums of squares associated with the polynomial decomposition are *not* orthogonal in the statistical sense, that is, *not* independent, because the observations within $\boldsymbol{y}_{hi}$ are in general correlated.

A more imaginative use of derived variables is to fit scientifically interpretable *non-linear* models to each time-sequence, $\boldsymbol{y}_{hi}$, and to use the parameter estimates from these fits as a natural set of derived variables. For example, we might decide that the individual time-sequences can be well described by a logistic growth model of the form

$$\mu_{hi}(t) = \alpha_{hi}[1 + \exp\{-\beta_{hi}(t - \delta_{hi})\}]^{-1},$$

in which case estimates of the asymptotes α_{hi}, slope parameters β_{hi} and points of inflection δ_{hi} form a natural set of derived variables. Note that non-linear least squares could be used to estimate the α_{hi}, β_{hi}, and δ_{hi}, and that no assumptions about the correlation structure are needed to validate the subsequent derived variable analysis. However, whilst validity is one thing, efficiency is quite another. The small-sample behaviour of ordinary least squares estimation in non-linear regression modelling is often poor, and this can only be exacerbated by correlations amongst the y_{hij}. Furthermore, the natural parametrization from a scientific point of view may not correspond to a good parametrization from a statistical point of view. In the simpler context of uncorrelated data, these issues are addressed in Bates and Watts (1980) and in Ratkowsky (1983).

Strictly, the derived variable approach breaks down when the array, y_{hij}, is incomplete because of missing values, or when the m times of observation are not common to all units. This is because the derived variable no longer satisfies the standard ANOVA assumption of a common variance for all observations. Also, we cannot simply weight the values of the derived variables according to the numbers of observations contained in the corresponding vectors, $\boldsymbol{y}_{hi}$, because of the unknown effects of the correlation structure in the original data. In practice, derived variables appear to be used somewhat optimistically with moderately incomplete data. The consequences of this are unclear, although the randomization justification for the ANOVA is available if we are prepared to assume that the mechanisms leading to the incompleteness of the data are independent of both the experimental treatments applied and the measurement process itself. As we shall see in Chapter 11, this is often not so.

Example 6.2. Growth of sitka spruce with and without ozone

To illustrate the use of derived variables we consider the 1989 growth data on Sitka spruce. Recall that these consist of 79 trees grown in four controlled environment chambers in which the first two chambers are treated with introduced ozone at 70 ppb. The respective numbers of trees in the four chambers are 27, 27, 12 and 13. The objective of the experiment was to investigate the effect of ozone on tree growth, and the response from each tree was a sequence of eight measurements of $y = \log(d^2h)$ on days 103, 130, 162, 190, 213, 247, 273 and 308 where day 1 is 1 January 1989, and d and h refer to stem diameter and height, respectively.

In Section 4.6 we saw that the dominant source of variation in these data was a random shift in the response profile for each tree. This suggests that the observed mean response would be a suitable derived variable. The appropriate analysis is then a one-way ANOVA estimating a separate mean response parameter, μ_h, in each of the four chambers, followed by estimation of the treatment versus control contrast,

$$c = (\mu_1 + \mu_2) - (\mu_3 + \mu_4).$$

The ANOVA table is as follows:

Source of variation	Sum of squares	df	Mean square
Between chambers	2.2448	3	0.7483
Residual	31.0320	75	0.4136
Total	33.2768	78	0.4266

Note that the F-statistic to test for chamber effects is not significant ($F_{3,75} = 1.81, p \simeq 0.15$). However, the estimate of the treatment versus control contrast is

$$\hat{c} = (5.89 + 5.87) - (6.17 + 6.29) = -0.70$$

with estimated standard error

$$\text{se}(\hat{c}) = \left\{ \sqrt{0.4136} \left(\frac{1}{27} + \frac{1}{27} + \frac{1}{12} + \frac{1}{13} \right) \right\} = 0.31$$

corresponding to a t-statistic of 2.26 on 75 df, which is significant ($p \simeq 0.03$). The conclusion from this analysis is that introduced ozone suppresses growth of sitka spruce, although the evidence is not overwhelming.

Example 6.3. Effect of pH on drying rate of holly leaves

This example concerns an experiment to investigate the effect of different pH treatments on the drying rate of holly leaves. The treatment was administered by intermittent spraying during the plant's life. Ten plants were allocated to each of four pH levels, 2.5, 3.5, 4.5 and 5.6, the last being a control. Leaves from the plant were then allowed to dry, and were weighed at each of 13 times, irregularly spaced over a three-day period. The recorded variable was the ratio of current to initial fresh weight. Thus, if $w(t)$ denotes the weight at time t, the recorded sequence of values from each plant is

$$x_j = w(t_j)/w(0), \; j = 1, \ldots, 13.$$

Note that $t_1 = 0$, and $x_1 = 1$ for every plant.

A plausible model for the drying process is one of exponential decay towards a lower bound which represents the dry weight of the plant material. If this lower bound is a small fraction of the initial weight, a convenient approximation is to assume exponential decay towards zero. This gives $w(t) = w(0)\exp(-\beta t)$ or, dividing by $w(0), x_j = \exp(-\beta t_j)$. Taking $y_j = \log(x_j)$, this in turn gives a linear regression through the origin,

$$y_j = -\beta t_j + \epsilon_j.$$

Figure 6.2 shows the 40 sequences of measurements, y_j, plotted against t_j. The linear model gives a good approximation, but with substantial variation in the slope, $-\beta$, for the different plants. This suggests using a least-squares estimate of β from each of the 40 plants as a derived variable. Note that β has a direct interpretation as a *drying rate*. Table 6.3 shows the resulting values, b_{hi}, $h = 1, \ldots, 4$; $i = 1, \ldots, 10$, the subscripts h and i referring to pH level and replicate within pH level, respectively.

To analyse this derived variable, we would assume that the b_{hi} are realizations of Gaussian random variables, with constant variance and means dependent on pH level. However, the pattern of the standard errors in Table 6.3 casts doubt on the constant variance assumption. For this reason, we transform the b_{hi} to $z_{hi} = \log(b_{hi})$ and assume that $z_{hi} \sim N(\mu_h, \sigma^2)$, where μ_h denotes the mean value of the derived variable within pH level h. Plots of b_{hi} against pH, and of z_{hi} against pH, show the effect of the log-transformation on the distribution of the derived variable. In particular, whilst there is a clear relationship between mean and variance for the b_{hi}, a constant variance model does seem reasonable for the z_{hi}.

We now consider three possible models for μ_h, the mean log-drying-rate within pH level, h. These are:

(1) $\mu_h = \mu$ (no pH effect);

(2) μ_h arbitrary;

(3) $\mu_h = \theta_0 + \theta_1 u_h$, where u_h denotes pH level in treatment group h.

A comparison between models 1 and 2 involves a one-way ANOVA of the data in Table 6.3. The F-statistic is 3.14 on 3 and 35 degrees of freedom,

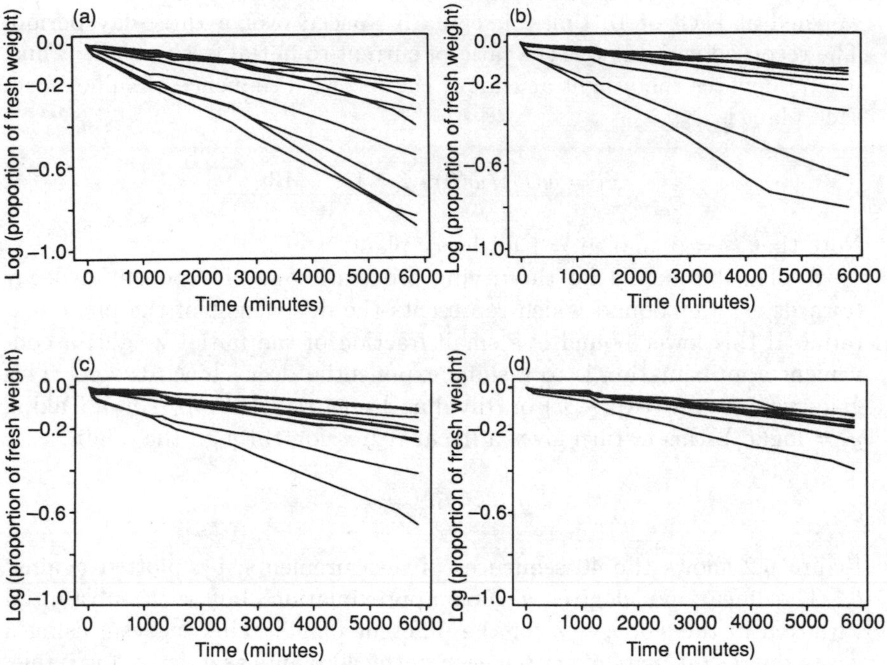

Fig. 6.2. Data on drying rates of holly leaves: (a) pH=2.5; (b) pH=3.5; (c) pH=4.5; (d) pH=5.6 (control).

corresponding to a p-value of 0.037 which is indicative of differences in mean response between pH levels. Model 3 is motivated by the observation that the observed rates of drying seem to depend on pH in a roughly linear fashion. The F-statistic to test the hypothesis $\theta_1 = 0$ within model 3, using the residual mean square from the saturated model 2, is 9.41 on 1 and 36 degrees of freedom. This corresponds to a p-value of 0.004, representing strong evidence against $\theta_1 = 0$. The F-statistic to test lack of fit of model 3 within model 2 is 0.31 on 2 and 36 degrees of freedom, which is clearly not significant ($p = 0.732$).

The conclusion from this analysis is that log-drying rate decreases, approximately linearly, with increasing pH.

6.4 Repeated measures

A *repeated measures* ANOVA can be regarded as a first attempt to provide a single analysis of a complete longitudinal dataset. The rationale for the analysis is to regard time as a factor on n levels in a hierarchial design with units as sub-plots. In agricultural research, this type of experiment

Table 6.3. Derived data (b_{hi}) for analysis of the drying rates of holly leaves

Replication	ph			
	2.5	3.5	4.5	5.6
1	0.33	0.27	0.76	0.23
2	0.30	0.35	0.43	0.68
3	0.91	0.26	0.18	0.28
4	1.41	0.10	0.21	0.32
5	0.41	0.24	0.17	0.21
6	0.41	1.51	0.32	0.43
7	1.26	1.17	0.35	0.31
8	0.57	0.26	0.43	0.26
9	1.51	0.29	1.11	0.28
10	0.55	0.41	0.36	0.28
Mean	0.77	0.54	0.44	0.33
Std. error	(0.15)	(0.14)	(0.09)	(0.04)

is usually called a *split-plot* experiment. However, the usual randomization justification for the split-plot analysis is not available because there is no sense in which the allocation of times to the n observations within each unit can be randomized. We therefore have to assume an underlying model for the data, namely,

$$y_{hij} = \beta_h + \gamma_{hj} + U_{hi} + Z_{hij}, \; j = 1, \ldots, n; \; i = 1, \ldots, m_h; \; h = 1, \ldots, g \tag{6.4.1}$$

in which the β_h represent main effects for treatments and the γ_{hj} interactions between treatments and times with the constraint that $\sum_{j=1}^{n} \gamma_{hj} = 0$, for all h. Also, the U_{hi} are mutually independent random effects for units and the Z_{hij} mutually independent random measurement errors.

In the model (6.4.1), $\mathrm{E}(Y_{hij}) = \beta_h + \gamma_{hj}$. If we assume that $U_{hi} \sim N(0, \nu^2)$ and $Z_{hij} \sim N(0, \sigma^2)$, then the resulting distribution of $\boldsymbol{Y}_{hi} = (Y_{hi1}, \ldots, Y_{hin})$ is multivariate Gaussian with covariance matrix

$$\boldsymbol{V} = \sigma^2 I + \nu^2 J,$$

where I is the identity matrix and J a matrix all of whose elements are 1. This implies a constant correlation, $\rho = \nu^2/(\nu^2 + \sigma^2)$, between any two observations on the same unit.

The split-plot ANOVA for the model (6.4.1) is given by the following table. In the table, we again use the dot notation for averaging, so that $y_{hi\cdot} = n^{-1} \sum_{j=1}^{n} y_{hij}$, $y_{h\cdot\cdot} = (m_h n)^{-1} \sum_{i=1}^{m_h} \sum_{j=1}^{n} y_{hij}$, and so on. Also,

Source of variation	Sum of squares	df
Between treatments	$BTSS_1 = m\sum_{h=1}^{g} m_h(y_{h\cdot\cdot} - y_{\cdot\cdot\cdot})^2$	$g-1$
Whole plot residual	$RSS_1 = TSS_1 - BTSS_1$	$m-g$
Whole plot total	$TSS_1 = m\sum_{h=1}^{g}\sum_{i=1}^{m_h}(y_{hi\cdot} - y_{\cdot\cdot\cdot})^2$	
Between times	$BTSS_2 = n\sum_{j=1}^{n}(y_{\cdot\cdot j} - y_{\cdot\cdot\cdot})^2$	$n-1$
Treatment by time interaction	$ISS_2 = \sum_{j=1}^{n}\sum_{h=1}^{g} n_h(y_{h\cdot j} - y_{\cdot\cdot})^2$ $-BTSS_1 - BTSS_2$	$(g-1)\times$ $(n-1)$
Split-plot residual	$RSS_2 = TSS_2 - ISS_2 - BTSS_2$ $-TSS_1$	$(m-g)\times$ $(n-1)$
Split-plot total	$TSS_2 = \sum_{h=1}^{g}\sum_{i=1}^{m_h}\sum_{j=1}^{n}(y_{hij} - y_{\cdot\cdot\cdot})^2$	$nm-1$

in the table, $m = \sum_{h=1}^{g} m_h$ denotes the total number of units. The first F-statistic associated with the table is

$$F_1 = \{BTSS_1/(g-1)\}/\{RSS_1/(m-g)\},$$

which tests the hypothesis that $\beta_h = \beta,\ h = 1\ldots,g$. The second is

$$F_2 = \{ISS_2/[(g-1)(n-1)]\}/\{RSS_2/[(m-g)(n-1)]\},$$

which tests the hypothesis that $\gamma_{hj} = \gamma_j,\ h = 1,\ldots,g$ for each of $j = 1,\ldots,n$, that is, that all g group mean response profiles are parallel.

The split-plot ANOVA strictly requires a complete data array. As noted in Section 6.1, recent developments in methodology for incomplete data relax this requirement. Alternatively, and in our view preferably, we can analyse incomplete data under the split-plot *model* (6.4.1) by the general likelihood-based approach given in Chapters 4 and 5 with the assumption of a constant correlation between any two measurements on the same unit and, incidentally, a general linear model for the mean response profiles.

Example 6.4. Growth of Sitka spruce with and without ozone (continued) The following table presents a split-plot ANOVA for the 1989 Sitka spruce data. We ignore possible chamber effects to ease comparison with the results from the robust analysis given in Section 4.6.

Source of variation	Sum of squares	df	Mean square
Between treatments	17.106	1	17.106
Whole plot residuals	249.108	77	3.235
Whole plot total	266.214	78	3.413
Between times	41.521	7	5.932
Treatment by time interaction	0.045	7	0.0064
Split-plot residual	5.091	539	0.0094
Split-plot total	312.871	631	

The F-statistic for differences between control and treated mean response is $F_1 = 5.29$ on 1 and 77 degrees of freedom, corresponding to a p-value of approximately 0.02. The F-statistic for departure from parallelism is $F_2 = 0.68$ on 7 and 539 degrees of freedom, corresponding to a p-value of approximately 0.69. The analysis therefore gives moderate evidence of a difference in mean response between control and treated trees, and no evidence of departure from parallelism.

6.5 Conclusions

The principal virtue of the ANOVA approach to longitudinal data analysis is its technical simplicity. The computational operations involved are elementary and there is an apparently reassuring familiarity in the solution of a complex class of problems by standard methods. However, we believe that this virtue is outweighed by the inherent limitations of the approach as noted in Section 6.1.

Provided that the data are complete, the method of derived variables can give a simple and easily interpretable analysis with a strong focus on particular aspects of the mean response profiles. Furthermore, it provides a feasible method of analysing inherently non-linear models for the mean response profiles, notwithstanding the statistical difficulties which arise with non-linear least-squares estimation in small samples. The method runs into trouble if the data are seriously incomplete, or if no single derived variable can address the relevant scientific questions. In the former case, the inference is strictly invalidated whilst in the latter there are difficulties in making a correct combined inference for the complete analysis. The method is of course not applicable if a key covariate changes within a subject.

Time-by-time ANOVA can be viewed as a special case of a derived variable analysis in which the problem of combined inference is seen in an acute form. Additionally, the implicitly cross-sectional view of the data fails to address the question of evolution over time which is usually of fundamental importance in longitudinal studies.

The split-plot ANOVA embodies strong assumptions about the covariance structure of the data. If these are reasonable, a model-based analysis under the assumed uniform correlation structure achieves the same ends, while coping naturally with missing values and allowing a structured linear model for the mean response profiles.

In summary, whilst ANOVA methods are undoubtedly useful in particular circumstances, they do not constitute a generally viable approach to longitudinal data analysis.

7 Generalized linear models for longitudinal data

This chapter surveys approaches to the analysis of discrete and continuous longitudinal data using extensions of generalized linear models (GLMs). We have shown in Chapter 4 how regression inferences using *linear* models can be made robust to assumptions about the correlation, especially when the number of observations per person, n_i, is small relative to the number of individuals, m. With linear models, although the *estimation* of the regression parameters must take into account the correlations in the data, their *interpretation* is essentially independent of the correlation structure. With non-linear models for discrete data, such as logistic regression, different assumptions about the source of correlation can lead to regression coefficients with distinct interpretations. The data analyst must therefore think even more carefully about the objectives of the analysis and the source of correlation in choosing an approach.

In this chapter we discuss three extensions of GLMs for longitudinal data: *marginal*, *random effects*, and *transition* models. The objective is to present the ideas underlying each model as well as their domains of application. We focus on the interpretation of regression coefficients. Chapters 8 to 10 present details about each method and examples of their use.

7.1 Marginal models

In a marginal model, the regression of the response on explanatory variables is modelled separately from within-person (within-unit) correlation. In the regression, we model the marginal expectation, $\mathrm{E}(Y_{ij})$, as a function of explanatory variables. By marginal expectation, we mean the average response over the sub-population that shares a common value of x. The marginal expectation is what we model in a cross-sectional study. Specifically, a marginal model has the following assumptions:

- the marginal expectation of the response, $\mathrm{E}(Y_{ij}) = \mu_{ij}$, depends on explanatory variables, $\boldsymbol{x}_{ij}$, by $h(\mu_{ij}) = \boldsymbol{x}_{ij}'\beta$ where h is a known *link* function such as the logit for binary responses or log for counts;

- the marginal variance depends on the marginal mean according to $\text{Var}(Y_{ij}) = v(\mu_{ij})\phi$ where v is a known variance function and ϕ is a scale parameter which may need to be estimated;
- the correlation between Y_{ij} and Y_{ik} is a function of the marginal means and perhaps of additional parameters, $\boldsymbol{\alpha}$, i.e. $\text{Corr}(Y_{ij}, Y_{ik}) = \rho(\mu_{ij}, \mu_{ik}; \boldsymbol{\alpha})$ where $\rho(\cdot)$ is a known function.

Marginal regression coefficients, $\boldsymbol{\beta}$, have the same interpretation as coefficients from a cross-sectional analysis. Marginal models are natural analogues for correlated data of GLMs for independent data.

Nearly all of the linear models introduced in Chapters 4 to 6 can be formulated as marginal models since they have as part of their specification an equation of the form $\text{E}(Y_{ij}) = \boldsymbol{x}'_{ij}\boldsymbol{\beta}$ for some set of explanatory variables, $\boldsymbol{x}_{ij}$. For example, consider a linear regression whose errors follow the exponential correlation model introduced in Section 4.2.2. Responses, Y_{ij}, are taken at integer times, $t_j = 1, 2, ..., n$, on each of m subjects. The mean responses are $\text{E}(Y_{ij}) = \boldsymbol{x}'_{ij}\boldsymbol{\beta}$. The covariance structure is given by $\text{Cov}(Y_{ij}, Y_{ik}) = \sigma^2 \exp(-\phi \mid t_j - t_k \mid)$ and the variance is assumed to be independent of the mean.

To illustrate a logistic marginal model, consider the problem of assessing the dependence of respiratory infection on vitamin A status in the Indonesian Children's Health Study. Let x_{ij} indicate whether or not child i is vitamin A deficient (1-yes; 0-no) at visit j. Let Y_{ij} denote whether the child has respiratory infection (1-yes; 0-no) and let $\mu_{ij} = \text{E}(Y_{ij})$. One marginal model is given by the following assumptions:

- $\text{logit}(\mu_{ij}) = \log \frac{\mu_{ij}}{1-\mu_{ij}} = \log \frac{\Pr(Y_{ij}=1)}{\Pr(Y_{ij}=0)} = \beta_0 + \beta_1 x_{ij}$
- $\text{Var}(Y_{ij}) = \mu_{ij}(1 - \mu_{ij})$
- $\text{Corr}(Y_{ij}, Y_{ik}) = \alpha$.

Here, the transformed regression coefficient $\exp(\beta_0)$ is the ratio of the frequency of infected to uninfected children among the sub-population that is not vitamin A deficient. The parameter $\exp(\beta_1)$ is the odds of infection among vitamin A deficient children divided by the odds among children replete with vitamin A. When the prevalence of the response is low as in the Indonesian example, the odds ratio is approximately the ratio of frequencies of infection among the vitamin A deficient and replete sub-groups. Note that $\exp(\beta_1)$ is a ratio of population frequencies so we refer to it as a *population-averaged* parameter. If all individuals with the same x have the same probability of disease, the population frequency is the same as the individual's probability. However, when there is heterogeneity in the risk of disease among subjects with a common x, the population frequency is the average of the individual risks.

The variance of a binary response is a known function of its mean as given in the second assumption above. The correlation between two observations for a child is assumed to be α regardless of the times or expectations of the observations. For continuous data, this form of correlation derives from a random intercept in the linear model. It can only be a first order approximation for binary outcomes. To see why, note that the correlation between two binary responses, Y_1 and Y_2 with means μ_1 and μ_2, is given by

$$\text{Corr}(Y_1, Y_2) = \frac{\Pr(Y_1 = 1, Y_2 = 1) - \mu_1\mu_2}{\{\mu_1(1-\mu_1)\mu_2(1-\mu_2)\}^{1/2}}. \tag{7.1.1}$$

The joint probability, $\Pr(Y_1 = 1, Y_2 = 1)$, is constrained to satisfy

$$\max(0, \mu_1 + \mu_2 - 1) < \Pr(Y_1 = 1, Y_2 = 1) < \min(\mu_1, \mu_2) \tag{7.1.2}$$

so that the correlation must satisfy a constraint which depends in a complicated way on the means μ_1 and μ_2 (Prentice, 1988). For this reason many authors, including Lipsitz (1989), Lipsitz *et al.* (1991) and Liang *et al.* (1992) have modelled the association among binary data using the odds ratio

$$\text{OR}(Y_1, Y_2) = \frac{\Pr(Y_1 = 1, Y_2 = 1)\Pr(Y_1 = 0, Y_2 = 0)}{\Pr(Y_1 = 1, Y_2 = 0)\Pr(Y_1 = 0, Y_2 = 1)} \tag{7.1.3}$$

which is not constrained by the means. For example, in the Indonesian study, we might assume that the odds ratios rather than correlations among all pairs of responses for a child are equal to an unknown constant, α. Specifying the means, variances and odds ratios fully determines the covariances among repeated outcomes for one child. Modelling odds ratios instead of correlations is further discussed in Chapter 8.

7.2 Random effects models

Section 5.2 introduced the linear random effects model where the response is assumed to be a linear function of explanatory variables with regression coefficients that vary from one individual to the next. This variability reflects natural heterogeneity due to unmeasured factors. An example is a simple linear regression for infant growth where the coefficients represent birth weight and growth rate. Children obviously are born at different weights and have different growth rates due to genetic and environmental factors which are difficult or impossible to quantify. A random effects model is a reasonable description if the set of coefficients from a population of children can be thought of as a sample from a distribution. Given the actual coefficients for a child, the linear random effects model further assumes that repeated observations for that person are independent. The correlation among repeated observations arises because we cannot observe the underlying growth curve, that is, the true regression coefficients, but have only imperfect measurements of weight on each infant.

This idea extends naturally to regression models for discrete and non-Gaussian continuous responses. It is assumed that the data for a subject are independent observations following a GLM, but that the regression coefficients can vary from person to person according to a distribution, F. To illustrate, once again consider a logistic model for the probability of respiratory infection in the Indonesian Children's Health Study. We might assume that the propensity for respiratory infection varies across children, reflecting their different genetic predispositions and unmeasured influences of environmental factors. The simplest model would assume that every child has their own propensity for respiratory disease but that the effect of vitamin A deficiency on this probability is the same for every child. This model takes the form

$$\text{logit}\,\Pr(Y_i = 1|U_i) = (\beta_0^* + U_i) + \beta_1^* x_{ij} \qquad (7.2.1)$$

where x_{ij} is 1 if child i is vitamin A deficient at visit j, and 0 otherwise. Given U_i, we further assume that the repeated observations for the ith child are independent of one another. Finally, the model requires an assumption about the distribution of the U_i across children in the population. Typically, a parametric model such as the Gaussian with mean zero and unknown variance, ν^2, is used.

In this example β_0^* is the log odds of respiratory infection for a typical child with random effect $U_i = 0$. The parameter β_1^* is the log odds ratio for respiratory infection when a child is deficient relative to when that same child is not. The variance ν^2 represents the degree of heterogeneity across children in the propensity for disease, not attributable to x.

The general specification of the random effects GLM is as follows.

- Given $\boldsymbol{U}_i$, the responses $Y_{i1}, \ldots, Y_{in_i}$ are mutually independent and follow a GLM with density $f(y_{ij}|\boldsymbol{U}_i) = \exp[\{(y_{ij}\theta_{ij} - \psi(\theta_{ij}))\}/\phi + c(y_{ij}, \phi)]$. The conditional moments, $\mu_{ij} = \text{E}(Y_{ij}|\boldsymbol{U}_i) = \psi'(\theta_{ij})$ and $v_{ij} = \text{Var}(Y_{ij}|\boldsymbol{U}_i) = \psi''(\theta_{ij})\phi$, satisfy $h(\mu_{ij}) = \boldsymbol{x}_{ij}'\boldsymbol{\beta}^* + \boldsymbol{d}_{ij}'\boldsymbol{U}_i$ and $v_{ij} = v(\mu_{ij})\phi$ where h and v are known link and variance functions, respectively, and $\boldsymbol{d}_{ij}$ is a subset of $\boldsymbol{x}_{ij}$.
- The random effects, $\boldsymbol{U}_i$, $i = 1, ..., m$, are mutually independent with a common underlying multivariate distribution, F.

To summarize, the basic idea underlying a random effects model is that there is natural heterogeneity across individuals in their regression coefficients and that this heterogeneity can be represented by a probability distribution. Correlation among observations for one person arises from their sharing unobservable variables, $\boldsymbol{U}_i$. A model of this type is sometimes referred to as a *latent variable* model (Bartholomew, 1987).

The random effects model is most useful when the objective is to make inference about individuals rather than the population average. In the MACS example, a random effects approach will allow us to estimate the CD4+ status of an individual man. In the Indonesian Children's Health Study, the random effects model would permit inference about the propensity for a particular child to have respiratory infection. The regression coefficients, $\boldsymbol{\beta}^*$, represent the effects of the explanatory variables on an individual child's chance of infection. This is in contrast to the marginal model coefficients which describe the effect of explanatory variables on the population average.

7.3 Transition (Markov) models

Under a transition model, correlation among $Y_{i1}, \ldots, Y_{in_i}$ exists because the past values, $Y_{i1}, ..., Y_{ij-1}$, explicitly influence the present observation, Y_{ij}. The past outcomes are treated as additional predictor variables.

The *marginal* model with an exponential autocorrelation function considered in Section 7.1 can be re-interpreted as a *transition* model by writing it as

$$Y_{ij} = \boldsymbol{x}'_{ij}\boldsymbol{\beta} + \epsilon_{ij}$$

where

$$\epsilon_{ij} = \alpha\epsilon_{ij-1} + Z_{ij},$$

$\alpha = \exp(-\phi)$, the Z_{ij} are mutually independent, $N(0, \tau^2)$ random variables and $\tau^2 = \sigma^2(1 - \alpha^2)$. By substituting $\epsilon_{ij} = Y_{ij} - \boldsymbol{x}'_{ij}\boldsymbol{\beta}$ into the second of these equations and re-arranging, we obtain the conditional distribution of Y_{ij}, given the preceding response, Y_{ij-1}, as

$$Y_{ij} \mid Y_{ij-1} \sim N\{\boldsymbol{x}'_{ij}\boldsymbol{\beta} + \alpha(Y_{ij-1} - \boldsymbol{x}'_{ij-1}\boldsymbol{\beta}), \tau^2\}.$$

This form treats both the explanatory variables and the prior response as explicit predictors of the current outcome.

The generalized linear transition model can be easily specified. We model the conditional distribution of Y_{ij} given the past as an explicit function of the q preceding responses. For example, in the Indonesian study, we might assume that the probability of respiratory infection for child i at visit j has a direct dependence on whether or not the child had infection at visit $j-1$, as well as on explanatory variables, $\boldsymbol{x}_{ij}$. One explicit formulation is

$$\operatorname{logit} \Pr(Y_{ij} = 1 | Y_{ij-1}, Y_{ij-2}, \ldots, Y_{i1}) = \boldsymbol{x}'_{ij}\boldsymbol{\beta}^{**} + \alpha Y_{ij-1}. \qquad (7.3.1)$$

Here, the chance of respiratory infection at time t_{ij} depends on explanatory variables but also on whether or not the child had infection three months earlier. The parameter $\exp(\alpha)$ is the ratio of the odds of infection among children who did and did not have infection at the prior visit. A coefficient in $\boldsymbol{\beta}^{**}$ can be interpreted as the change per unit change in x in the log odds of infection, among children who were free of infection at the previous visit.

This transition model for respiratory infection is a first order Markov chain (Feller, 1968, volume 1, p. 372). With a dichotomous response, Y_{ij}, observed at equally-spaced intervals, the process is described by the 2×2 transition matrix whose elements are $\Pr(Y_{ij} = y_{ij} | Y_{ij-1} = y_{ij-1})$ where each of y_{ij-1} and y_{ij} may take values 0 or 1. The logistic regression equation above specifies the transition matrix as

$$
\begin{array}{ccc}
 & & y_{ij} \\
 & 0 & 1 \\
 & 0 \quad \frac{1}{1+\exp(\boldsymbol{x}_{ij}'\boldsymbol{\beta}^{**})} & \frac{\exp(\boldsymbol{x}_{ij}'\boldsymbol{\beta}^{**})}{1+\exp(\boldsymbol{x}_{ij}'\boldsymbol{\beta}^{**})} \\
y_{ij-1} & & \\
 & 1 \quad \frac{1}{1+\exp(\boldsymbol{x}_{ij}'\boldsymbol{\beta}^{**}+\alpha)} & \frac{\exp(\boldsymbol{x}_{ij}'\boldsymbol{\beta}^{**}+\alpha)}{1+\exp(\boldsymbol{x}_{ij}'\boldsymbol{\beta}^{**}+\alpha)} .
\end{array}
\qquad (7.3.2)
$$

Note that because the transition probabilities depend on explanatory variables, the transition matrix can vary across individuals.

To specify the general transition model, let $H_{ij} = \{y_{i1}, \ldots, y_{ij-1}\}$ represent the past responses for the ith subject. Also, let $\mu_{ij}^c = \mathrm{E}(Y_{ij} \mid H_{ij})$ and $v_{ij}^c = \mathrm{Var}(Y_{ij} \mid H_{ij})$ be the conditional mean and variance of Y_{ij} given past responses and the explanatory variables. Analogous to the GLM for independent data, we assume:

- $h(\mu_{ij}^c) = \boldsymbol{x}_{ij}'\boldsymbol{\beta}^{**} + \sum_{r=1}^{s} f_r(H_{ij}; \boldsymbol{\alpha})$
- $v_{ij}^c = v(\mu_{ij}^c)\phi$.

Here, we are modelling the transition from the prior state as represented by the functions f_r, to the present response. The past outcomes, after transformation by the known functions, f_r, are treated as additional explanatory variables. It may also be important to include interactions among prior responses when predicting the current value. If the model for the conditional mean is correctly specified, we can treat the repeated transitions for a person as independent events and use standard statistical methods. Korn and Whittemore (1979), Ware *et al.* (1988), Wong (1986), Zeger and Qaqish (1988), and Kaufmann (1987) discuss specific examples of transition models for binary and count data. Transition models are reviewed in more detail in Chapter 10.

7.4 Contrasting approaches

To contrast further the three modelling approaches, we consider the hypothetical linear model for infant growth and the logistic model for the respiratory disease data from the ICHS.

In the linear case, it is possible to formulate the three regression approaches to have coefficients with the same interpretation. That is, coefficients from random effects and transition linear models can have marginal interpretations as well. To illustrate, consider the simple linear regression model for infant growth

$$Y_{ij} = \beta_0 + \beta_1 t_{ij} + \epsilon_{ij}$$

where t_{ij} is the age, in months, of child i at visit j, Y_{ij} is the weight at age t_{ij} and ϵ_{ij} is a mean-zero deviation. Because children do not all grow at the same rate, the residuals, $\epsilon_{i1}, \ldots, \epsilon_{in_i}$, for child i will likely be correlated with one another. The marginal modelling approach is to assume:

(1) $\mathrm{E}(Y_{ij}) = \beta_0 + \beta_1 t_{ij}$;

(2) $\mathrm{Corr}(\epsilon_{ij}, \epsilon_{ik}) = \rho(t_{ij}, t_{ik}; \boldsymbol{\alpha})$.

Assumption (1) is that the average weight for all infants in the population at any time t is $\beta_0+\beta_1 t$. The parameter β_1 is therefore the change per month in the population-average weight. Assumption (2) specifies the nature of the autocorrelation; a specific example might be that

$$\rho(t_{ij}, t_{ik}; \boldsymbol{\alpha}) = \begin{matrix} \alpha_0, & |t_{ij} - t_{ik}| < 6 \text{ months} \\ \alpha_1, & |t_{ij} - t_{ik}| \geq 6 \text{ months} \end{matrix} .$$

That is, residuals within 6 months of each other have correlation α_0, while those further apart in time have correlation α_1. In the marginal approach, we separate the modelling of the regression and the correlation; either can be changed without necessarily changing the other.

A linear random effects model for infant growth can be written

$$Y_{ij} = \beta_0^* + U_{i0} + (\beta_1^* + U_{i1})t_{ij} + Z_{ij} \qquad (7.4.1)$$

where Z_{ij}, $j = 1, \ldots, n_i$ are independent, mean-zero deviations with variance σ^2 and (U_{i0}, U_{i1}) are independent realizations from a mean-zero distribution with covariance matrix whose elements are $\mathrm{Var}(U_{i0}) = G_{11}$, $\mathrm{Var}(U_{i1}) = G_{22}$ and $\mathrm{Cov}(U_{i0}, U_{i1}) = G_{12}$. For mathematical convenience, we typically assume that all random variables follow Gaussian distributions. Figure 7.1(a) shows a sample of 10 linear growth curves whose intercepts and slopes are simulated from a bivariate Gaussian distribution with means $\beta_0^* = 8$ lbs, $\beta_1^* = 10$ lbs/year, variances $G_{11} = 2.0$ lbs^2,

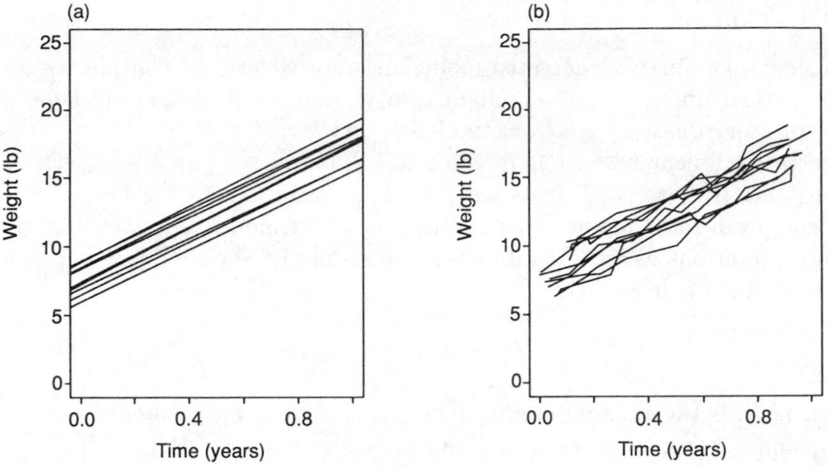

Fig. 7.1. Simulation of a random effects growth model: (a) underlying linear growth curves; (b) observed data.

$G_{22} = 0.4$ (lbs/year)^2 and correlation $\rho = 0.2$. Of course, we do not observe these curves, but only imprecise values, y_{ij}. One realization of the data is shown in Fig. 7.1(b) where the residual standard deviation is $\sigma = 0.5$.

In the linear random effects model, the regression coefficients also have a marginal interpretation since $\mathrm{E}(Y_{ij}) = \beta_0^* + \beta_1^* t_{ij}$. This is because the average of the rates of growth for individuals is the same as the change in the population-average weight across time in a linear model.

The correlation for the linear growth model is given by

$$\mathrm{Corr}(Y_{ij}, Y_{ik}) = \frac{G_{11} + G_{22} t_{ij} t_{ik} + (t_{ij} + t_{ik}) G_{12}}{\{(G_{11} + G_{22} t_{ij}^2 + \sigma^2)(G_{11} + G_{22} t_{ik}^2 + \sigma^2)\}^{1/2}}.$$

If there is no heterogeneity in the growth rates ($G_{22} = 0$), then $\mathrm{Corr}(Y_{ij}, Y_{ik}) = G_{11}/(G_{11} + \sigma^2)$. That is, the correlation is the same for any two observation times reflecting the fact that it derives from their sharing a common random intercept.

A transition model for infant growth might have the form

$$Y_{ij} = \beta_0^{**} + \beta_1^{**} t_{ij} + \epsilon_{ij} \tag{7.4.2}$$

$$\epsilon_{ij} = \alpha \epsilon_{ij-1} + Z_{ij} \tag{7.4.3}$$

where the Z_{ij} are independent mean zero innovations with variance σ^2. This model can be re-expressed as

$$Y_{ij} = \beta_0^{**} + \beta_1^{**} t_{ij} + \alpha(Y_{ij-1} - \beta_0^{**} - \beta_1^{**} t_{ij-1}) + Z_{ij}. \tag{7.4.4}$$

Hence, $E(Y_{ij}|y_{ij-1},\ldots,y_{i1}) = \beta_0^{**} + \beta_1^{**}t_{ij} + \alpha(y_{ij-1} - \beta_0^{**} - \beta_1^{**}t_{ij-1})$. The expectation for the present response given the past depends explicitly on the previous observation. Note that (7.4.2) and (7.4.3) imply also that $E(Y_{ij}) = \beta_0^{**} + \beta_1^{**}t_{ij}$ so that this form of a transition model has coefficients which also have a marginal interpretation.

Econometricians use autoregressive (transition) models for prediction. Some prefer the form

$$Y_{ij} = \boldsymbol{x}'_{ij}\boldsymbol{\beta}^{+} + \alpha Y_{ij-1} + Z_{ij}. \tag{7.4.5}$$

Here, the response is regressed on the covariates and on the previous outcome itself without adjusting for its expectation. The predictions from (7.4.4) and (7.4.5) are identical. However, the interpretation of $\boldsymbol{\beta}^{**}$ and $\boldsymbol{\beta}^{+}$ differ. Equation (7.4.5) implies that $E(Y_{ij}) = \sum_{r=0}^{\infty} \alpha^r \boldsymbol{x}'_{ij-r}\boldsymbol{\beta}^{+}$ so that $\boldsymbol{\beta}^{+}$ does not have a marginal interpretation as the coefficient for $\boldsymbol{x}_{ij}$. The formula for the marginal mean will change as the model for the correlation changes, for example by adding additional lagged values of Y in the model. Hence, the interpretation of the regression coefficients $\boldsymbol{\beta}^{+}$ depends on the assumed form of the autocorrelation model. When the dependence of Y on x is the focus, we believe the formulation in (7.4.4) is preferable.

We have shown for the linear model that regression coefficients can have a marginal interpretation for each of the three approaches. With non-linear link functions, such as the logit, this is not the case. To expand on this point, we now consider the logistic regression model for the ICHS example of vitamin A and respiratory disease.

The random effects logistic model

$$\text{logit}\,\Pr(Y_{ij} = 1|U_i) = \beta_0^* + U_i + \beta_1^* x_{ij}$$

states that each child has their own baseline risk of infection $\exp(\beta_0^* + U_i)/\{1 + \exp(\beta_0^* + U_i)\}$ and that a child's odds of infection are multiplied by $\exp(\beta_1^*)$ if they become vitamin A deficient. To illustrate this idea, Fig. 7.2 shows the risk of infection for 100 hypothetical children whose U_is were chosen at random from a Gaussian distribution. Each child is represented by a vertical line connecting their risk with and without vitamin A deficiency plotted at the child's value for U_i. The Gaussian curve of the U_i is shown at the bottom. We used the parameter values $\beta_0 = -2.0$, $\beta_1 = 0.4$ and $\text{Var}(U_i) = \nu^2 = 2.0$, so that a child with random effect $U_i = 0.0$ has a 17 per cent chance of infection when deficient and a 12 per cent chance when not. Since the random intercepts follow a Gaussian distribution, 95 per cent of the children who are not deficient have risks in the range 0.8 per cent to 68 per cent. Note that the logistic model assumes that the odds ratio for vitamin A deficiency is the same for every child, equal to $\exp(0.4) = 1.5$. But the corresponding change in absolute risk differs depending on the

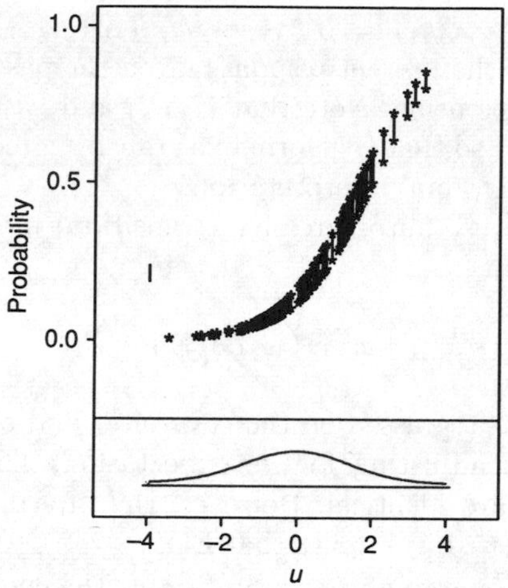

Fig. 7.2. Simulation of risk of infection with and without dichotomous risk factor, for 100 children in a logistic model with random intercept. Population average risks with and without infection are indicated by the vertical line on the left. The Gaussian density for the random intercepts is shown below.

baseline rate. Children with a lower propensity for infection ($U_i < 0$) at the left side of the figure have a smaller change in absolute risk than those with propensities near 0.5.

The population rate of infection is the average risk, which is given by

$$\begin{aligned}\Pr(Y_{ij} = 1) &= \int \Pr(Y_{ij} = 1 \mid U_i) dF(U_i) \\ &= \int \frac{\exp(\beta_0^* + U_i + \beta_1^* x_{ij})}{1 + \exp(\beta_0^* + U_i + \beta_1^* x_{ij})} f(U_i; \nu^2) dU_i\end{aligned}$$

where $f(.)$ is the Gaussian density function. The marginal rates, $\Pr(Y_{ij} = 1)$, are 0.18 and 0.23 for the sub-groups that are vitamin A replete and deficient respectively under the parameter values above.

In the marginal model, we ignore the differences among children and model the population-average, $\Pr(Y_{ij} = 1)$, rather than $\Pr(Y_{ij} = 1 \mid U_i)$, by assuming that

$$\operatorname{logit} \Pr(Y_{ij} = 1) = \beta_0 + \beta_1 x_{ij}.$$

Here, the infection rate in the sub-group that has sufficient vitamin A is $\exp(\beta_0)/\{1 + \exp(\beta_0)\} = 0.18$ so that $\beta_0 = -1.23$. The odds ratio for vitamin A deficiency is $\exp(\beta_1) = \{0.23/(1.0 - 0.23)\}/\{0.18/(1.0 - 0.18)\} = 1.36$ so that $\beta_1 = 0.31$. We have now established an important point. The

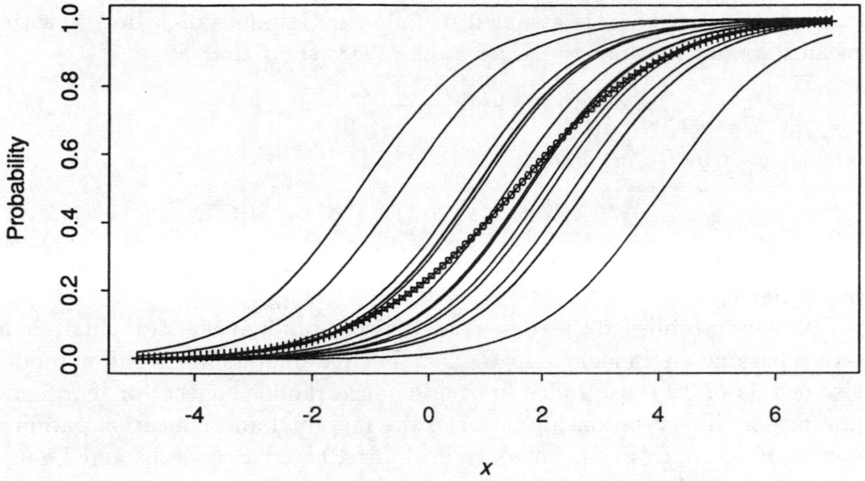

Fig. 7.3. Simulation of the probability of a positive response in a logistic model with random intercept and a continuous covariate. —— : sample curves; ++++++ : average curve.

marginal and random effects model parameters differ in the logistic model. The former describes the ratio of population odds; the latter describes the ratio of an individual's odds. Note that the marginal parameter values are smaller in absolute value than their random effects analogues. The vertical line on the left side of Fig. 7.2 shows the change in the population rates with and without vitamin A for comparison with the person-specific rates which change size as a function of the random effect.

This attenuation of the parameter values in the marginal model also occurs if x is a continuous, rather than discrete, variable. Figure 7.3 displays $\Pr(Y_{ij} = 1|U_i)$ as a function of x_{ij} in the simple model, $\text{logit}\Pr(Y_{ij} = 1|U_i) = (\beta_0^* + U_i) + \beta_1^* x_{ij}$, for a number of randomly generated U_i from a mean-zero Gaussian distribution with $\nu^2 = 2.0$. Also shown is the average curve at each value of x. Notice how the steep rate of increase as a function of x in the individual curves is attenuated in the average curve. This phenomenon is well-known in the related errors-in-variables regression problem; see for example Stefanski and Carroll (1985).

Neuhaus *et al.* (1991) show that if $\text{Var}(U_i) > 0$, then the elements of the marginal ($\boldsymbol{\beta}$) and random effects ($\boldsymbol{\beta}^*$) regression vectors satisfy

(1) $| \beta_k | \leq | \beta_k^* |$, $k = 1, \ldots, p$;

(2) equality holds if and only if $\beta_k^* = 0$;

(3) the discrepancy between β_k and β_k^* increases with $\text{Var}(U_i)$.

In particular, if U_i is assumed to follow a Gaussian distribution with mean zero and variance ν^2, Zeger *et al.* (1988) show that

$$\text{logitE}(Y_{ij}) \approx (c^2\nu^2 + 1)^{-1/2}\boldsymbol{x}'_{ij}\boldsymbol{\beta}^*, \tag{7.4.6}$$

where $c = 16\sqrt{3}/(15\pi)$ so that

$$\boldsymbol{\beta} \approx (c^2\nu^2 + 1)^{-1/2}\boldsymbol{\beta}^*, \tag{7.4.7}$$

where $c^2 \approx 0.346$. Note that (7.4.7) is consistent with the three properties listed above.

We have established the connection between random effects and marginal model parameters in logistic regression. The parameters in transition models such as (7.3.1) also differ from either the random effects or marginal parameters. The relationship between the marginal and transition parameters can be established, but only in limited cases. See Zeger and Liang (1992) for further discussion.

In log-linear models for counted data, random effects and marginal parameters can be equivalent in some important special cases. Consider the random effects model

$$\log \text{E}(Y_{ij}|\boldsymbol{U}_i) = \boldsymbol{x}'_{ij}\boldsymbol{\beta}^* + \boldsymbol{d}'_{ij}\boldsymbol{U}_i \tag{7.4.8}$$

where $\boldsymbol{d}_{ij}$ is a vector containing a subset of the variables in $\boldsymbol{x}_{ij}$. The marginal expectation has the form

$$\text{E}(Y_{ij}) = \int \exp(\boldsymbol{x}'_{ij}\boldsymbol{\beta}^*)\exp(\boldsymbol{d}'_{ij}\boldsymbol{U}_i)f(\boldsymbol{U}_i)d\boldsymbol{U}_i. \tag{7.4.9}$$

In the Gaussian random intercept case with $d_{ij} = 1$, note that $\text{E}(Y_{ij}) = \exp(\boldsymbol{x}'_{ij}\boldsymbol{\beta}^* + \nu^2/2)$, so that the marginal expectation has the same exponential form apart from an additive constant in the exponent. Hence, if we fit a marginal model which assumes that $\text{E}(Y_{ij}) = \exp(\boldsymbol{x}'_{ij}\boldsymbol{\beta})$, all of the parameters except the intercept will have the same value and interpretation as in the random effects model.

7.5 Inferences

With random effects and transitional extensions of the GLM, it is possible to estimate unknown parameters using traditional maximum likelihood methods. Starting with random effects models, the likelihood of the data, expressed as a function of the unknown parameters, is given by

$$L(\boldsymbol{\beta}^*, \boldsymbol{\alpha}; \boldsymbol{y}) = \prod_{i=1}^{m} \int \prod_{j=1}^{n_i} f(y_{ij}|\boldsymbol{U}_i)f(\boldsymbol{U}_i; \boldsymbol{\alpha})d\boldsymbol{U}_i \tag{7.5.1}$$

where $\boldsymbol{\alpha}$ represents the parameters of the random effects distribution. The likelihood is the integral over the unobserved random effects of the joint distribution of the data and the random effects. With Gaussian data, the integral has a closed form and relatively simple methods exist for maximizing the likelihood or restricted likelihood, as discussed in Chapter 5. With non-linear models, numerical integration techniques are often necessary to evaluate the likelihood. Techniques for finding maximum likelihood estimates are discussed further in Chapter 9.

Transition models can also be fitted using maximum likelihood. The joint distribution of the responses, $Y_{i1}, \ldots, Y_{in_i}$, can be written in the form

$$f(y_{i1}, \ldots, y_{in_i}) = f(y_{in_i} | y_{in_i-1}, \ldots, y_{i1}) f(y_{in_i-1} | y_{in_i-2}, \ldots, y_{i1}) \ldots \\ \ldots f(y_{i2}|y_{i1}) f(y_{i1}). \qquad (7.5.2)$$

In a first order Markov model,

$$f(y_{ij}|y_{ij-1}, \ldots, y_{i1}; \boldsymbol{\beta}^{**}, \boldsymbol{\alpha}) = f(y_{ij}|y_{ij-1}; \boldsymbol{\beta}^{**}, \boldsymbol{\alpha})$$

so that the likelihood contribution from person i simplifies to

$$f(y_{i1}, \ldots, y_{in_i}; \boldsymbol{\beta}^{**}, \boldsymbol{\alpha}) = f(y_{i1}; \boldsymbol{\beta}^{**}, \boldsymbol{\alpha}) \prod_{j=2}^{n_i} f(y_{ij}|y_{ij-1}; \boldsymbol{\beta}^{**}, \boldsymbol{\alpha}). \qquad (7.5.3)$$

Again, in the linear model this likelihood is easy to evaluate and maximize. For GLMs with non-linear link functions, one difficulty is that the marginal distribution of Y_{i1} often can not be determined from the conditional distributions $f(y_{ij}|y_{ij-1})$ without additional assumptions. A simple alternative is to maximize the *conditional likelihood* of $Y_{i2}, \ldots, Y_{in_i}$ given Y_{i1}, which is obtained by omitting $f(y_{i1})$ from the equation above. Conditional maximum likelihood estimates can be found using standard GLM software, treating functions of the previous responses as explanatory variables. Inferences conditional on Y_{i1} are less efficient than mles, but are all that is available without additional assumptions about $f(Y_{i1})$. Inferences about transition models are discussed in more detail in Chapter 10.

In our marginal modelling approach, we specify only the first two moments of the responses for each person. With Gaussian data, the first two moments fully determine the likelihood, but this is not the case with other members of the GLM family. To specify the entire likelihood, additional assumptions about higher order moments are also necessary. Examples for binary data are given by Prentice and Zhao (1991), Fitzmaurice *et al.* (1993) and Liang *et al.* (1992).

Even if additional assumptions are made, the likelihood is often intractable and involves many nuisance parameters in addition to $\boldsymbol{\alpha}$ and $\boldsymbol{\beta}$ that must be estimated. For this reason, a reasonable approach in many problems is to use *generalized estimating equations* or GEE, a multivariate analogue of quasi-likelihood which we now outline. The reader can find details in Liang and Zeger (1986), Zeger and Liang (1986) and Prentice (1988).

In the absence of a convenient likelihood function to work with, it is sensible to estimate $\boldsymbol{\beta}$ by solving a multivariate analogue of the quasi-score function (Wedderburn, 1974):

$$\boldsymbol{S}_\beta(\boldsymbol{\beta}, \boldsymbol{\alpha}) = \sum_{i=1}^{m} \left(\frac{\partial \boldsymbol{\mu}_i}{\partial \boldsymbol{\beta}} \right)' \operatorname{Var}(\boldsymbol{Y}_i)^{-1} (\boldsymbol{Y}_i - \boldsymbol{\mu}_i) = \boldsymbol{0}. \tag{7.5.4}$$

In the multivariate case, there is the additional complication that $\boldsymbol{S}_\beta$ depends on $\boldsymbol{\alpha}$ as well as on $\boldsymbol{\beta}$ since $\operatorname{Var}(\boldsymbol{Y}_i) = \operatorname{Var}(\boldsymbol{Y}_i; \boldsymbol{\beta}, \boldsymbol{\alpha})$. This can be overcome by replacing $\boldsymbol{\alpha}$ in the equation above by an $m^{1/2}$-consistent estimate, $\hat{\boldsymbol{\alpha}}(\hat{\boldsymbol{\beta}})$. Liang and Zeger (1986) and Gourieroux *et al.* (1984) show that the solution of the resulting equation is asymptotically as efficient as if $\boldsymbol{\alpha}$ were known.

The correlation parameters $\boldsymbol{\alpha}$ may be estimated by simultaneously solving $\boldsymbol{S}_\beta = \boldsymbol{0}$ and

$$\boldsymbol{S}_\alpha(\boldsymbol{\beta}, \boldsymbol{\alpha}) = \sum_{i=1}^{m} \left(\frac{\partial \boldsymbol{\eta}_i}{\partial \boldsymbol{\alpha}} \right)' H_i^{-1} (\boldsymbol{W}_i - \boldsymbol{\eta}_i) = \boldsymbol{0} \tag{7.5.5}$$

where $\boldsymbol{W}_i = (Y_{i1}Y_{i2}, Y_{i1}Y_{i3}, \ldots, Y_{in_i-1}Y_{in_i}, Y_{i1}^2, Y_{i2}^2, \ldots, Y_{in_i}^2)'$, the set of all products of pairs of responses and squared responses, and $\boldsymbol{\eta}_i = \mathrm{E}(\boldsymbol{W}_i; \boldsymbol{\beta}, \boldsymbol{\alpha})$ (Prentice, 1988). The choice of the weight matrix, H_i, depends on the type of responses. For binary response, the last n_i components of $\boldsymbol{W}_i$ can be ignored since the variance of a binary response is determined by its mean. In this case, we can use

$$H_i = \begin{pmatrix} \operatorname{Var}(Y_{i1}Y_{i2}) & & & 0 \\ & \operatorname{Var}(Y_{i1}Y_{i3}) & & \\ & & \ddots & \\ 0 & & & \operatorname{Var}(Y_{in_i-1}Y_{in_i}) \end{pmatrix}$$

which ensures that $\boldsymbol{S}_\alpha$ depends on $\boldsymbol{\beta}$ and $\boldsymbol{\alpha}$ only. For counted data, we suggest the use of the $n_i^* \times n_i^*$ identity matrix for H_i when solving $\boldsymbol{S}_\alpha = \boldsymbol{0}$, where $n_i^* = \binom{n_i}{2} + n_i$ is the number of elements of $\boldsymbol{W}_i$.

The suggested H_i are not optimal (Godambe, 1960). However, our experience in applications has been that the choice of H_i has small impact on inference for $\boldsymbol{\beta}$ when m is large. Furthermore, to use the most suitable weight, namely $\operatorname{Var}(\boldsymbol{W}_i)$, we would need to make further assumptions about the third and the fourth joint moments of $\boldsymbol{Y}_i$. See Fitzmaurice *et al.* (1993) and Liang *et al.* (1992) for details in the case of binary responses.

The solution, $(\hat{\boldsymbol{\beta}}, \hat{\boldsymbol{\alpha}})$, of $\boldsymbol{S}_\beta = \mathbf{0}$ and $\boldsymbol{S}_\alpha = \mathbf{0}$ is asymptotically Gaussian (Liang and Zeger, 1986), with variance consistently estimated by

$$\left(\sum_{i=1}^{m} C_i' B_i^{-1} D_i\right)^{-1} \left(\sum_{i=1}^{m} C_i' B_i^{-1} V_{0i} B_i^{-1} C_i\right) \left(\sum_{i=1}^{m} D_i' B_i^{-1} C_i\right)^{-1} \qquad (7.5.6)$$

evaluated at $(\hat{\boldsymbol{\beta}}, \hat{\boldsymbol{\alpha}})$, where

$$C_i = \begin{pmatrix} \frac{\partial \boldsymbol{\mu}_i}{\partial \boldsymbol{\beta}} & 0 \\ 0 & \frac{\partial \boldsymbol{\eta}_i}{\partial \boldsymbol{\alpha}} \end{pmatrix}, B_i = \begin{pmatrix} \mathrm{Var}(\boldsymbol{Y}_i) & 0 \\ 0 & H_i \end{pmatrix}, D_i = \begin{pmatrix} \frac{\partial \boldsymbol{\mu}_i}{\partial \boldsymbol{\beta}} & \frac{\partial \boldsymbol{\mu}_i}{\partial \boldsymbol{\alpha}} \\ \frac{\partial \boldsymbol{\eta}_i}{\partial \boldsymbol{\beta}} & \frac{\partial \boldsymbol{\eta}_i}{\partial \boldsymbol{\alpha}} \end{pmatrix}$$

and

$$V_{0i} = \begin{pmatrix} \boldsymbol{y}_i - \boldsymbol{\mu}_i \\ \boldsymbol{w}_i - \boldsymbol{\eta}_i \end{pmatrix} \begin{pmatrix} \boldsymbol{y}_i - \boldsymbol{\mu}_i \\ \boldsymbol{w}_i - \boldsymbol{\eta}_i \end{pmatrix}'.$$

This variance estimate is consistent as the number of individuals contributing to each element of the matrix goes to infinity. For example, in an analysis of variance problem, the number of units in each treatment group must get large.

The proposed method of estimation enjoys two properties. First, $\hat{\boldsymbol{\beta}}$ is nearly efficient relative to the maximum likelihood estimates of $\boldsymbol{\beta}$ in many practical situations provided that $\mathrm{Var}(\boldsymbol{Y}_i)$ has been reasonably approximated (e.g. Liang and Zeger, 1986; Liang *et al.*, 1991). In fact, the GEE is the maximum likelihood score equation for multivariate Gaussian data and for binary data from a log-linear model when $\mathrm{Var}(\boldsymbol{Y}_i)$ is correctly specified (Fitzmaurice *et al.*, 1993). Second, $\hat{\boldsymbol{\beta}}$ is consistent as $m \to \infty$, even if the covariance structure of $\boldsymbol{Y}_i$ is incorrectly specified. When regression coefficients are the scientific focus as in the examples here, one should invest the lion's share of time in modelling the mean structure, while using a reasonable approximation to the covariance. The robustness of the inferences about $\boldsymbol{\beta}$ can be checked by fitting a final model using different covariance assumptions and comparing the two sets of estimates and their robust standard errors. If they differ substantially, a more careful treatment of the covariance model may be necessary.

In this section, we have focused on inference for marginal models. Transition models can also be fitted using GEE. The reader is referred to Korn and Whittemore (1979), Ware *et al.* (1988), and references therein. Further discussion is provided in Chapter 10.

8

Marginal models

8.1 Introduction

This chapter describes regression methods that characterize the marginal expectation of a discrete or continuous response, Y, as a function of explanatory variables. The methods are designed to permit separate modelling of the regression of Y on $\boldsymbol{x}$, and the association among repeated observations of Y for each individual. Marginal models are appropriate when inferences about the population-average are the focus. For example, in a clinical trial the average difference between control and treatment is most important, not the difference for any one individual. Marginal models are also useful for epidemiological studies. For example, in the Indonesian Children's Health Study, a marginal model can be used to address questions such as:

- what is the age-specific prevalence of respiratory infection in children?
- is the prevalence of respiratory infection greater in the sub-population of children with vitamin A deficiency?
- how does the association of vitamin A deficiency and respiratory infection change with age?

Note that the scientific objectives are to characterize and contrast populations of children.

In addition to modelling the effects of covariates on the marginal expectations, we must also specify a model for the association among observations from each subject. This is to be contrasted with the random effects and transitional models where the covariate effects and within-subject association are modelled through a single equation. As discussed in Chapter 7, the three approaches all lead to the same class of linear models for Gaussian data. But in the discrete data case, different (non-linear) models can lead to different interpretations for the regression coefficients. The choice of model should therefore depend on the scientific question being addresssed.

This chapter discusses marginal models for categorical and count longitudinal data. Similar ideas apply to the analysis of skewed continuous

(Paik, 1992) and censored data which will not be covered here. Section 8.2 focuses on binary data contrasting traditional log-linear and marginal models. In Subsection 8.2.2, we compare three distinct parametrizations of the covariance among repeated binary observations, in terms of correlations, conditional odds ratios, and marginal odds ratios. Subsection 8.2.3 considers *generalized estimating equations* (GEE) for estimating regression and association parameters without specifying the entire likelihood. Examples of logistic models for binary longitudinal data are presented in Section 8.3. Discussion of ordered categorical and multinomial responses is delayed until Chapter 10. Section 8.4 treats models for count data.

8.2 Binary responses

8.2.1 *The log-linear model*

This section focuses on probability models for $\Pr(Y_{i1}, \ldots, Y_{in_i})$. To simplify the notation, we suppress for the moment the individual's subscript, i, as well as dependence on covariates, $\boldsymbol{x}_{ij}$.

The most widely used probability model for multivariate binary data is the log-linear model (Bishop *et al.* 1975)

$$\Pr(\boldsymbol{Y} = \boldsymbol{y}) = c(\boldsymbol{\theta}) \exp\left(\sum_{j=1}^{n} \theta_j y_j + \sum_{j_1 < j_2} \theta_{j_1 j_2} y_{j_1} y_{j_2} + \ldots + \theta_{1\ldots n} y_1 \ldots y_n\right)$$

where the $(2^n - 1)$-vector of canonical parameters is

$$\boldsymbol{\theta} = (\theta_1, \ldots, \theta_n, \theta_{12}, \ldots, \theta_{n-1n}, \ldots, \theta_{1\ldots n}).$$

The function $c(\boldsymbol{\theta})$ normalizes the probability distribution to sum to one. The equation above represents a saturated model in which the only constraint on the 2^n cell probabilities is that they add to one.

The canonical parameters facilitate calculation of cell probabilities but are less useful for describing $\Pr(\boldsymbol{Y} = \boldsymbol{y})$ as a function of the explanatory variables $\boldsymbol{x}$, because, for example, the parameter θ_{jk} describes the association between Y_j and Y_k *conditional* on all the other responses, Y_l, $l \neq j, k$. To establish this point, consider the quadratic exponential model (Zhao and Prentice, 1990; Gourieroux *et al.* 1984), a log-linear model with third and higher order terms equal to 0, so that

$$\Pr(\boldsymbol{Y} = \boldsymbol{y}) = c(\boldsymbol{\theta}) \exp\left(\sum_{j=1}^{n} \theta_j y_j + \sum_{j<k} \theta_{jk} y_j y_k\right).$$

For this model,

$$\log\left\{\frac{\Pr(Y_j = 1 \mid Y_k = y_k, Y_l = 0, l \neq j, k)}{\Pr(Y_j = 0 \mid Y_k = y_k, Y_l = 0, l \neq j, k)}\right\} = \theta_j + \theta_{jk} y_k.$$

Here, θ_j is the log odds for $Y_j = 1$ given that the remaining responses Y_l, $l \neq j$ are all zero. Similarly, θ_{jk} is the log odds ratio describing the association between Y_j and Y_k given that all the other responses are fixed, here set equal to zero. The log-linear canonical parameters, $\boldsymbol{\theta}$, may be undesirable if we now want to formulate a model in which they depend on explanatory variables by letting $\boldsymbol{\theta} = \boldsymbol{\theta}(\boldsymbol{x})$. Suppose that $Y_1, \ldots, Y_n$ indicate whether a child has respiratory infection each month and the key explanatory variable is an indicator of whether the child's mother smokes, $x = 1$ for smokers. If we now assume that the quadratic exponential model above has different parameter values for the smokers and non-smokers, we have a regression model. But the difference between $\theta_j(x = 1)$ and $\theta_j(x = 0)$ represents the effect of smoking on the conditional probability of respiratory infection at visit j given that there was no infection at any of the other visits. Since mother's smoking is likely to influence all visits, a part of the smoking effect will be conditioned away when we control for the other outcomes. A better formulation of the log-linear model is with parameters that describe the marginal probability $\Pr(Y_j = 1 \mid \boldsymbol{x})$ as a function of the smoking variable (Neuhaus and Jewell, 1990).

A second limitation of the canonical parameters is that their interpretation depends on the number of responses, n. Hence, if we add or delete an observation for an individual, the interpretation and value of the canonical parameters will change. In longitudinal studies, the number of observations commonly differs across subjects so that the applicability of models formulated in terms of $\boldsymbol{\theta}$ is limited.

8.2.2 *Log-linear models for marginal means*

We can still build a log-linear model by starting with the marginal parameters,

$$\mu_j = \Pr(Y_j = 1), \quad j = 1, \ldots, n. \tag{8.2.1}$$

The saturated log-linear model for an n-vector $\boldsymbol{Y}$ has $2^n - 1$ free parameters. If we start with the n μ_js, we have $2^n - 1 - n$ parameters to be specified. This can be done in a number of useful ways; we briefly consider three of these.

The first is to use the μ_js, plus the second and higher order canonical parameters, $\theta_{12}, \ldots, \theta_{n-1n}, \ldots, \theta_{12\ldots n}$, as proposed by Fitzmaurice *et al.* (1993). If we let $\boldsymbol{w} = (y_1 y_2, y_1 y_3, \ldots, y_{n-1} y_n, y_1 y_2 y_3, \ldots, y_1 y_2 \cdots y_n)$, $\boldsymbol{\theta}_1 = (\theta_1, \ldots, \theta_n)$ and $\boldsymbol{\theta}_2 = (\theta_{12}, \theta_{13}, \ldots, \theta_{n-1n}, \ldots, \theta_{1\ldots n})$ then the log-linear model can be written as

$$\Pr(\boldsymbol{Y} = \boldsymbol{y}) = c(\boldsymbol{\theta}_1, \boldsymbol{\theta}_2) \exp(\boldsymbol{y}' \boldsymbol{\theta}_1 + \boldsymbol{w}' \boldsymbol{\theta}_2).$$

The parameters of interest are $\boldsymbol{\mu} = (\mu_1, \ldots, \mu_n) = \boldsymbol{\mu}(\boldsymbol{\theta}_1, \boldsymbol{\theta}_2)$, so we can make a transformation from the canonical parameters, $(\boldsymbol{\theta}_1, \boldsymbol{\theta}_2)$, to $(\boldsymbol{\mu}, \boldsymbol{\theta}_2)$. In the regression context, we assume that the marginal means, μ_j, satisfy a model such as $\text{logit}\mu_j = \boldsymbol{x}_j'\boldsymbol{\beta}$. Interestingly, the score equation for $\boldsymbol{\beta}$ under this parametrization takes the generalized estimation equation (GEE) form

$$\left(\frac{\partial \boldsymbol{\mu}}{\partial \boldsymbol{\beta}}\right)' \text{Var}(\boldsymbol{Y})^{-1}(\boldsymbol{Y} - \boldsymbol{\mu}) = 0$$

where the variance matrix of Y is determined by both $\boldsymbol{\beta}$ and $\boldsymbol{\theta}_2$. This demonstrates that the solution of the GEE is the maximum likelihood estimate under the log-linear model when the covariance assumption, $\text{Var}(\boldsymbol{Y}_i)$, is correct for all $i = 1, \ldots, m$.

The limitation of this parametrization is that the interpretation of $\boldsymbol{\theta}_2$ as a conditional odds ratio depends on the number of other responses in a cluster. Hence, this formulation is most useful when the number of observations per person is the same, at least by design, as for example is often the case in clinical trials. Also, the conditional odds ratios are not easily interpreted when the association among responses is itself a focus of the study.

The second parametrization of the log-linear model that uses marginal means was proposed by Bahadur (1961). Here, second order moments are specified in terms of correlations. If we let $R_j = (Y_j - \mu_j)/\{\mu_j(1 - \mu_j)\}^{1/2}$, $\rho_{jk} = \text{Corr}(Y_j, Y_k) = \text{E}(R_j R_k)$, $\rho_{jkl} = \text{E}(R_j R_k R_l)$ and so on, up to $\rho_{1\ldots n} = \text{E}(R_1 \ldots R_n)$, then we can write the probability distribution $\Pr(\boldsymbol{Y} = \boldsymbol{y})$ as

$$\prod_{j=1}^{n} \mu_j^{y_j}(1 - \mu_j)^{(1-y_j)}$$
$$\times \left(1 + \sum_{j<k} \rho_{jk} r_j r_k + \sum_{j<k<l} \rho_{jkl} r_j r_k r_l + \ldots + \rho_{1\ldots n} r_1 r_2 \ldots r_n\right).$$

Here, the joint distribution is expressed in terms of the marginal means, pairwise correlations, and higher moments of the standardized variables R_j.

The Bahadur representation is attractive because it uses marginal probabilites and correlations, which are familiar parameters from the analysis of continuous responses. The serious drawback, however, is that, unlike for Gaussian data, the correlations among binary responses are constrained in complicated ways by the marginal means. Hence, if we assume that the marginal means depend on covariates, $\boldsymbol{x}$, it may not be correct to assume that the correlations and higher order moments are independent of $\boldsymbol{x}$, as would be convenient.

A compromise between conditional odds ratios which are unconstrained by the means but which have interpretations that depend on n, and correlations which are seriously constrained, is to parametrize the likelihood in terms of marginal odds ratios. These have weaker constraints and their interpretations are independent of n. The marginal odds ratio is defined as

$$\gamma_{jk} = \frac{\Pr(Y_j = 1, Y_k = 1)\Pr(Y_j = 0, Y_k = 0)}{\Pr(Y_j = 1, Y_k = 0)\Pr(Y_j = 0, Y_k = 1)}. \tag{8.2.2}$$

It takes values in $(0, \infty)$; a value greater than one indicates positive association. The odds ratio is a popular measure of association because it is unconstrained on the logarithmic scale and is in common use as the parameter in a logistic regression.

The full distribution of $\boldsymbol{Y}$ may be specified in terms of the means $\boldsymbol{\mu}$, the odds ratios $\boldsymbol{\gamma} = (\gamma_{12}, \ldots, \gamma_{n-1n})$ and contrasts of odds ratios, the first two given by:

$$\begin{aligned}
\zeta_{j_1 j_2 j_3} &= \log \mathrm{OR}(y_{j_1}, y_{j_2} | y_{j_3} = 1) - \log OR(y_{j_1}, y_{j_2} | y_{j_3} = 0), \\
\zeta_{j_1 j_2 j_3 j_4} &= \log \mathrm{OR}(y_{j_1}, y_{j_2} | y_{j_3} = 1, y_{j_4} = 1) \\
&\quad - \log \mathrm{OR}(y_{j_1}, y_{j_2} | y_{j_3} = 1, y_{j_4} = 0) \\
&\quad - \log \mathrm{OR}\,(y_{j_1}, y_{j_2} | y_{j_3} = 0, y_{j_4} = 1) \\
&\quad + \log \mathrm{OR}(y_{j_1}, y_{j_2} | y_{j_3} = 0, y_{j_4} = 0).
\end{aligned}$$

The general term has the somewhat complicated expression

$$\zeta_{j_1, \ldots, j_n} = \sum_{y_{j_3}, \ldots, y_{j_n} = 0,1} (-1)^{b(\boldsymbol{y})} \log\mathrm{OR}(y_{j_1}, y_{j_2} | y_{j_3}, \cdots, y_{j_n})$$

where $b(\boldsymbol{y}) = \sum_{\ell=3}^{n} y_{j\ell} + n - 2$. Liang *et al.* (1992) discuss the evaluation of the likelihood function in terms of these odds ratios and their contrasts. This is not a simple matter except in cases with a small number of observations per person. For example, with two observations the joint distribution, $\Pr(Y_1 = y_1, Y_2 = y_2)$, can be written as

$$\mu_1^{y_1}(1 - \mu_1)^{1-y_1}\mu_2^{y_2}(1 - \mu_2)^{1-y_2} + (-1)^{(y_1 - y_2)}(\mu_{11} - \mu_1\mu_2),$$

where $\mu_{11} = \Pr(Y_1 = Y_2 = 1)$ is given in terms of μ_j and γ_{12} by

$$\mu_{11} = \begin{cases} \frac{1-(\mu_1+\mu_2)(1-\gamma_{12}) - [\{1-(\mu_1+\mu_2)(1-\gamma_{12})\}^2 - 4(\gamma_{12}-1)\gamma_{12}\mu_1\mu_2]^{1/2}}{2(\gamma_{12}-1)}, & \gamma_{12} \neq 1 \\ \mu_1\mu_2, & \gamma_{12} = 1. \end{cases}$$

We note that there are $\binom{n_i}{2}$ pairs of odds ratios per subject, and this number becomes substantial when n_i increases. The problem of having a large number of nuisance parameters can be alleviated by using a regression model for the odds ratios as is done for the marginal expectations. The simplest model is $\gamma_{ijk} = \gamma$ for all i, j, k which says that the degree of association is the same for all pairs of observations from the same subject. Alternatively, we might assume that

$$\log \gamma_{ijk} = \alpha_0 + \alpha_1 \mid t_{ij} - t_{ik} \mid^{-1},$$

that is, the degree of association is inversely proportional to the time between observations. In general, we write

$$\gamma_{ijk} = \gamma(\boldsymbol{\alpha}) \tag{8.2.3}$$

where $\boldsymbol{\alpha}$ is a vector of q parameters to be estimated.

We often do not have sensible, simple models for third and higher order moments regardless of which formulation we adopt. Even when a probability model is fully specified, the likelihood can be complicated to evaluate except with small and constant n_i. For these reasons, we now apply the generalized estimating equation approach discussed in Section 7.5 to logistic regression.

8.2.3 *Generalized estimating equations*

In the logistic regression model, we specify that the marginal expectations, $\mathrm{E}(Y_{ij}) = \mu_{ij}$, satisfy $\mathrm{logit}\mu_{ij} = \boldsymbol{x}_{ij}'\boldsymbol{\beta}$. We have seen that the GEE estimating function introduced in Section 7.5 is the score equation for $\boldsymbol{\beta}$ when the data follow a log-linear probability distribution and we correctly specify $\mathrm{Var}(\boldsymbol{Y}_i)$. Even when we must model the covariance, it seems sensible to estimate $\boldsymbol{\beta}$ by solving the GEE

$$S_{\boldsymbol{\beta}}(\boldsymbol{\beta}, \boldsymbol{\alpha}) = \sum_{i=1}^{m} \left(\frac{\partial \boldsymbol{\mu}_i}{\partial \boldsymbol{\beta}} \right)' \mathrm{Var}(\boldsymbol{Y}_i)^{-1} (\boldsymbol{Y}_i - \boldsymbol{\mu}_i) = 0. \tag{8.2.4}$$

The quantity $S_{\boldsymbol{\beta}}(\boldsymbol{\beta}, \boldsymbol{\alpha})$ can be viewed as a multivariate version of the quasi-score function first proposed by Wedderburn (1974) with an additional complication that it depends not only on $\boldsymbol{\beta}$ but also on $\boldsymbol{\alpha}$, since $\mathrm{Var}(\boldsymbol{Y}_i) = \mathrm{Var}(\boldsymbol{Y}_i; \boldsymbol{\beta}, \boldsymbol{\alpha})$. The dependence on $\boldsymbol{\alpha}$ can be resolved by replacing $\boldsymbol{\alpha}$ in the GEE above with a $m^{1/2}$-consistent estimate, $\hat{\boldsymbol{\alpha}}(\hat{\boldsymbol{\beta}})$. Liang and Zeger (1986) and Gourieroux *et al.* (1984) have shown that the solution of the resulting equation is asymptotically as efficient as if $\boldsymbol{\alpha}$ were known.

With binary responses, the association parameters, $\boldsymbol{\alpha}$, may be formulated and estimated in a number of ways. Liang and Zeger (1986) parametrize $\mathrm{Var}(\boldsymbol{Y}_i)$ in terms of correlations and use moment estimators for the unknown parameters. Fitzmaurice *et al.* (1993) use conditional odds ratios for data sets with block designs and estimate the association parameters using maximum likelihood. Lipsitz *et al.* (1991), Liang *et al.* (1992) and Carey *et al.* (1993) formulate $\text{Var}(\boldsymbol{Y}_i)$ in terms of marginal odds ratios.

Following Prentice (1988), we can estimate association parameters by adding a second set of estimating equations $S_{\boldsymbol{\alpha}}(\boldsymbol{\beta}, \boldsymbol{\alpha}) = 0$ and simultaneously solving the expanded equations for $\hat{\boldsymbol{\beta}}$ and $\hat{\boldsymbol{\alpha}}$. The $\boldsymbol{\alpha}$ equations take the form

$$S_{\boldsymbol{\alpha}}(\boldsymbol{\beta}, \boldsymbol{\alpha}) = \sum_{i=1}^{m} \left(\frac{\partial \boldsymbol{\eta}_i}{\partial \boldsymbol{\alpha}}\right)^t H_i^{-1}(\boldsymbol{W}_i - \boldsymbol{\eta}_i) = 0 \qquad (8.2.5)$$

where

$$\boldsymbol{W}_i = (R_{i1}R_{i2}, R_{i1}R_{i3}, \ldots, R_{in_i-1}R_{in_i}),$$

$H_i = \text{diag}\{\text{Var}(R_{i1}R_{i2}), \ldots, \text{Var}(R_{in_i-1}R_{in_i})\}$, $R_{ij} = \{Y_{ij} - \mu_{ij}\}/\{\mu_{ij}(1 - \mu_{ij})\}^{1/2}$ and $\boldsymbol{\eta}_i = E(\boldsymbol{W}_i)$. With binary responses, $\boldsymbol{\eta}_i$ and H_i are fully determined by the mean and correlation models without additional assumptions about higher order moments. One simple example is if we assume that $\text{Corr}(Y_{ij}, Y_{ik}) = \alpha$ for all i and all $j \neq k$. Here, if we use the simpler identity weighting matrix in (8.2.5), we can estimate α by

$$\hat{\alpha} = (1/N^*) \sum_{i=1}^{m} \sum_{j<k} r_{ij} r_{ik}$$

where $N^* = \sum_i \binom{n_i}{2}$ and $r_{ij} = (y_{ij} - \hat{\mu}_{ij})/\{\hat{\mu}_{ij}(1 - \hat{\mu}_{ij})\}^{1/2}$.

When marginal odds ratios are used to model association, $\boldsymbol{\alpha}$ can be estimated by the approach introduced by Carey *et al.* (1993) and developed by Carey (1992). They estimate the odds ratios using an offset logistic regression. Let α_{ijk} be the log odds ratio for responses Y_{ij} and Y_{ik} and let $\mu_{ijk} = \Pr(Y_{ij} = Y_{ik} = 1)$. Then, it can easily be shown that

$$\text{logit}\Pr(Y_{ij} = 1 \mid Y_{ik} = y_{ik}) = \alpha_{ijk} y_{ik} + \log\left(\frac{\mu_{ij} - \mu_{ijk}}{1 - \mu_{ij} - \mu_{ik} + \mu_{ijk}}\right).$$

Suppose that we now impose the model that all the odds ratios are the same, that is, $\alpha_{ijk} = \alpha$. Then α can be estimated by a logistic regression of y_{ij} on y_{ik}, $1 \leq j < k \leq n_i$, $i = 1, \ldots, m$, using the second term on the right hand side of the equation above as an offset. Note that the offset depends on both $\boldsymbol{\beta}$ and $\boldsymbol{\alpha}$ so that iteration is required. More generally, we can assume a model $\alpha_{ijk} = \boldsymbol{d}_{ijk}{}'\boldsymbol{\alpha}$, where $\boldsymbol{d}_{ijk}$ is a set of covariates that characterizes the log odds ratio between observations j and k. We then estimate the vector $\boldsymbol{\alpha}$ by a logistic regression of y_{ij} on the product $\boldsymbol{d}_{ijk} y_{ik}$. Carey *et al.* (1993) give more details of this implementation of GEE, which they call 'alternating logistic regressions', or ALR.

Table 8.1. Data from a 2×2 crossover trial on cerebrovascular deficiency adapted from Jones and Kenward (1989, p.90), where treatments A and B are active drug and placebo, respectively; the outcome indicates whether an electrocardiagram was judged abnormal (0) or normal (1).

	Responses					Period	
Group	(1,1)	(0,1)	(1,0)	(0,0)	Total	1	2
AB	22	0	6	6	34	28	22
BA	18	4	2	9	33	20	22

8.3 Examples

In this section we demonstrate the use of marginal models for binary data from a few biomedical studies. The first two examples are from clinical trials with crossover designs. Crossovers are commonly used in biomedical studies to compare treatments for chronic diseases such as asthma and hypertension. The third example is from the observational study of the association between vitamin A deficiency and respiratory infection in Indonesian children.

Example 8.1. A 2×2 crossover trial

The data shown in Table 8.1 are adapted from Jones and Kenward (1989, p.90) who reported results from a crossover trial on cerebrovascular deficiency in which an active drug (A) and a placebo (B) were compared. Only data from centre 2 are used for illustration. Thirty-four patients received the active drug (A) followed by placebo (B); another 33 patients were treated in the reverse order. The response variable is defined to be 0 for an abnormal and 1 for a normal electrocardiogram reading. Table 8.1 shows, for example, that 22 patients in the AB group respond positively on both treatments, whereas 4 patients in the BA group respond negatively to the first treatment (B) but positively to the second treatment (A).

Also shown in Table 8.1 are the marginal distributions of responses at the two periods for both the AB and BA groups. Focusing on the data at period 1 for the moment, we observe that 82 per cent (28/34) of patients receiving the active drug (A) were normal as opposed to 61 per cent (20/33) of the patients receiving the placebo (B). This gives $(13 \times 28)/(20 \times 6) = 3.0$ as an estimate of the odds ratio comparing the chance of being normal for the active drug versus the placebo. The standard error for $\log 3.0 = 1.1$ is estimated as $(28^{-1} + 6^{-1} + 20^{-1} + 13^{-1})^{1/2} = 0.57$. For an explanation of these calculations, see Appendix A, Example A.1. The estimate of the treatment effect, measured by the odds ratio, is greater than one, indicating that the active drug produces a higher proportion of normal readings. However, the treatment effect is not statistically significant at the 0.05

level due to the large standard error. It is therefore desirable to combine the odds ratio estimated from the first period with that from second period, $22 \times 12/(11 \times 22) = 1.1$, to reduce the sampling error. Two problems arise. First, this approach ignores the possible treatment-by-period interaction known as the *carry-over effect* whereby the effect of treatment in period 1 influences the response in the next period. Secondly, the two responses obtained from the same subject are likely to be dependent. In fact, both groups show strong degrees of within-subject dependence. The odds ratio defined in (8.2.2) to measure within-subject association, is estimated to be $22 \times 6/(6 \times 0.5) = 44$ and $18 \times 9/(4 \times 2) = 20.3$ for groups AB and BA, respectively.

We now use the GEE method, together with a sensible choice of model, to make inferences about the treatment effect, combining the data from both periods. Conceptually, 2×2 crossover trials can be viewed as longitudinal studies with $n_i = n = 2$. In this example, $m = 67$. The two major covariates, period (x_1) and treatment (x_2), are both time-dependent, and are coded as

$$x_1 = \begin{cases} 1 & \text{period 2} \\ 0 & \text{period 1,} \end{cases} \quad x_2 = \begin{cases} 1 & \text{active drug (A)} \\ 0 & \text{placebo (B).} \end{cases}$$

We first fit a logistic regression model

$$\operatorname{logit} \Pr(Y_{ij} = 1) = \beta_0 + \beta_1 x_{ij1} + \beta_2 x_{ij2} + \beta_3 x_{ij1} x_{ij2},$$

under the assumption of a constant odds ratio, γ, across subjects. Results from Table 8.2 show little support for a treatment-by-period interaction ($\hat{\beta}_3 = 1.02 \pm 0.98$) and a strong within-subject association ($\hat{\gamma} = \exp(3.54) = 34.5$). The association parameter indicates that subjects with normal responses at the first visit have odds of normal readings at the next visit that are almost 35 times higher than those whose first response was abnormal.

When dropping the interaction term from the model, we estimate a statistically significant treatment effect ($\hat{\beta}_2 = 0.57 \pm 0.23$). Thus, the overall odds of a normal electrocardiogram reading are estimated to be 77 per cent higher ($0.77 = \exp(0.57) - 1$) when using the active drug as compared to the placebo. We note that if we incorrectly assume that there is no within-subject dependence, as was done in the last column of Table 8.2, the treatment effect is erroneously assessed to be statistically insignificant ($\hat{\beta}_2 = 0.56 \pm 0.38$). However, this potential pitfall can be avoided by using the robust standard error (0.23) instead of the model-based standard error (0.38).

Example 8.2. A three-treatment, three-period crossover trial

This example demonstrates how marginal models can be easily extended to more complicated designs. It also serves to contrast the results from

Table 8.2. Logistic regression coefficients (standard errors) from GEE for the 2×2 crossover trial on cerebrovascular deficiency. Models 1 and 2 are fitted using alternating logistic regressions (Carey *et al.* 1993).

	Model		
Variable	1	2	3
Intercept	1.54	1.24	1.22
	(0.45)	(0.30)	(0.30)*
			[0.34]**
Period (x_1)	-0.85	-0.30	-0.27
	(0.58)	(0.23)	(0.23)
			[0.38]
Treatment (x_2)	1.11	0.57	0.56
	(0.57)	(0.23)	(0.23)
			[0.38]
$x_1 * x_2$	-1.02	-	-
	(0.98)	-	-
$\log\gamma$	3.54	3.56	-
	(0.82)	(0.81)	-

*robust s.e.

** model-based s.e.

using marginal and conditional parameters. Table 8.3 shows the data drawn from a crossover trial in which two treatments were compared with placebo for the relief of primary dysmenorrhea (Jones and Kenward, 1987). The placebo is labelled treatment A, whilst treatments B and C are analgesic with low and high dose, respectively. The response variable, measured at the end of each period, is coded one for some relief and zero for no relief. Table 8.3 shows, for example, that among 15 patients randomized to the group receiving treatments A, B, and C in order, 9 patients had the outcome (011) indicating no relief in period 1 followed by some relief for both the second and third periods.

Restricting attention to the data from the first period only, we notice that the fraction of patients with relief is considerably lower, 23 per cent = 7/31, in the placebo group than in the low, 74 per cent =20/27, or high, 75 per cent =21/28, analgesic groups. As in the 2 × 2 crossover example, we want to pool information across periods taking account of the correlation among repeated responses for each individual. To do so, we use a marginal model where the primary covariates are defined as follows:

Table 8.3. Data from a 3×3 crossover trial on primary dysmenorrhea from Jones and Kenward (1987).

	Response								
Group	000	100	010	001	110	101	011	111	Total
ABC	0	0	2	2	1	0	9	1	15
ACB	2	1	0	0	0	0	9	4	16
BAC	0	1	1	1	0	8	3	1	15
BCA	0	1	1	1	8	0	0	1	12
CAB	3	0	0	0	1	7	2	1	14
CBA	1	5	0	0	4	3	1	0	14

$$x_1(x_2)\left\{= \begin{array}{ll} 1 & \text{period } 2(3) \\ 0 & \text{otherwise,} \end{array}\right. \quad x_3(x_4) = \left\{ \begin{array}{ll} 1 & \text{treatment B(C)} \\ 0 & \text{otherwise} \end{array}\right. ,$$

$$x_5(x_6) = \left\{ \begin{array}{ll} 1 & \text{the previous assignment is B(C)} \\ 0 & \text{otherwise} \end{array}\right. .$$

We first examine the carry-over effect by fitting a model

$$\operatorname{logit} \Pr(Y_{ij} = 1) = \beta_0 + \beta_1 x_{ij1} + \beta_2 x_{ij2} + \beta_3 x_{ij3} + \beta_4 x_{ij4} + \beta_5 x_{ij5} + \beta_6 x_{ij6} \tag{8.3.1}$$

for $j = 1, 2, 3$ and $i = 1, \ldots, 86$. We considered two different models for the association: that the marginal odds ratios were constant for all 3 pairs of times (Model 1); and that they took three distinct values (Model 2). The inferences for $\boldsymbol{\beta}$ did not change materially as we changed the assumption about the association.

Results for Models 1 and 2 reported in Table 8.4 suggest that there is moderate evidence of a carry-over effect for the high dose analgesic treatment. The odds of dysmenorrhea relief in periods 2 or 3 are reduced by a factor of 60 per cent ($= 1 - \exp(-0.919)$) among patients who received treatment C in the previous period rather than A, irrespective of the current treatment assignment.

We next examine the data for possible treatment-by-period interactions by adding x_1x_3, x_1x_4, x_2x_3 and x_2x_4 to the previous model. It appears, as shown for Models 3 and 4 of Table 8.4, that the assumption that treatment effects are constant over time is adequate. We therefore return to the first model, and conclude that the odds of dysmenorrhea relief for the group receiving the low (high) dose of analgesic is $\exp(2.09) = 8.1$ ($\exp(2.07) =$

Table 8.4. Coefficients and (standard errors) from logistic regression analyses using GEE for the 3×3 crossover trial on primary dysmenorrhea. Models were fitted using the alternating logistic regression implementation of GEE.

Variable	Model 1	Model 2	Model 3	Model 4
Intercept	-1.08	-1.10	-1.21	-1.19
	(0.32)	(0.32)	(0.42)	(0.42)
Period$_2(x_1)$	0.42	0.38	0.96	0.88
	(0.41)	(0.41)	(0.72)	(0.72)
Period$_3(x_2)$	0.59	0.55	0.56	0.53
	(0.46)	(0.45)	(0.79)	(0.80)
Treatment$_B(x_3)$	2.09	2.09	2.74	2.66
	(0.42)	(0.42)	(0.69)	(0.67)
Treatment$_C(x_4)$	2.07	2.12	2.00	2.01
	(0.42)	(0.42)	(0.57)	(0.57)
Carryover$_B(x_5)$	-0.15	-0.091	-0.32	-0.34
	(0.51)	(0.50)	(0.55)	(0.55)
Carryover$_C(x_6)$	-0.92	-0.86	-0.86	-0.81
	(0.44)	(0.43)	(0.49)	(0.49)
x_1x_3			-1.55	-1.42
			(0.87)	(0.84)
x_1x_4			-0.052	0.006
			(0.89)	(0.90)
x_2x_3			-0.39	-0.33
			(0.92)	(0.89)
x_2x_4			0.42	0.44
			(0.91)	(0.92)
$\log \gamma_{12}$	-0.22	-0.96	-0.19	-0.87
	(0.38)	(0.62)	(0.35)	(0.90)
$\log \gamma_{13}$	-0.22	0.11	-0.19	0.18
	(0.38)	(0.72)	(0.35)	(0.75)
$\log \gamma_{23}$	-0.22	0.29	-0.19	0.20
	(0.38)	(0.70)	(0.35)	(0.68)

7.9) times higher than when using the placebo and that the two doses of analgesic have similar effects.

Jones and Kenward (1987) use canonical parameters in a log-linear model to analyse three-period crossover trials. They used a model in which $\text{logit}\,\Pr(Y_{ij} \mid Y_{ik},\ k \neq j, \boldsymbol{x}_{ij})$ is a linear function of the other responses, $Y_{ik},\ k \neq j$, and of the covariates defined above. When applied to the same data, their model found little evidence of carry-over effects. The lack of statistical significance in their analysis is likely due to the fact that the variables x_5 and x_6 were modelled simultaneously with Y_{ij-1}. When there are treatment effects, as is the case here, the carry-over effects of variables x_5 and x_6 will be partly attributed to Y_{ij-1}, which has also been affected by the treatment during the last period. By conditioning on the prior outcome, we attribute some of the effect of the last-period treatment to the last-period outcome. By the analogous argument, the estimated treatment effects will also be smaller in the conditional model as applied by Jones and Kenward (1987). The conceptual difficulties in conditioning on other outcomes with longitudinal and other studies that generate correlated data have been addressed in more detail by Liang *et al.* (1992).

Example 8.3. Respiratory infection in Indonesian preschool children

Two hundred and seventy-five preschool children in Indonesia were examined for up to six consecutive quarters for the presence of respiratory infection. This is a subset of a cohort studied by Sommer *et al.* (1984). A primary question of interest is whether the prevalence of respiratory infection is higher among children who suffer xerophthalmia, an ocular manifestation of chronic vitamin A deficiency. Also of interest is the change in the prevalence of respiratory infection with age. It is worth noting that either question can be addressed using both cross-sectional and longitudinal data. In what follows, we will look specifically at the ageing question.

The prevalence of respiratory infection in six consecutive quarters, as shown in Table 8.5, reveals a possible seasonal trend with a summer maximum. The prevalence of xerophthalmia also indicates some seasonality, with a maximum in winter.

We begin our analysis by considering only data from the first visit. The results are summarized in Table 8.6. Model 1 in Table 8.6 is a logistic regression of respiratory infection on xerophthalmia and age, adjusting for gender and height for age as a percentage of the United States National Center for Health Statistics standard. The results suggest a strong cross-sectional age effect on the prevalence of respiratory infection. The effect is non-linear on the logit scale, as the quadratic term for age is statistically significant and negative in sign. As shown in Fig. 8.1, the cross-sectional analysis suggests that the prevalence of respiratory infection increases from age 12 months and reaches its peak at 20 months before starting to decline. If we now include the data from all 6 visits (Model 2 in Table 8.6), this

Table 8.5. Prevalence (in per cent) of respiratory infection and xerophthalmia by visit.

	Visit (season)					
Prevalence (per cent)	1(Su*)	2(A)	3(W)	4(S)	5(Su)	6(A)
Respiratory infection	12.6	4.6	7.3	3.8	14.9	9.4
Xerophthalmia	3.9	6.1	6.2	5.5	3.1	3.0
No. of children	230	214	177	183	195	201

*Su=Summer; A=Autumn; W=Winter; S=Spring

concave relationship is preserved qualitatively, as shown in Fig. 8.1. We note that an annual sine and cosine have been included in Model 2 to adjust for seasonality but that this has very little impact on the age coefficients. The discrepancy among the age coefficients in Models 1 and 2 may be explained by Fig. 8.2 in which the cross-sectional age effects on respiratory infection are displayed graphically for each of six visits. The age coefficients in Model 2 can be interpreted as weighted averages of the cross-sectional age coefficients from each visit.

The association between respiratory infection and xerophthalmia is positive, although not statistically significant at the 5 per cent level. There is limited information about the association with xerophthalmia in this partial data set which includes only 52 events of vitamin A deficiency. See Sommer *et al.* (1984) for an analysis of the complete data set of over 23000 observations.

Table 8.6. Logistic regressions of the prevalence of respiratory function on age and xerophthalmia adjusting for gender, season, and height for age. Models 1 and 2 estimate cross-sectional effects; Models 3 and 4 distinguish cross-sectional from longitudinal effects. Models 2–4 are fitted using the alternating logistic regression implementation of GEE.

	Model			
Variable	1	2	3	4
Intercept	-1.47	-2.05	-1.76	-2.21
	(0.36)	(0.21)	(0.25)	(0.32)
Gender	-0.66	-0.49	-0.53	-0.53
	(0.44)	(0.24)	(0.24)	(0.24)
Height for age	-0.11	-0.042	-0.051	-0.048
	(0.041)	(0.023)	(0.025)	(0.024)
Seasonal cosine	-	-0.59	-	-0.54
		(0.17)		(0.21)
Seasonal sine	-	-0.16	-	-0.016
	-	(0.14)	-	(0.18)
Xerophthalmia	0.44	0.50	0.53	0.64
	(1.15)	(0.44)	(0.45)	(0.44)
Age	-0.089	-0.030	-	-
	(0.027)	(0.008)	-	-
Age^2	-0.0026	-0.0010	-	-
	(0.0011)	(0.0004)	-	-
Age at entry	-	-	-0.053	-0.053
	-	-	(0.013)	(0.013)
$(\text{Age at entry})^2$	-	-	-0.0013	-0.0013
	-	-	(0.0005)	(0.0005)
Follow-up time	-	-	-0.19	-0.082
	-	-	(0.071)	(0.099)
$(\text{Follow-up})^2$	-	-	0.013	0.007
	-	-	(0.004)	(0.007)
$\log(\gamma)$		0.49	0.46	0.49
		(0.27)	(0.26)	(0.26)

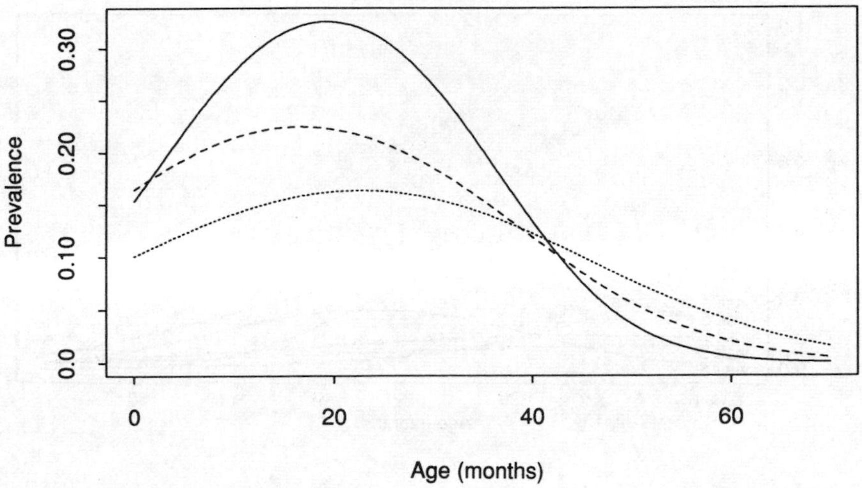

Fig. 8.1. Prevalence of respiratory infection as a function of age for three different models. —— : Model 1; .... : Model 2; - - - : Model 3.

We now want to distinguish the contributions of cross-sectional and longitudinal information to the estimated relationship of respiratory infection and age. That is, we want to separate differences among sub-populations of children at different ages at a fixed time (cross-sectional) from changes in children over time (longitudinal). To do so, we first decompose the variable age_{ij}, the age of the ith child at the jth visit, where $j = 1, \ldots, 6$, as the sum

$$age_{i1} + (age_{ij} - age_{i1}).$$

If we allow separate regression coefficients β_C and β_L for age_{i1} and $age_{ij} - age_{i1}$, respectively, the regression coefficient β_L describes the change of the risk of respiratory infection as the children grow older. The parameter β_C describes the age effect which would be estimated from purely cross-sectional data. The reader is referred to a more detailed discussion on this distinction in Sections 1.2 and 2.2. Note that the distinction between β_C and β_L defined for linear models holds only approximately for logistic and other non-linear models.

Results for Model 3 in Table 8.6 providc a different picture of how the risk of respiratory infection may be associated with age. The cross-sectional parameters suggest that the risk of respiratory infection climbs steadily in the first 20 months of life before declining; see Fig. 8.1. The longitudinal age parameters suggest otherwise. The risk of respiratory infection declines

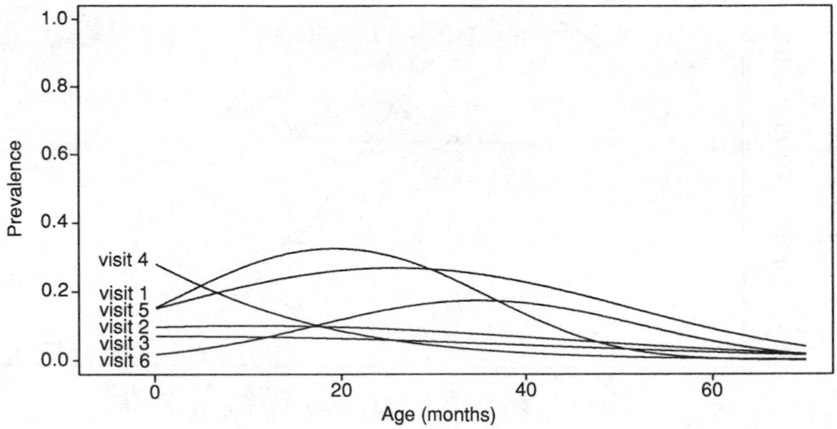

Fig. 8.2. Prevalence of respiratory infection as a function of age estimated separately for each of the six visits.

in the first 7 to 8 months of follow-up before rising later in life; see Fig. 8.3. This pattern is consistent even if we subdivide the study population into five cohorts according to the age at entry, namely, 0-12, 13-24, 25-36, 37-48, and 49 months or older. This pattern of a convex relationship between age and the risk of respiratory infection appears to coincide with the pattern of seasonality noted earlier. To determine whether seasonality is responsible, we add the annual harmonic in Model 4. The results are shown in the last column of Table 8.6. In fact, the longitudinal age parameters are now all insignificant. It makes sense that we can learn little about the effects of ageing from data collected over 18 months if we restrict our attention to longitudinal information. This is especially true in the presence of a substantial seasonal signal. On the other hand, much can be learned by comparing children of different ages so long as we can assume that there are not cohort effects confounding the inferences about age.

8.4 Counted responses

8.4.1 *Parametric modelling for count data*

Count data are increasingly common in the biological sciences. Examples include the number of panic attacks occuring during a six month interval after receiving treatment; the number of sexual partners in a three month period recorded in an HIV prevention programme, and the number of infant deaths per month before and after introduction of a prenatal care programme. Traditionally, the Poisson distribution has been the most commonly used model for count data. It has the form

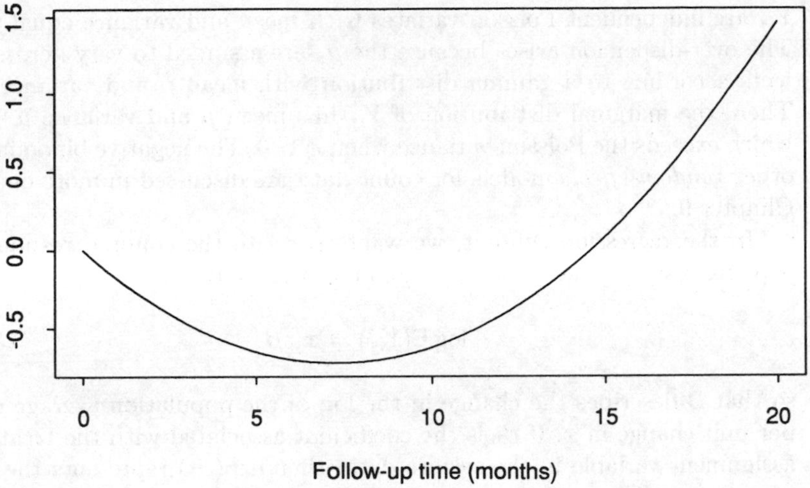

Fig. 8.3. The logarithm of the risk of respiratory infection as a function of follow-up time relative to the risk at an individual's first visit.

Table 8.7. Variance to mean ratios for the epileptic seizure data.

	Visit			
Treatment	1	2	3	4
Treated	38.7	16.8	23.8	18.8
Placebo	10.8	7.5	24.5	7.3

$$\Pr(Y = y) = \mu^y e^{-\mu}/y!, \quad y = 0, 1, 2, \ldots$$

The Poisson distribution is completely specified by the parameter μ, which is both the mean and the variance of the distribution, that is, $\mu = \mathrm{E}(Y) = \mathrm{Var}(Y)$. Unfortunately, the Poisson assumption that the mean and variance are equal is often inconsistent with empirical evidence.

To illustrate the problem, consider the seizure data in Table 1.5, which represent the number of epileptic seizures in each of four two-week intervals, for treatment and control groups with a total of 59 individuals. Table 8.7 gives the ratio of the sample variances to the means of the counts for each treatment-by-visit combination. A high degree of extra-Poisson variation is evident as the variance-to-mean ratios range from 7 to 39, whereas ratios of one correspond to the Poisson model. A commonly used model for *over-dispersed* count data, where the variance exceeds the mean, is the negative-binomial distribution. Here, we assume that given a rate μ_i, the

Y_{ij} are independent Poisson variates with mean and variance equal to μ_i. The over-dispersion arises because the μ_i are assumed to vary across subjects according to a gamma distribution with mean μ and variance $\phi\mu^2$. Then, the marginal distribution of Y_{ij} has mean μ and variance $\mu + \phi\mu^2$ which exceeds the Poisson variance when $\phi > 0$. The negative binomial and other *random-effects* models for count data are discussed in more detail in Chapter 9.

In the regression context, we want to relate the counted response to explanatory variables. The most common assumption is

$$\log \mathrm{E}(Y_{ij}) = \boldsymbol{x}'_{ij}\boldsymbol{\beta} \tag{8.4.1}$$

so that $\boldsymbol{\beta}$ describes the change in the log of the population average count per unit change in x. If β_1 is the coefficient associated with the treatment assignment variable in the seizure data, then $\exp(\beta_1)$ represents the ratio of average seizure rates, measured as the number of seizures per two-week period, for the treated patients compared to that among the control patients. A negative value of β_1 is evidence that the treatment is effective relative to the placebo in controlling the seizure rate.

To account for the over-dispersion which is not prescribed by the Poisson distribution, one can assume instead that

$$\mathrm{Var}(Y_{ij}) = \phi_{ij}\mathrm{E}(Y_{ij}) \tag{8.4.2}$$

with $\phi_{ij} > 1$. To control the number of over-dispersion parameters, a regression model for the ϕ_{ij} is needed. For this, we can assume $\phi_{ij} = \phi(\boldsymbol{\alpha}_1)$, where α_1 is a vector of q_1 parameters. A simple version of $\phi(\boldsymbol{\alpha}_1)$ would be $\phi_{ij} = \phi_1$ if treated and ϕ_2 if control. Alternatively, we may allow a different ϕ at different time points, constant across subjects.

Until now, we have assumed that the interval lengths, t_{ij}, during which the events are observed, are the same for each subject, and for each visit. This is appropriate for the seizure data where the intervals after treatment are all two weeks. A special problem may emerge if the observations are collected at irregular times and Y_{ij} represents the number of events between successive visits. This problem, however, can be easily corrected by decomposing the marginal mean, $\mu_{ij} = \mathrm{E}(Y_{ij})$, into the product of t_{ij}, the known observation period, and λ_{ij}, an unknown parameter representing the rate of the counting process per unit time. The log-linear model (8.4.1) can then be applied to λ_{ij}, that is,

$$\log \lambda_{ij} = \boldsymbol{x}'_{ij}\boldsymbol{\beta},$$

so that a regression coefficient in $\boldsymbol{\beta}$ is a logarithm of the ratio of rates per unit interval of time. We note that

$$\log \mathrm{E}(Y_{ij}) = \log t_{ij} + \boldsymbol{x}_{ij}'\boldsymbol{\beta}.$$

This equation shows that we can take account of different interval lengths by introducing an *offset*, $\log t_{ij}$, into the log-linear model as explained in more detail by McCullagh and Nelder (1989).

Thus far, we have ignored parametric models for the correlation among repeated observations on a unit. The most natural models involve random effects, which will be discussed in Chapter 9. Taking a marginal approach, we can specify a parametric model for the correlation coefficient $\rho = \rho(\boldsymbol{\alpha}_2)$, where $\boldsymbol{\alpha}_2$ is a q_2-vector of parameters. As with binary responses, the ranges of correlations for pairs of counts are restricted by their means. Sensible models in addition to random effects models for the association among counts need further development.

8.4.2 *Generalized estimating equation approach*

We have suggested the use of $S_{\boldsymbol{\beta}}$ in (8.2.4) to estimate the regression coefficient $\boldsymbol{\beta}$ for binary responses. The same estimating function can be used here, except that $S_{\boldsymbol{\beta}}$ depends not only on $\boldsymbol{\beta}$ and $\boldsymbol{\alpha}_2$, but also on $\boldsymbol{\alpha}_1$ because of the need to account for over-dispersion. Let $\boldsymbol{\alpha} = (\boldsymbol{\alpha}_1, \boldsymbol{\alpha}_2)$ be a vector of $q = q_1 + q_2$ parameters. We can now solve simultaneously the equations $S_{\boldsymbol{\beta}} = 0$ and

$$S_{\boldsymbol{\alpha}} = \sum_{i=1}^{m} \left(\frac{\partial \boldsymbol{\eta}_i}{\partial \boldsymbol{\alpha}}\right)' (\boldsymbol{W}_i - \boldsymbol{\eta}_i^*) = 0,$$

where $\boldsymbol{W}_i' = (Y_{i1}Y_{i2}, \ldots, Y_{in_i-1}Y_{in_i}, Y_{i1}^2, \ldots, Y_{in_n}^2)$ and $\boldsymbol{\eta}_i = E(\boldsymbol{W}_i)$.

There are two minor discrepancies between this procedure and the one suggested in Section 8.2.3. Firstly, the n_i squared terms of the Y_{ij}'s have been added to $\boldsymbol{W}_i$ to estimate the overdispersion parameters, $\boldsymbol{\phi} = \boldsymbol{\phi}(\boldsymbol{\alpha}_1)$. Secondly, the diagonal matrix H_i used in (8.2.5) is set to be the identity matrix, so that $\left(S_{\boldsymbol{\beta}}, S_{\boldsymbol{\alpha}}\right)$ depends on $\boldsymbol{\beta}$ and $\boldsymbol{\alpha}$ only. For binary data, $\mathrm{Cov}(Y_{ij}, Y_{ik})$ for $j \neq k$ is completely specified by (8.2.1) and (8.2.3), and no additional parameters are introduced. However, this is not the case for count data. By forcing H_i to be the identity matrix we potentially lose efficiency in estimating $\boldsymbol{\alpha}$. Our experience in the past has suggested that this loss of efficiency has very little impact on the $\boldsymbol{\beta}$ estimation. Using $H_i = I$ avoids estimating additional higher order parameters, and this reduces sampling variation.

8.4.3 *An example*

We now revisit the seizure count data set considered briefly in Section 8.4.1. Fifty-nine epileptics suffering from partial seizures were randomized

Table 8.8. Summary statistics for the epileptic seizure data.

Group	Age	Baseline seizure counts in 8 weeks
Treatment	27.7 ± 1.19	31.6 ± 5.03
Placebo	29.0 ± 1.13	30.8 ± 4.93

Table 8.9. Averaged seizure rates (per two weeks) by treatment and visit.

Treatment	Visit	Seizure rate
Progabide	Baseline	7.90 (6.91)*
	1-4	7.96 (5.71)*
Placebo	Baseline	7.70
	1-4	8.60
Cross-product ratio		0.90 † (0.74)*

*summary statistics when patient #207 is deleted
† 0.90 = (7.96/7.90)/(8.60/7.70)

to receive either the anti-epileptic drug progabide ($m_1 = 31$) or a placebo ($m_2 = 28$). Patients from the two groups appear to be comparable in terms of baseline age (in years) and eight-week baseline seizure counts as shown in Table 8.8. The main objective of this study was to determine whether progabide reduces the rate of seizures.

Table 8.9 gives the mean seizure rates per two weeks stratified by treatment group and time (baseline versus visits 1–4). Overall, there is very little change in two-week seizure counts for the treated group (7.96/7.90 = 101 per cent) and a small increase in the placebo group (8.60/7.70 = 112 per cent). The treatment effect, measured by the cross-product ratio, is reported in the last row of Table 8.9. Patient number 207 appears to be very unusual. He had an extremely high seizure count (151 in eight weeks) at baseline and his count doubled after treatment to 302 seizures in eight weeks. If his data are set aside, the cross-product ratio drops from 0.90 to 0.74, giving some indication of a treatment benefit.

For more formal inference we now use a log-linear regression fitted by the GEE method as described in Sections 8.4.1 and 8.4.2. To estimate the overall treatment effect, we use the following model:

$$\log \mathrm{E}(Y_{ij}) = \log t_{ij} + \beta_0 + \beta_1 x_{i1} + \beta_2 x_{i2} + \beta_3 x_{i1} x_{i2}, \quad \begin{array}{l} j = 0, 1, 2, 3, 4 \\ i = 1, \ldots, 59. \end{array}$$

Here, $t_{ij} = 8$ if $j = 0$ and $t_{ij} = 2$ if $j = 1, 2, 3, 4$. The $\log t_{ij}$ term is needed to account for different observation periods. The covariates are defined as

$$x_{i1} = \begin{cases} 1 \text{ if visit 1,2,3 or 4} \\ 0 \text{ if baseline} \end{cases}, \quad x_{i2} = \begin{cases} 1 \text{ if progabide} \\ 0 \text{ if placebo.} \end{cases}$$

The variable x_{i2} is included in the model to allow different baseline seizure counts between the treated and placebo groups. The parameter $\exp(\beta_1)$ is the ratio of the average seizure rate after treatment to before treatment for the placebo group. The coefficient of interest, β_3, represents the difference in the logarithm of the post- to pre-treatment ratio between the progabide and placebo groups. A negative coefficient corresponds to a greater reduction (or smaller increase) in the seizure counts for the progabide group.

The results given in Table 8.10 suggest that, overall, there is very little difference between the treatment and placebo groups in the change of seizure counts before and after the randomization ($\hat{\beta}_3 = -0.10 \pm 0.21$) if patient number 207 is included. If he is set aside, there is modest evidence that progabide is favoured over the placebo. Note also that different conclusions would be drawn if the strong over-dispersion ($\hat{\phi} = 19.44$) was ignored since the standardized test statistic, $Z = \hat{\beta}_3/\sqrt{\{\mathrm{Var}(\hat{\beta}_3)\}}$ would have been inflated by $(19.4)^{1/2} = 4.4$, and would be statistically significant at the 5 per cent level.

8.5 Further reading

This chapter has discussed the fitting of marginal models to discrete longitudinal data. We have emphasized the problem of describing the relationship between the marginal expectation of the response and covariates at each visit. In contrast to Chapter 5, we have given very little attention to the mechanisms by which the within-subject association may have arisen. Readers interested in this issue are referred to Zeger *et al.* (1988) and Thall and Vail (1990).

The use of generalized estimating equations to estimate regression coefficients specified by the marginal models has been studied extensively over the last five years. For more detailed treatments, see Liang and Zeger (1986), Zeger and Liang (1986), Prentice (1988), Zhao and Prentice (1990), Thall and Vail (1990), Liang *et al.* (1992), and Fitzmaurice *et al.* (1993).

Table 8.10. Log-linear regression coefficients and robust standard errors (in parentheses) for analysis of seizure rates. The model in 8.4.3 was fitted using GEE assuming exchangeable correlation, with and without patient number 207 who had unusual pre- and post-randomization seizure counts.

Variable	Complete data	Patient 207 deleted
Intercept	1.35	1.35
	(.16)	(.16)
Time (x_1)	0.11	0.11
	(.12)	(.12)
Treatment(x_2)	0.027	-0.11
	(.22)	(.19)
x_1x_2	-0.10	-0.30
	(.21)	(.17)
Over-dispersion parameter	19.4	10.4
Correlation coefficient	0.78	0.60

It is not uncommon in biomedical studies to encounter response variables which are categorical with more than two categories. A typical example is when the response measures the severity of injury. Regression models for polytomous responses, ordinal or nominal, are available. See, for example, McCullagh (1980), Anderson (1982), and particularly Dale (1986) who studied marginal models for bivariate responses. Extensions of GEE to this situation can be found in Stram *et al.* (1988), and Liang *et al.* (1992).

9

Random effects models

9.1 Introduction

Chapter 8 has dealt with marginal models whose regression parameters have population average interpretations. In this chapter we consider random effects models in which the regression coefficients measure the more direct influence of explanatory variables on the responses for heterogeneous individuals. Linear models with random effects were discussed in Section 5.2 in the context of a general framework for parametric modelling of correlation in longitudinal data. The use of random effects in the generalized linear model (GLM) family was introduced in Section 7.2 where the coefficients were contrasted with those from marginal and transition models. In Chapter 7, we used the notation $\boldsymbol{\beta}, \boldsymbol{\beta}^*$, and $\boldsymbol{\beta}^{**}$ to distinguish marginal, random effects, and transition coefficients, respectively. For the remainder of the book, we will use $\boldsymbol{\beta}$ for any regression coefficients as long as the type of model is clear from the context.

As discussed in Section 7.2, the basic premise of the random effects model is that there is natural heterogeneity among subjects in a subset of the regression coefficients, for example in the intercepts. Using the GLM framework, we assume that conditional on unobservable variables $\boldsymbol{U}_i$, we have independent responses from a distribution in the exponential family (see Appendix A, equation (A.5.1) such that

$$h\{\mathrm{E}(Y_{ij} \mid \boldsymbol{U}_i)\} = \boldsymbol{x}'_{ij}\boldsymbol{\beta} + \boldsymbol{d}'_{ij}\boldsymbol{U}_i. \tag{9.1.1}$$

Here, h is the link function and $\boldsymbol{x}_{ij}$ and $\boldsymbol{d}_{ij}$ are p and q-vectors of covariates, respectively.

To illustrate briefly, consider the 2×2 crossover trial data in Example 8.1 and suppose that a person's response follows the logistic model

$$\operatorname{logit} \Pr(Y_{ij} = 1 \mid U_i) = \beta_0 + U_i + \beta_1 x_{ij} \tag{9.1.2}$$

where x_{ij} indicates whether the person received placebo ($x = 0$) or drug ($x = 1$). This model states that each person on placebo has their own probability of a normal response ($Y = 1$) given by $\exp(\beta_0 + U_i)/\{1 +$

$\exp(\beta_0 + U_i)\}$. It further states that a person's odds of a normal response are multiplied by $\exp(\beta_1)$ when taking the drug, regardless of the initial risk.

Another simple case is when $\boldsymbol{x}_{ij} = \boldsymbol{d}_{ij}$ so that each individual can be thought to have their own regression coefficients $\boldsymbol{\beta} + \boldsymbol{U}_i$. With a large number of observations for a subject, we could estimate their individual coefficients from $\boldsymbol{y}_i$. But in practice, we have limited data and must borrow strength across subjects to make inferences about either $\boldsymbol{\beta}$ or the $\boldsymbol{U}_i$. This is accomplished by assuming that the $\boldsymbol{U}_i$ are independent realizations from a distribution.

There are two distinct approaches to inference about random effects models. The first is appropriate when we are interested in a particular subset of regression coefficients, none of which are assumed to vary across subjects. For example, suppose we want to estimate the drug effect in the crossover trial and believe that only the intercept, not the drug-effect, varies across individuals. Here, we can treat the $\boldsymbol{U}_i$ as nuisance variables and condition them out of the problem. That is, we can attempt to use only that part of the data which does not contain information about the $\boldsymbol{U}_i$ when making inference about particular coefficients. The second approach is appropriate when the subject-specific coefficients are themselves of interest or when conditioning away the information about the random effects discards too much information about an important regression coefficient. Here, we operate as if the $\boldsymbol{U}_i$ are an independent sample from some distribution and estimate both the fixed effects, $\boldsymbol{\beta}$, and the random effects $\boldsymbol{U}_i$ under this working model.

In random intercept models, the choice between these two approaches relates to the distinction between cross-sectional and longitudinal information discussed in Sections 1.1 and 1.4, and illustrated in Fig. 1.1. With the conditional likelihood approach, we use only longitudinal information, that is, comparisons within subjects, to estimate $\boldsymbol{\beta}$. When we assume that the $\boldsymbol{U}_i$ follow a distribution, we combine both longitudinal and cross-sectional information. The relative weight given to each source is determined by the variability among the $\boldsymbol{U}_i$. When there is large variability across subjects, the analysis should weight the longitudinal information more heavily since comparisons within a subject are likely to be more precise than comparisons among subjects.

Note that a fundamental assumption of the random effects model is that the $\boldsymbol{U}_i$ are independent of the explanatory variables. If this assumption is incorrect in the random intercept case, the conditional analysis will still give a consistent estimate of $\boldsymbol{\beta}$. The random effects analysis will not. Econometricians (e.g. Hausman, 1978) have developed what they call *specification tests* to check the random effects assumption. The model for longitudinal data in Section 1.4, which includes both cross-sectional and longitudinal parameters, is one way to allow random intercepts to depend on explanatory variables.

The remainder of the chapter is organized as follows. Section 9.2 discusses estimation of $\boldsymbol{\beta}$ in the random effects GLM using both conditional and full maximum likelihood. The latter is computationally expensive to implement in general, and we shall review an approximate method which works reasonably well for estimation of $\boldsymbol{\beta}$. We consider logistic models for binary data in Section 9.3 and Poisson regression models for count data in Section 9.4.

9.2 Estimation for generalized linear mixed models

In the random effects GLM we assume:

(1) the conditional distribution of Y_{ij} given $\boldsymbol{U}_i$ follows a distribution from the exponential family with density $f(y_{ij} \mid \boldsymbol{U}_i; \boldsymbol{\beta})$;

(2) given $\boldsymbol{U}_i$, the repeated measurements, $Y_{i1}, \ldots, Y_{n_i}$, are independent;

(3) the $\boldsymbol{U}_i$ are independent and identically distributed with density function $f(\boldsymbol{u}_i; G)$.

Let $\boldsymbol{U} = (\boldsymbol{U}_1, \ldots, \boldsymbol{U}_m)$. In the subsection below on conditional likelihood, we will treat the random effects as nuisance variables – that is, as if they were fixed parameters to be removed from the problem so that we need not rely on the third assumption above. In the subsection on maximum likelihood estimation, we will treat $\boldsymbol{U}$ as a set of unobserved variables which we then integrate out of the likelihood. We will adopt the assumption that the random effects distribution is Gaussian with mean zero and variance matrix G.

9.2.1 *Conditional likelihood*

In this subsection, we review conditional maximum likelihood estimation for $\boldsymbol{\beta}$. McCullagh and Nelder (1989, Section 7.2) present a more general treatment for GLMs. The main idea is to treat the random effects, $\boldsymbol{U}_i$, as a set of nuisance parameters and to estimate $\boldsymbol{\beta}$ using the conditional likelihood of the data given the sufficient statistics for the $\boldsymbol{U}_i$.

Treating $\boldsymbol{U}$ as fixed, the likelihood function for $\boldsymbol{\beta}$ and $\boldsymbol{U}$ is

$$\prod_{i=1}^{m}\prod_{j=1}^{n_i} f(y_{ij} \mid \boldsymbol{\beta}, \boldsymbol{U}_i) \propto \prod_{i=1}^{m}\prod_{j=1}^{n_i} \exp\{\theta_{ij} y_{ij} - \psi(\theta_{ij})\}, \tag{9.2.1}$$

where $\theta_{ij} = \theta_{ij}(\boldsymbol{\beta}, \boldsymbol{U})$. To simplify the discussion, we restrict attention to canonical link functions (McCullagh and Nelder, 1989, p.32) for which $\theta_{ij} = \boldsymbol{x}_{ij}'\boldsymbol{\beta} + \boldsymbol{d}_{ij}'\boldsymbol{U}_i$. Then the likelihood above can be written

$$\exp\left\{\boldsymbol{\beta}'\sum_{i,j}\boldsymbol{x}_{ij}y_{ij}+\sum_{i}\boldsymbol{U}_i{}'\sum_{j}\boldsymbol{d}_{ij}y_{ij}-\sum_{i,j}\psi(\theta_{ij})\right\}. \tag{9.2.2}$$

Hence, the sufficient statistics for $\boldsymbol{\beta}$ and $\boldsymbol{U}_i$ are $\sum_{i,j}\boldsymbol{x}_{ij}y_{ij}$ and $\sum_j \boldsymbol{d}_{ij}y_{ij}$ respectively, and $\sum_j \boldsymbol{d}_{ij}y_{ij}$ is sufficient for $\boldsymbol{U}_i$ for fixed $\boldsymbol{\beta}$.

The conditional likelihood is proportional to the conditional distribution of the data given the sufficient statistics for the $\boldsymbol{U}_i$. The contribution from subject i has the form

$$f(\boldsymbol{y}_i \mid \Sigma_j \boldsymbol{d}_{ij}y_{ij}=\boldsymbol{b}_i;\boldsymbol{\beta})=\frac{f(\boldsymbol{y}_i;\boldsymbol{\beta},\boldsymbol{U}_i)}{f(\Sigma_j\boldsymbol{d}_{ij}y_{ij}=\boldsymbol{b}_i;\boldsymbol{\beta},\boldsymbol{U}_i)} \tag{9.2.3}$$

$$=\frac{f(\Sigma_j\boldsymbol{x}_{ij}y_{ij}=\boldsymbol{a}_i,\ \Sigma_j\boldsymbol{d}_{ij}y_{ij}=\boldsymbol{b}_i;\ \boldsymbol{\beta},\boldsymbol{U}_i)}{f(\Sigma_j\boldsymbol{d}_{ij}y_{ij}=\boldsymbol{b}_i;\ \boldsymbol{\beta},\boldsymbol{U}_i)}.$$

For a discrete GLM this expression can be written as

$$\frac{\Sigma_{R_{i1}}\exp(\boldsymbol{\beta}'\boldsymbol{a}_i+\boldsymbol{U}_i'\boldsymbol{b}_i)}{\Sigma_{R_{i2}}\exp(\boldsymbol{\beta}'\Sigma_j\boldsymbol{x}_{ij}y_{ij}+\boldsymbol{U}_i'\boldsymbol{b}_i)}$$

where R_{i1} is the set of possible values for $\boldsymbol{y}_i$ such that $\Sigma_j\boldsymbol{x}_{ij}y_{ij}=\boldsymbol{a}_i$ and $\Sigma_j\boldsymbol{d}_{ij}y_{ij}=\boldsymbol{b}_i$, and R_{i2} is the set of values for $\boldsymbol{y}_i$ such that $\Sigma_j\boldsymbol{d}_{ij}y_{ij}=\boldsymbol{b}_i$. The conditional likelihood for $\boldsymbol{\beta}$ given the data for all m individuals simplifies to

$$\prod_{i=1}^{m}\frac{\Sigma_{R_{i1}}\exp(\boldsymbol{\beta}'\boldsymbol{a}_i)}{\Sigma_{R_{i2}}\exp(\boldsymbol{\beta}'\sum_{j=1}^{n_i}\boldsymbol{x}_{ij}y_{ij})}. \tag{9.2.4}$$

For simple cases such as the random intercept model, the conditional likelihood is reasonably easy to maximize (Breslow and Day, 1980). Random intercept models for binary and count data are considered in more detail below.

9.2.2 *Maximum likelihood estimation*

Here, we will treat the $\boldsymbol{U}_i$ as a sample of independent unobservable variables from a random effects distribution. Qualitatively, this assumption implies that we can learn about one individual's coefficients by understanding the variability in coefficients across the population. When there is little variability, we should rely on the population average coefficients to estimate those for an individual. When there is substantial variation, we must rely more heavily on the data from each individual in estimating their own coefficients. This idea is illustrated in the CD4+ example in Section 5.6.

The likelihood function for the unknown parameter $\boldsymbol{\delta}$, which is defined to include both $\boldsymbol{\beta}$ and the elements of G, is

$$L(\boldsymbol{\delta};\boldsymbol{y}) = \prod_{i=1}^{m} \int \prod_{j=1}^{n_i} f(y_{ij} \mid \boldsymbol{U}_i;\boldsymbol{\beta}) f(\boldsymbol{U}_i;G)d\boldsymbol{U}_i. \tag{9.2.5}$$

This is just the marginal distribution of $\boldsymbol{Y}$ obtained by integrating the joint distribution of $\boldsymbol{Y}$ and $\boldsymbol{U}$ with respect to $\boldsymbol{U}$. In some special cases such as the Gaussian linear model (Chapters 4–6), the integral above has a closed form, but for most non-Gaussian models, numerical methods are required for its evaluation.

To find the maximum likelihood estimate, we can solve the score equations obtained by setting to zero the derivative with respect to $\boldsymbol{\delta}$ of the log likelihood. If we imagine that the 'complete' data for an individual comprise $(\boldsymbol{y}_i, \boldsymbol{U}_i)$ and if we restrict attention for the moment to canonical link functions, then the *complete data* score function for $\boldsymbol{\beta}$ has a particularly simple form

$$S_{\boldsymbol{\beta}}(\boldsymbol{\delta} \mid \boldsymbol{y}, \boldsymbol{U}) = \sum_{i=1}^{m}\sum_{j=1}^{n_i} \boldsymbol{x}_{ij}\{y_{ij} - \mu_{ij}(\boldsymbol{U}_i)\} = 0 \tag{9.2.6}$$

where $\mu_{ij}(\boldsymbol{U}_i) = \mathrm{E}(y_{ij} \mid \boldsymbol{U}_i) = h^{-1}(\boldsymbol{x}_{ij}'\boldsymbol{\beta} + \boldsymbol{d}_{ij}'\boldsymbol{U}_i)$. The observed data score equations are obtained by taking the expectation of the complete data equations with respect to the conditional distribution of the unobserved random effects given the data. That is, we define the observed data score functions, $S_{\boldsymbol{\beta}}(\boldsymbol{\delta} \mid \boldsymbol{y})$, as the expectations of the complete data score functions, $S_{\boldsymbol{\beta}}(\boldsymbol{\delta} \mid \boldsymbol{y}, \boldsymbol{U})$, with respect to the conditional distribution of $\boldsymbol{U}$ given $\boldsymbol{y}$. This gives

$$S_{\boldsymbol{\beta}}(\boldsymbol{\delta} \mid \boldsymbol{y}) = \sum_{i=1}^{m}\sum_{j=1}^{n_i} \boldsymbol{x}_{ij}\{y_{ij} - \mathrm{E}(\mu_{ij}(\boldsymbol{U}_i) \mid \boldsymbol{y}_i)\} = 0. \tag{9.2.7}$$

The score equations for G can similarly be obtained as

$$S_G(\boldsymbol{\delta} \mid \boldsymbol{y}) = \tfrac{1}{2}G^{-1}\left\{\sum_{i=1}^{m} \mathrm{E}(\boldsymbol{U}_i\boldsymbol{U}_i' \mid \boldsymbol{y}_i)\right\}G^{-1} - \frac{m}{2}G^{-1} = 0. \tag{9.2.8}$$

To solve for the maximum likelihood estimate of $\boldsymbol{\delta}$, a common strategy is to use the EM algorithm (Dempster *et al.* 1977). This algorithm iterates between an E-step, which involves evaluating the expectations in the score equations above using the current values of the parameters, and an M-step, in which we solve the score equations to give updated parameter estimates.

The dimension of the integration involved in the conditional expectations is q, the dimension of $\boldsymbol{U}_i$. When q is one or two, numerical integration techniques can be implemented reasonably easily (e.g. Crouch and Spiegelman, 1990). For higher dimensional problems, Monte Carlo integration methods can be used. See for example, the application of Gibbs sampling in Zeger and Karim (1991).

An alternative strategy is to approximate the score equations in such a way that the integrations can be avoided. This approach has been used by Stiratelli *et al.* (1984) for logistic models with Gaussian random effects, Karim (1991), Schall (1991), and Breslow and Clayton (1993) for random effects GLMs, and Lindstrom and Bates (1990) for non-linear regression models with Gaussian random effects and errors. The central idea is to use conditional modes rather than conditional means in the score equation for $\boldsymbol{\beta}$. This is equivalent to approximating the conditional distribution of $\boldsymbol{U}_i$ given $\boldsymbol{y}_i$ by a Gaussian distribution with the same mode and curvature. By using modes rather than means, we replace the integration with an optimization that can be incorporated into the M-step.

To specify the algorithm more completely, let $v_{ij} = \mathrm{Var}(y_{ij} \mid \boldsymbol{U}_i)$ and $Q_i = \mathrm{diag}\{v_{ij}h'(\mu_{ij})^2\}$. Let $\boldsymbol{z}_i$ be the surrogate response defined to have elements $z_{ij} = h(\mu_{ij}) + (y_{ij} - \mu_{ij})h'(\mu_{ij})$, $j = 1, \ldots, n_i$ and define the $n_i \times n_i$ matrix $V_i = Q_i + D_i G D_i'$ where D_i is the $n_i \times q$ matrix whose jth row is $\boldsymbol{d}_{ij}$. For a fixed G, updated values of $\boldsymbol{\beta}$ and $\boldsymbol{U}$ are obtained by iteratively solving

$$\hat{\boldsymbol{\beta}} = \left(\sum_{i=1}^{m} X_i' V_i^{-1} X_i \right)^{-1} \sum_{i=1}^{m} X_i' V_i^{-1} \boldsymbol{z}_i \qquad (9.2.9)$$

and

$$\hat{\boldsymbol{U}}_i = G D_i V_i^{-1} (\boldsymbol{z}_i - X_i \boldsymbol{\beta}).$$

These equations are an application of Harville's (1977) method for linear random effects models to a linearized version of the possibly non-linear estimating equations in the GLM extension.

To estimate G, note that the score equation (9.2.8) implies that

$$\hat{G} = m^{-1} \sum_{i=1}^{m} \mathrm{E}(\boldsymbol{U}_i \boldsymbol{U}_i' \mid \boldsymbol{y}_i)\} \qquad (9.2.10)$$

$$= m^{-1} \sum_{i=1}^{m} \mathrm{E}(\boldsymbol{U}_i \mid \boldsymbol{y}_i) \mathrm{E}(\boldsymbol{U}_i \mid \boldsymbol{y}_i)' + m^{-1} \sum_{i=1}^{m} \mathrm{Var}(\boldsymbol{U}_i \mid \boldsymbol{y}_i). \qquad (9.2.11)$$

The quantity $\hat{\boldsymbol{U}}_i$ is an estimate of $\mathrm{E}(\boldsymbol{U}_i \mid \boldsymbol{y}_i)$. An estimate of the conditional variance is

$$m^{-1}\sum_{i=1}^{m}(D_i'Q_i^{-1}D_i+G^{-1})^{-1}.$$

Note that the parameters appear on both sides of equations (9.2.9) and (9.2.10) so that the algorithm proceeds by iteratively updating first the estimates of the regression coefficients and the random effects, and then the variance of the random effects, until the parameter estimates converge. A variety of slightly different algorithms for estimation of G have been proposed. See Breslow and Clayton (1993) for one specific implementation and an evaluation of its performance.

This approximate method gives reasonable estimates of $\boldsymbol{\beta}$ in many problems. The estimates of $\boldsymbol{U}_i$ and G are more sensitive to the Gaussian approximation to the conditional distribution. The approximation breaks down when there are few observations per subject and the GLM is far from the Gaussian. Karim (1991) and Breslow and Clayton (1993) have evaluated this approximate method for some specific random effects GLMs.

9.3 Logistic regression for binary responses

9.3.1 *Conditional likelihood approach*

In this subsection, we consider the random intercept logistic model for binary data given by

$$\operatorname{logit}\Pr(Y_{ij}=1\mid U_i)=\beta_0+U_i+\boldsymbol{x}_{ij}'\boldsymbol{\beta}. \tag{9.3.1}$$

To simplify the discussion, we will write $\gamma_i=\beta_0+U_i$ and assume that $\boldsymbol{x}_{ij}$ does not include an intercept term. The joint likelihood function for $\boldsymbol{\beta}$ and the γ_i is proportional to

$$\prod_{i=1}^{m}\exp\left[\gamma_i\sum_{j=1}^{n_i}y_{ij}+\left(\sum_{j=1}^{n_i}y_{ij}\boldsymbol{x}_{ij}'\right)\boldsymbol{\beta}-\sum_{j=1}^{n_i}\log\left\{1+\exp(\gamma_i+\boldsymbol{x}_{ij}'\boldsymbol{\beta})\right\}\right]. \tag{9.3.2}$$

The conditional likelihood for β given the sufficient statistics for the γ_i has the form

$$\prod_{i=1}^{m}\frac{\exp(\sum_{j=1}^{n_i}y_{ij}\boldsymbol{x}_{ij}'\boldsymbol{\beta})}{\sum_{R_i}\exp(\sum_{\ell=1}^{y_{i.}}\boldsymbol{x}_{i\ell}'\boldsymbol{\beta})} \tag{9.3.3}$$

where $y_{i.}=\sum_{j=1}^{n_i}y_{ij}$ and the index set R_i contains all the $\binom{n_i}{y_{i.}}$ ways of choosing $y_{i.}$ positive responses out of n_i repeated observations.

The conditional likelihood above is equivalent to the one derived in stratified case-control studies (Breslow and Day, 1980). In that context, there are m strata; in the ith stratum, there are $y_{i.}$ cases and $n_i-y_{i.}$ controls. This connection is important in that any statistical package suitable

Table 9.1. Notation for a 2×2 crossover trial.

Group	(1,1)	(0,1)	(1,0)	(0,0)
AB	a_1	b_1	c_1	d_1
BA	a_2	c_2	b_2	d_2

for the analysis of the stratified case-control studies can be used to fit a random intercept logistic model to binary longitudinal data with little or no modification.

Example 9.1. The 2×2 crossover trial
Let a, b, c and d denote the numbers of response pairs for each of the four possible combinations of outcomes in a 2×2 crossover trial, as shown in Table 9.1. For example, b_1 is the number of subjects in the first group, who received the active treatment (A) followed by placebo (B) with outcomes (1,0), that is with a normal response ($Y = 1$) at the first visit and an abnormal response ($Y = 0$) at the second.

For the logistic model (9.3.1) which includes only the treatment variable x_1, the conditional likelihood (9.2.4) reduces to

$$\left\{\frac{\exp(\beta_1)}{1+\exp(\beta_1)}\right\}^{b_1+b_2}\left\{\frac{1}{1+\exp(\beta_1)}\right\}^{c_1+c_2}.$$

The estimate of β_1 which maximizes this conditional likelihood is

$$\hat{\beta}_1 = \log\{(b_1+b_2)/(c_1+c_2)\}.$$

Its variance can be estimated by $(b_1+b_2)^{-1} + (c_1+c_2)^{-1}$.

If the period effect (x_2) is now added to the model, the conditional likelihood function becomes

$$\frac{\exp\{(\beta_1+\beta_2)b_1\}}{\{1+\exp(\beta_1+\beta_2)\}^{b_1+c_1}}\frac{\exp\{(\beta_1-\beta_2)b_2\}}{\{1+\exp(\beta_1-\beta_2)\}^{b_2+c_2}}.$$

The maximum conditional likelihood estimate of β_1 and the corresponding variance estimate are, respectively

$$\hat{\beta}_1 = \tfrac{1}{2}\log\left(\frac{b_1 b_2}{c_1 c_2}\right), \quad \hat{\mathrm{Var}}(\hat{\beta}_1) = \tfrac{1}{4}\left(b_1^{-1}+c_1^{-1}+b_2^{-1}+c_2^{-1}\right).$$

For the 2×2 crossover data on cerebrovascular deficiency, we have $b_1 = 6$, $c_1 = 0$, $b_2 = 4$, and $c_2 = 2$. For the model including the treatment variable only, the treatment effect, β_1, is estimated as $\log\{(6+4)/(0+2)\} = 1.61$

Table 9.2. Results for a conditional likelihood analysis of data from a 3×3 crossover trial

Variable	Treatment		Period		Carryover	
	B	C	2	3	B	C
Coefficient	1.98	1.71	0.69	0.85	-0.14	-1.24
Standard Error	(0.45)	(0.41)	(0.56)	(0.58)	(0.60)	(0.65)

with estimated standard error 0.77. A similar result is obtained when the period effect is included in the model. The treatment effect is now estimated as $\frac{1}{2}\log\{6 \times 4/(0.5 \times 2)\} = 1.59$ with estimated standard error 0.85. Note that we have used an ad hoc convention of replacing the zero cell with 0.5 in this calculation. Nevertheless, the data indicate that the odds of a normal electrocardiogram for the treated patients are about five-fold (exp(1.6)) greater than the odds for patients receiving the placebo. This finding from the conditional inference is to be contrasted with the results from fitting a marginal model in Section 8.3. For the marginal model, we estimated a roughly two-fold increase in the odds for the treated group. The smaller value from the marginal analysis is consistent with the theoretical inequality stated in Section 7.4.

Example 9.2. The 3×3 crossover trial

Table 9.2 gives the results of fitting a random intercept logistic regression model by conditional likelihood to the crossover data from Example 8.2. The table reports the estimated regression coefficients and their standard errors.

Strong treatment effects are evident after adjusting for period and carryover effects. The chance of dysmenorrhea relief for a patient is increased by a factor of $\exp(1.98) = 7.3$ if her treatment is switched from placebo to a low dose of analgesic and by a factor of $\exp(1.71) = 5.5$ if treatment is switched from placebo to a high dose.

The principal advantage of the conditional likelihood approach is that we remove the random effects from the likelihood by which we estimate $\boldsymbol{\beta}$ without having to make the assumption that they are a sample from a particular probability distribution. The disadvantage is that we rely entirely on within-subject comparisons. So persons with $y_{i.} = n_i$ or $y_{i.} = 0$ provide no information about the regression coefficients. In the 2×2 crossover trials, this means that $a_1+d_1+a_2+d_2$ pairs are uninformative. In the example considered, this accounts for 82 per cent (55/67) of the subjects under observation. Consequently, standard errors of regression estimates tend to be larger than in a marginal or random effects analysis. For example, the standard error of the regression coefficient for the treatment variable is here

0.91, as opposed to a value of 0.23 obtained from the marginal model. At the extreme, the conditional analysis provides no information about coefficients of explanatory variables which do not vary over time. This can be seen by examining (9.3.3). The product of any time-independent covariate and its coefficient will factor out of the sum in both the numerator and denominator and cancel from the conditional likelihood. This is sensible since we have conditioned away all information about each subject's intercept and thus cannot use estimates which are based entirely upon comparisons across subjects.

We now turn to the situation in which the $\boldsymbol{U}_i$ are treated as an independent sample from a random effects distribution. We begin by reviewing the traditional but simpler random effects models for binary data and then consider the logistic-Gaussian model more specifically.

9.3.2 *Random effects models for binary data*

Historically, the motivation for random effects models has been the observation that the variability among clustered binary responses exceeds what would be expected due to binomial variation alone. Random effects models were introduced to account for this so-called *extra-binomial variation.* The beta-binomial distribution (Skellam, 1948) was one of the earliest. Let $\{Y_{i1}, \ldots, Y_{in_i}\}$ represent the n_i binary responses from cluster i. Here the cluster could be a litter in a teratology experiment, a family or household in a genetic study or an individual in a longitudinal study. The beta-binomial distribution assumes that:

(1) conditional on μ_i, the responses $Y_{i1}, \ldots, Y_{in_i}$ are independent with common probability μ_i,

(2) the μ_i follow a beta distribution with mean μ and variance $\delta\mu(1-\mu)$.

Unconditionally, the total number of positive responses for a cluster, $Y_{i.} = Y_{i1} + \ldots + Y_{in}$, has a beta-binomial distribution with

$$\mathrm{E}(Y_{i.}) = n_i \mu_i$$

and

$$\mathrm{Var}(Y_{i.}) = n_i \mu_i (1 - \mu_i)\{1 + (n_i - 1)\delta\}.$$

The over-dispersion parameter δ is the correlation for each pair of binary responses from the same cluster.

The beta-binomial distribution has been used to model the incidence of non-infectious diseases in a household (Griffiths, 1973), the number of malformed fetuses in a litter (Williams, 1975), and the number of chromosomal aberrant cells among repeated samples for an individual (Prentice,

1986). Prentice (1986) points out that the correlation coefficient, δ, need not be positive in the beta-binomial model as previously thought, but that its lower bound is

$$\delta_0 = \max\{-\mu/(n-\mu-1), -(1-\mu)/(n+\mu)\}.$$

The beta-binomial framework can be extended so that a parametric model may be imposed on the cluster-specific means, μ_i. For example, μ_i might be assumed to depend on cluster-level explanatory variables, $\boldsymbol{x}_i$, through a logistic function, $\text{logit}(\mu_i) = \boldsymbol{x}_i'\boldsymbol{\beta}$.

Originally, it was assumed that the beta-binomial distribution required each response from the same cluster to have a common probability, μ_i. In the regression set-up, this required the covariates to be the same for all observations within a cluster, that is, $\boldsymbol{x}_{i1} = \ldots = \boldsymbol{x}_{in_i} = \boldsymbol{x}_i$. However, Rosner (1984) has extended the beta-binomial to allow the covariates to vary within clusters. His model for one cluster is formally equivalent to the following n_i logistic regressions:

$$\begin{aligned} &\text{logit}\Pr(Y_{ij} = 1 \mid y_{i1}, \ldots, y_{ij-1}, y_{ij+1}, \ldots, y_{in_i}, \boldsymbol{x}_{ij}) \\ &= \log\left(\tfrac{\theta_{i1} + w_{ij}\theta_{i2}}{1-\theta_{i1}+(n_i-1-w_{ij})\theta_{i2}}\right) + \boldsymbol{x}_{ij}'\boldsymbol{\beta}^*, \; j = 1, \ldots, n_i \end{aligned} \tag{9.3.4}$$

where $w_{ij} = y_{i.} - y_{ij}$, θ_{i1} is an intercept parameter and θ_{i2} characterizes the association between pairs of responses for the same cluster. This clever extension of the beta-binomial does have some important limitations. First, its regression coefficients, $\boldsymbol{\beta}^*$, measure the effect of $\boldsymbol{x}_{ij}$ on Y_{ij} which cannot first be explained by the other responses in the cluster. Hence, the effects of cluster-level covariates may often be attributed to the other observations within the cluster, rather than to the covariate itself. This drawback is particularly severe when the cluster sizes vary so that different numbers of other responses are conditioned upon in the different clusters. This is a particular problem for longitudinal studies where the number of observations per person often varies. In addition, in longitudinal studies, it may be awkward to model the probability for the first response as a function of responses which come later in time. See, for example, Jones and Kenward (1987) who consider models for crossover trials.

The logistic model introduced in Section 9.1 adds the random effects on the same scale as the fixed effects. To our knowledge, this approach was first considered in the biostatistical literature by Pierce and Sands (1975) in an unpublished Oregon State University report. They assumed a Gaussian distribution for a univariate random intercept. Since then, the logistic model with Gaussian random effects has been studied extensively, including work by Williams (1982), Stiratelli *et al.* (1984), Anderson and Aitkin (1985), Gilmour *et al.* (1985), Zeger *et al.* (1988), Zeger and Karim

(1991), Breslow and Clayton (1993), and Waclawiw and Liang (1993). One exception is Conoway (1990), who used a log-log link and a log-gamma distribution for the random intercepts.

9.3.3 *Examples of logistic models with Gaussian random effects*

The approach to fitting the random effects model within the GLM framework has been covered in Section 9.2. No particular computational simplification arises when we focus on the logistic model with Gaussian random effects. The likelihood function for $\boldsymbol{\beta}$ and G is

$$L(\boldsymbol{\beta}, G; \boldsymbol{y}) = \prod_{i=1}^{m} \int \prod_{j=1}^{n_i} \{\mu_{ij}(\boldsymbol{\beta}, \boldsymbol{U}_i)\}^{y_{ij}} \{1 - \mu_{ij}(\boldsymbol{\beta}, \boldsymbol{U}_i)\}^{1-y_{ij}} f(\boldsymbol{U}_i; G) d\boldsymbol{U}_i \tag{9.3.5}$$

where $\mu_{ij}(\boldsymbol{\beta}, \boldsymbol{U}_i) = \mathrm{E}(Y_{ij} \mid \boldsymbol{U}_i; \boldsymbol{\beta})$. With the logit link and Gaussian assumption on the $\boldsymbol{U}_i$, this reduces to

$$\prod_{i=1}^{m} \int \exp\left[\boldsymbol{\beta}' \sum_j \boldsymbol{x}_{ij} y_{ij} + \boldsymbol{U}_i{}' \sum_j \boldsymbol{d}_{ij} y_{ij} - \sum_j \log\{1 + \exp(\boldsymbol{x}_{ij}'\boldsymbol{\beta} + \boldsymbol{d}_{ij}'\boldsymbol{U}_i)\}\right]$$
$$\times (2\pi)^{-1} \mid G \mid^{-q/2} \exp(-\boldsymbol{U}_i' G^{-1} \boldsymbol{U}_i / 2) d\boldsymbol{U}_i$$

where G is the $q \times q$ variance matrix of each $\boldsymbol{U}_i$. Crouch and Spiegelman (1990) present numerical integration methods tailored for the logistic-Gaussian integral above. Zeger and Karim (1991) have used a Gibbs sampling Monte Carlo algorithm to simulate from a posterior distribution similar to this likelihood function. In the examples below, we use numerical integration to obtain maximum likelihood estimates.

Example 9.1 (continued)

For the 2×2 crossover trial on cerebrovascular deficiency, we assume a logistic regression model with additive effects of treatment, period and a random intercept $\gamma_i = \beta_0 + U_i$, assumed to follow a Gaussian distribution with variance G. Table 9.3 gives maximum likelihood estimates of $\boldsymbol{\beta}$ and $G^{1/2}$. For comparison, the table also presents regression coefficients obtained by fitting a marginal model.

Focusing first on the maximum likelihood estimates, there is clear evidence of substantial heterogeneity among subjects. The standard deviation of the random intercept distribution is estimated to be 4.9 with a standard error of 1.8. By the Gaussian assumption for the intercepts on the logit scale, roughly 95 per cent of subjects would fall within 9.8 logit units of the overall mean. But this range on the logit scale translates into probabilities which range from essentially 0 to 1. Hence, the data suggest that some people have little chance and others very high chance of a normal

Table 9.3. Regression estimates and standard errors (in parentheses) of random effects and marginal models fitted to the 2×2 crossover data for cerebrovascular deficiency adapted from Jones and Kenward (1989) and presented in Table 8.1.

	Random effects model	Marginal model	Ratio of random effects to marginal
Intercept	4.1	1.2	3.4
	(1.7)	(0.30)	
Treatment	1.9	0.57	3.3
	(0.91)	(0.23)	
Period	-1.0	-0.30	3.3
	(0.81)	(0.23)	
$G^{1/2}$	4.9	-	
	(1.8)		

reading given either treatment. Assuming a constant treatment effect for all persons, the odds of a normal response for a subject are estimated to be 6.7 (exp(1.9)) times higher on the active drug than on the placebo.

The last column of Table 9.3 presents the ratios of regression estimates obtained from the random effects model and from the marginal model. The three ratios are all close to $(0.346\hat{G}+1)^{1/2} = 3.1$, the theoretical value discussed in Section 7.4 and in Zeger *et al.* (1988).

Example 9.3 (continued)

Fitting a random effects model to the data from the Indonesian study allows us to address the question of how an individual child's risk for respiratory infection will change if their vitamin A status were to change. This is accomplished by allowing each child to have a distinct intercept which represents their propensity to infection. Table 9.4 gives regression estimates from models analogous to models 2 to 4 in Table 8.6. Here, we have accounted for correlation by including random intercepts which are assumed to follow a Gaussian distribution with variance G.

Table 9.4. Regression estimates and standard errors (in parentheses) of random effects models for the Indonesian study on respiratory infection (Sommer *et al.*, 1984).

Variable	Model 1	Model 2	Model 3
Intercept	-2.2 (0.24)	-1.9 (0.30)	-2.4 (0.37)
Sex	-0.51 (0.25)	-0.56 (0.26)	-0.55 (0.26)
Height for age	-0.044 (0.022)	-0.052 (0.022)	-0.049 (0.022)
Seasonal cosine	-0.61 (0.17)	-	-0.56 (0.22)
Seasonal sine	-0.17 (0.17)	-	-0.019 (0.22)
Xerophthalmia	0.54 (0.48)	0.57 (0.48)	0.69 (0.49)
Age	-0.031 (0.0079)	-	-
Age^2	-0.0011 (0.00040)	-	-
Age at entry	-	-0.054 (0.011)	-0.055 (0.011)
Age at entry^2	-	-0.0014 (0.00049)	-0.0014 (0.00050)
Follow-up time	-	-0.20 (0.074)	-0.085 (0.10)
Follow-up time^2	-	0.014 (0.0048)	0.0069 (0.0068)
$G^{1/2}$	0.72 (0.23)	0.71 (0.24)	0.72 (0.24)

Note that the estimates of the random effects standard deviation are around 0.7, statistically significantly different from zero but smaller than in the crossover trial of Example 9.1. Among children with linear predictor equal to the intercept, -2.2, in Model 1 (average age, weight, female, vitamin A sufficient), about 95 per cent would have a probability of infection between 0.03 and 0.31. This still represents considerable heterogeneity in the propensity for infection. The relative odds of infection associated with xerophthalmia (vitamin A deficiency) are estimated to be $\exp(0.54) = 1.7$ in Model 1 and are not significantly different from 1. The lack of significance of the effect is due to the small number of xerophthalmia cases (52) in this illustrative subset of the original data. Finally, the longitudinal age effect on the risk of respiratory infection seen in Model 2 can be explained, to a large extent, by the seasonal trend as shown by fitting Model 3.

Because there is less heterogeneity among subjects, the regression estimates seen above are similar to the marginal model coefficients in Table 8.6. Again, the ratios of the marginal and random effects coefficients are close to $(0.346\hat{G} + 1)^{1/2}$ as discussed in Section 7.4.

9.4 Counted responses

9.4.1 *Conditional likelihood method*

We now consider conditional maximum likelihood estimation of the random intercept log-linear model for count data. Specifically, we assume that conditional on $\gamma_i = \beta_0 + U_i$,

(1) Y_{ij} follows a Poisson distribution such that

$$\log\mathrm{E}(Y_{ij} \mid \gamma_i) = \gamma_i + \boldsymbol{x}_{ij}'\boldsymbol{\beta} + \log(t_{ij}), \quad j = 1, \ldots, n_i; \quad \text{and}$$

(2) $Y_{i1}, \ldots, Y_{in_i}$ are independent.

Under these assumptions, the likelihood function for $\boldsymbol{\beta}$ and $\gamma_1, \ldots, \gamma_m$ is completely specified and is proportional to

$$\prod_{i=1}^{m} \exp\left\{\gamma_i \sum_{j=1}^{n_i} y_{ij} + \boldsymbol{\beta}' \sum_{j=1}^{n_i} y_{ij}\boldsymbol{x}_{ij} + \sum_{j=1}^{n_i} y_{ij}\log(t_{ij}) - \sum_{j=1}^{n_i} t_{ij}\exp(\gamma_i + \boldsymbol{x}_{ij}'\boldsymbol{\beta})\right\}. \tag{9.4.1}$$

By conditioning on $y_{i.} = \Sigma_{j=1}^{n_i} y_{ij}$ as was done in Section 9.2.1, we obtain the following conditional likelihood which depends on $\boldsymbol{\beta}$ only:

$$\prod_{i=1}^{m} \binom{y_{i.}}{y_{i1}, \ldots, y_{in_i}} \prod_{j=1}^{n_i} \left(\frac{t_{ij}e^{\boldsymbol{x}_{ij}'\boldsymbol{\beta}}}{\sum_{\ell=1}^{n_i} t_{i\ell}e^{\boldsymbol{x}_{i\ell}'\boldsymbol{\beta}}}\right)^{y_{ij}}. \tag{9.4.2}$$

The contribution in (9.4.2) for subject i is a multinomial probability in which

$$\pi_{ij} = t_{ij}\exp(\boldsymbol{x}'_{ij}\boldsymbol{\beta})/\sum_{\ell=1}^{n_i} t_{i\ell}\exp(\boldsymbol{x}'_{i\ell}\boldsymbol{\beta})$$

represents the probability that each of the $y_{i.}$ events will fall into 'category j', $j = 1, \ldots, n_i$. We now use the conditional likelihood approach to continue the analysis of the seizure data.

Example 9.4 (epileptic seizures)
Returning to Example 8.4, we first consider the model

$$\log \mathrm{E}(Y_{ij} \mid \gamma_i) = \gamma_i + \beta_1 x_{ij1} + \beta_2 x_{ij2} + \beta_3 x_{ij1} x_{ij2} + \log(t_{ij})\begin{matrix} j = 0, 1, \ldots, 4, \\ i = 1, \ldots, 59 \end{matrix}$$

where

$$x_{ij1} = \begin{cases} 1 \text{ if the } ith \text{ subject is assigned to the progabide group} \\ 0 \text{ if the } ith \text{ subject is assigned to the placebo group,} \end{cases}$$

$$x_{ij2} = \begin{cases} 1 \text{ if } j = 1, 2, 3, \text{ or } 4 \\ 0 \text{ if } j = 0. \end{cases}$$

Here, e^{γ_i} is the expected baseline seizure count for the ith subject, $i = 1, \ldots, 59$. The coefficient β_2 represents the log ratio of the seizure rates post versus pre-randomization for the placebo group. Note that this is assumed to take the same value for each subject. Similary, $\beta_2 + \beta_3$ represents the log ratio of rates for the treated group so that β_3 is the treatment effect coefficient.

Because $\boldsymbol{x}_{i1} = \boldsymbol{x}_{i2} = \boldsymbol{x}_{i3} = \boldsymbol{x}_{i4}$ and $t_{i0} = 8 = t_{i1} + \cdots + t_{i4}$, the conditional likelihood for $\boldsymbol{\beta}$ reduces to

$$\prod_{i=1}^{28} \binom{y_{i.}}{y_{i0}} \left(\frac{\exp(\beta_2)}{1 + \exp(\beta_2)}\right)^{y_{i.} - y_{i0}} \left(\frac{1}{1 + \exp(\beta_2)}\right)^{y_{i0}} \times$$

$$\prod_{i=29}^{59} \binom{y_{i.}}{y_{i0}} \left(\frac{\exp(\beta_2 + \beta_3)}{1 + \exp(\beta_2 + \beta_3)}\right)^{y_{i.} - y_{i0}} \left(\frac{1}{1 + \exp(\beta_2 + \beta_3)}\right)^{y_{i0}}.$$

Thus, the conditional likelihood reduces to a comparison of two sets of binomial observations with 'sample sizes' 28 and 31. In the placebo group, each subject contributes a statistic $y_{i0}/y_{i.}$ for estimating the parameter $\pi_1 = 1/\{1 + \exp(\beta_2)\}$, which is the common probability that an individual's seizure occurred before rather than after the randomization. Similarly, the common probability in the progabide group is $\pi_2 = 1/\{1 + \exp(\beta_2 + \beta_3)\}$.

Table 9.5. Results of conditional likelihood analysis (coefficient ± standard error) of seizure data.

	Complete data	Patient # 207 deleted
β_2	0.11 ± 0.047	0.11 ± 0.047
β_3	-0.10 ± 0.065	-0.30 ± 0.070
Pearson's χ^2	289.9	227.1

Thus, a negative value for β_3 indicates that a relatively larger fraction of the total seizures in the treatment group occurred before rather than after randomization as compared to the placebo group. In other words, a negative β_3 indicates that the treatment is effective.

Table 9.5 gives the conditional maximum likelihood estimates of β_2 and β_3 and their standard errors. With the full data set, there is modest evidence that progabide is more effective than the placebo in reducing the occurrence of seizures ($\hat{\beta}_3 = -0.10 \pm 0.065$). With possible outlier patient number 207 deleted, a stronger treatment effect is suggested, ($\hat{\beta}_3 = -0.30 \pm 0.070$). However, the calculation of the Pearson χ^2 statistic

$$\sum_{i=1}^{28} \frac{(y_{i0} - y_{i.}\hat{\pi}_1)^2}{y_{i.}\hat{\pi}_1(1-\hat{\pi}_1)} + \sum_{i=29}^{59} \frac{(y_{i0} - y_{i.}\hat{\pi}_2)^2}{y_{i.}\hat{\pi}_2(1-\hat{\pi}_2)},$$

where $\hat{\pi}_1 = 1/\{1 + \exp(\hat{\beta}_2)\}$ and $\hat{\pi}_2 = 1/\{1 + \exp(\hat{\beta}_2 + \hat{\beta}_3)\}$, also reveals that the fitted model is grossly inadequate. For example, with 57 degrees of freedom in the full data, the fitted model is rejected at the 0.01 level. The same conclusion is reached when the outlying individual is set aside. An important implication of this observation is that the estimated standard errors for the elements of $\hat{\boldsymbol{\beta}}$ may be too small. This may be due to the inadequacy of the assumption that the change in seizure rate is common to everyone within a treatment group.

One way to address this possibility is to introduce a random effect U_{i2} for the pre-post explanatory variable x_2. However, the conditional likelihood method would no longer be appropriate since all relevant information about β_3 will be conditioned away. We must instead use the random effects approach which is the topic of the following section.

9.4.2 *Random effects models for counts*

The Poisson distribution has a long tradition as a model for count data, but in biomedical applications it is rarely the case that $\mathrm{Var}(Y) = \mathrm{E}(Y)$ as is implied by the Poisson assumption. Typically, the variance exceeds the mean (Breslow, 1984). As discussed in Section 9.3.2 for binomial data, this over-dispersion can be explained by assuming that there is natural heterogeneity among the expected responses across observations. If the means are assumed to follow a gamma distribution, the marginal distribution of the counts is the *negative binomial* distribution. Specifically, this distribution arises from the assumptions that

(1) conditional on μ_i, the response variable Y_{ij} has a Poisson distribution with mean μ_i,

(2) the μ_i are independent gamma random variables with mean μ and variance $\phi\mu^2$.

Then, the unconditional distribution of Y_{ij} is negative binomial with

$$\mathrm{E}(Y_{ij}) = \mu \quad \text{and} \quad \mathrm{Var}(Y_{ij}) = \mu + \phi\mu^2. \tag{9.4.3}$$

The use of the negative binomial model dates back at least to the work of Greenwood and Yule (1920) who modelled over-dispersed accident counts.

The simplest extension of the negative binomial model is to assume that the μ_i depend on covariates $\boldsymbol{x}_i$ through some parametric function. The most common is the log-linear model for which

$$\log(\mu_i) = \boldsymbol{x}_i'\boldsymbol{\beta}. \tag{9.4.4}$$

One important limitation of this model for application to longitudinal data is that the explanatory variables in the regression above do not vary within subjects. Morton (1987) proposed a solution to this problem. In the longitudinal context, if we once again let $Y_{i1}, \ldots, Y_{in_i}$ denote the counted responses from the *ith* subject, Morton (1987) assumed that:

(1) Conditional on an independent unobserved variable ϵ_i, $\mathrm{E}(Y_{ij} \mid \epsilon_i) = \exp(\boldsymbol{x}_{ij}'\boldsymbol{\beta})\epsilon_i$, $j = 1, \ldots, n_i$;

(2) $\mathrm{Var}(Y_{ij} \mid \epsilon_i) = \phi\mathrm{E}(Y_{ij} \mid \epsilon_i)$;

(3) $\mathrm{E}(\epsilon_i) = 1$ and $\mathrm{Var}(\epsilon_i) = \sigma^2$.

Note that assumptions (1), (2), and (3) imply that $\mathrm{E}(Y_{ij}) = \exp(\boldsymbol{x}_{ij}'\boldsymbol{\beta}) = \mu_{ij}$ and $\mathrm{Var}(Y_{ij}) = \phi\mu_{ij} + \sigma^2\mu_{ij}^2$. Morton (1987) extended this approach to include more complicated nesting structures as well. He used a quasi-likelihood estimation approach, which is similar to GEE, for estimating $\boldsymbol{\beta}, \phi$, and σ^2. An attractive feature of this model is that it is not necessary to specify the complete distribution of ϵ_i, only its first two moments.

The model that is the focus of the remainder of this chapter adds the random effects on the same scale as the fixed effects as follows:

(1′) $\log \mathrm{E}(Y_{ij} \mid \boldsymbol{U}_i) = \boldsymbol{x}'_{ij}\boldsymbol{\beta} + \boldsymbol{d}'_{ij}\boldsymbol{U}_i$;

(2′) Given $\boldsymbol{U}_i$, the responses $Y_{i1}, \ldots, Y_{in_i}$ are independent Poisson variables with mean $\mathrm{E}(Y_{ij} \mid U_i)$;

(3′) the $\boldsymbol{U}_i$ are independent realizations from a distribution with density function $f(\boldsymbol{u}_i; G)$.

This second approach allows the contribution of the random effects to vary within a subject, that is, $\boldsymbol{d}_{ij}$ need not be constant for a given i. If this flexibility is needed, the second approach is to be preferred. In fact, the expression (1) is readily seen as a special case of (1′) with $d_{ij} = 1$ and $\boldsymbol{U}_i = \log \epsilon_i$. On the other hand, in order to make inferences about $\boldsymbol{\beta}$ and G, one needs to specify a distribution for the $\boldsymbol{U}_i$. The following section illustrates the use of Poisson regression with Gaussian random effects by re-analysing the seizure data.

9.4.3 *Poisson–Gaussian random effects models*

Example 9.4 (continued)

In this section we fit two models to the progabide data which differ only on how the random effects are incorporated. Model 1 is a log-linear model with a random intercept. In Model 2, we add a second random effect for the pre/post-treatment indicator (x_2) so that

$$\log \mathrm{E}(Y_{ij}|\boldsymbol{U}_i) = \beta_0 + \beta_1 x_{ij1} + \beta_2 x_{ij2} + \beta_3 x_{ij1} x_{ij2} + U_{i1} + x_{ij2} U_{i2} + \log(t_{ij}),$$

where $\boldsymbol{U}_i = (U_{i1}, U_{i2})$ is assumed to follow a Gaussian distribution with mean (0,0) and variance matrix G with elements

$$\begin{pmatrix} G_{11} & G_{12} \\ G_{21} & G_{22} \end{pmatrix}.$$

The inclusion of U_{i2} allows us to address the concern raised at the end of Section 9.4.1, that there might be heterogeneity among subjects in the ratio of the expected seizure counts before and after the randomization. The degree of heterogeneity can be measured by the magnitude of G_{22}, the variance of U_{i2}. Both models were fitted using the approximate maximum likelihood algorithm outlined in Section 9.2.2.

Table 9.6 presents results from fitting Models 1 and 2, with and without patient number 207. As expected, the results from Model 1 are in close agreement with those from the conditional approach given in Section 9.4.1. However, this model is refuted by the statistical significance of G_{22}, which in Model 2 using the complete data we estimate to be 0.24 with standard error 0.062.

Table 9.6. Estimates and standard errors (in parentheses) for random effects Poisson regression models fitted to the progabide data with and without patient number 207.

	Model 1		Model 2	
Variable	complete data	without 207	complete data	without 207
Intercept	1.0 (0.15)	1.0 (0.14)	1.1 (0.14)	1.1 (0.13)
Treatment	-0.023 (0.20)	-0.009 (0.19)	0.050 (0.18)	-0.029 (0.19)
Time	0.11 (0.047)	0.11 (0.047)	0.002 (0.11)	0.010 (0.11)
Treatment-by-time	-0.10 (0.065)	-0.30 (0.070)	-0.31 (0.15)	-0.34 (0.15)
G_{11}	0.62 (0.12	0.53 (0.10)	0.51 (0.10)	0.46 (0.10)
G_{12}	-	-	0.054 (0.056)	0.014 (0.053)
G_{22}	-	-	0.24 (0.062)	0.22 (0.059)

Focusing on the results for Model 2 fitted to the complete data, subjects in the placebo group have expected seizure rates after treatment which are estimated to be roughly the same as before treatment ($\exp(\hat{\beta}_2) = \exp(0.002) = 1.002$). For the progabide group, the seizure rates are reduced after treatment by about 27 per cent ($1 - \exp(0.002 - 0.31) = 0.27$). Hence, the treatment seems to have a modest effect: the estimated effect is $\hat{\beta}_3 = -0.31$ with a standard error of 0.15. Finally, if we set aside for the moment patient number 207, who had unusually high seizure rates, then the evidence that progabide is effective is somewhat stronger, with ($\hat{\beta}_3 = -0.34 \pm 0.15$). The analysis without patient 207 is only exploratory, and is carried out in order to understand this patient's influence on the overall results. Patient number 207 has been identified because of unusual seizure counts and perhaps has special medical problems.

9.5 Further reading

Throughout this chapter, we have used the Gaussian distribution as a convenient model for the random effects. When the regression coefficients are of primary interest, the specific form of the random effects distribution is less important. However, when the random effects are themselves the focus, as in the CD4+ example in Chapter 5, inferences are more dependent on the assumptions about their distribution. Lange and Ryan (1989) suggest a graphical way to test the Gaussian assumption when the response variables are continuous. When the response variables are discrete, the same task becomes more difficult. Davidian and Gallant (1992) have recently developed a non-parametric approach to estimating the random effects distribution with non-linear models.

The statistical literature on random effects GLMs has grown enormously in the last decade. Key papers in the biostatistics literature include: Laird and Ware (1982); Stiratelli *et al.* (1984); Gilmour *et al.* (1985); Schall (1990); Zeger and Karim (1991); Waclawiw and Liang (1993); Breslow and Clayton (1993); Fahrmeir (1992). These papers also provide useful additional references.

10
Transition models

This chapter considers extensions of generalized linear models (GLMs) for describing the conditional distribution of each response y_{ij} as an explicit function of past responses $y_{ij-1}, \ldots, y_{i1}$ and covariates $\boldsymbol{x}_{ij}$. We will focus on the case where the observation times t_{ij} are equally spaced. To simplify notation, we denote the history for subject i at visit j by $H_{ij} = \{y_{ik},\ k = 1, \ldots, j-1\}$. As above, we will continue to condition on the past and present values of the covariates without explicitly listing them.

The most useful transition models are Markov chains for which the conditional distribution of y_{ij} given H_{ij} depends only on the q prior observations $y_{ij-1}, \ldots, y_{ij-q}$. The integer q is referred to as the model *order*. Sections 10.1 and 10.2 provide a general treatment of Markov GLMs and fitting procedures. Section 10.3 deals with categorical data. Section 10.4 briefly discusses models for counted responses which are in an earlier stage of development than the corresponding models for categorical data.

10.1 General

As discussed in Section 7.3, a transition model specifies a GLM for the conditional distribution of Y_{ij} given the past responses, H_{ij}. The form of the conditional GLM is

$$f(y_{ij} \mid H_{ij}) = \exp\{[y_{ij}\theta_{ij} - \psi(\theta_{ij})]/\phi + c(y_{ij}, \phi)\} \qquad (10.1.1)$$

for known functions $\psi(\theta_{ij})$ and $c(y_{ij}, \phi)$. The conditional mean and variance are

$$\mu_{ij}^c = \mathrm{E}(Y_{ij} \mid H_{ij}) = \psi'(\theta_{ij}) \quad \text{and} \quad v_{ij}^c = \mathrm{Var}(Y_{ij} \mid H_{ij}) = \psi''(\theta_{ij})\phi.$$

We will consider transition models where the conditional mean and variance satisfy the equations

$$h(\mu_{ij}^c) = \boldsymbol{x}_{ij}{}'\boldsymbol{\beta} + \sum_{r=1}^{s} f_r(H_{ij}; \boldsymbol{\alpha})$$

for suitable functions $f_r(\cdot)$, and

$$v_{ij}^c = g(\mu_{ij}^c)\phi \tag{10.1.2}$$

where h and g are known link and variance functions determined from the specific form of the density function above. Appendix A gives additional details on GLMs.

In words, the transition model expresses the conditonal mean μ_{ij}^c as a function of both the covariates $\boldsymbol{x}_{ij}$ and of the past responses $y_{ij-1}, \ldots, y_{ij-q}$. Past responses or functions thereof are simply treated as additional explanatory variables. We assume that the past affects the present through the sum of s terms, each of which may depend on the q prior values. The following examples with different link functions illustrate the range of transition models which are available.

Linear link – a linear regression with autoregressive errors for Gaussian data (Tsay, 1984) is a Markov model. It has the form

$$Y_{ij} = \boldsymbol{x}_{ij}'\boldsymbol{\beta} + \sum_{r=1}^{q} \alpha_r (Y_{ij-r} - \boldsymbol{x}_{ij-r}'\boldsymbol{\beta}) + Z_{ij}$$

where the Z_{ij} are independent, mean-zero, Gaussian innovations. This is a transition model with $h(\mu_{ij}^c) = \mu_{ij}^c$, $g(\mu_{ij}^c) = 1$ and $f_r = \alpha_r(y_{ij-r} - \boldsymbol{x}_{ij-r}'\boldsymbol{\beta})$. Note that the present observation, Y_{ij}, is a linear function of $\boldsymbol{x}_{ij}$ and of the earlier deviations $Y_{ij-r} - \boldsymbol{x}_{ij-r}'\boldsymbol{\beta}$, $r = 1, \ldots, q$.

Logit link – an example of a logistic regression model for binary responses that comprises a first order Markov chain (Cox, 1970; Korn and Whittemore, 1979 and Zeger *et al.* 1985) is

$$\text{logit } \Pr(Y_{ij} = 1 \mid H_{ij}) = \boldsymbol{x}_{ij}'\boldsymbol{\beta} + \alpha y_{ij-1},$$

previously given as equation (7.3.1). Here

$$h(\mu_{ij}^c) = \text{logit}(\mu_{ij}^c) = \log\left(\frac{\mu_{ij}^c}{1 - \mu_{ij}^c}\right), \quad g(\mu_{ij}^c) = \mu_{ij}^c(1 - \mu_{ij}^c)$$

and

$$f_r(H_{ij}, \alpha) = \alpha_r y_{ij-r}, \quad s = q = 1.$$

A simple extension to a model of order q has the form

$$\text{logit } \Pr(Y_{ij} = 1 \mid H_{ij}) = \boldsymbol{x}_{ij}'\boldsymbol{\beta}_q + \sum_{r=1}^{q} \alpha_r y_{ij-r}.$$

The notation $\boldsymbol{\beta}_q$ indicates that the value and interpretation of the regression coefficients changes with the Markov order, q.

Log-link – with count data we can assume a log-linear model where Y_{ij} given H_{ij} follows a Poisson distribution. Zeger and Qaqish (1988) discussed a first order Markov chain with $f_1 = \alpha\{\log(y^*_{ij-1}) - \boldsymbol{x}_{ij-1}{}'\boldsymbol{\beta}\}$, where $y^*_{ij} = \max(y_{ij}, c), 0 < c < 1$. This leads to

$$\mu^c_{ij} = \mathrm{E}(Y_{ij} \mid H_{ij}) = \exp(\boldsymbol{x}_{ij}{}'\boldsymbol{\beta}) \left(\frac{y^*_{ij-1}}{\exp(\boldsymbol{x}'_{ij-1}\boldsymbol{\beta})} \right)^{\alpha}.$$

The constant c prevents $y_{ij-1} = 0$ from being an absorbing state whereby $y_{ij-1} = 0$ forces all future responses to be 0. Note when $\alpha > 0$, we have an increased expectation, μ^c_{ij}, when the previous outcome, y_{ij-1}, exceeds $\exp(\boldsymbol{x}'_{ij-1}\boldsymbol{\beta})$. When $\alpha < 0$, a higher value at t_{ij-1} causes a lower value at t_{ij}.

In the linear regression model, the transition model can be formulated with $f_r = \alpha_r(y_{ij-r} - \boldsymbol{x}_{ij-1}{}'\boldsymbol{\beta})$ so that $\mathrm{E}(Y_{ij}) = \boldsymbol{x}_{ij}{}'\boldsymbol{\beta}$ for different values of q. In the logistic and log-linear cases, it is difficult to formulate models in such a way that $\boldsymbol{\beta}$ has the same meaning for different assumptions about the time-dependence. When $\boldsymbol{\beta}$ is the scientific focus, the careful data analyst should examine the sensitivity of the substantive findings to the choice of time-dependence model. This issue is discussed below by way of example.

Section 7.5 briefly discussed the method of conditional maximum likelihood for fitting the simplest logistic transition model. We now consider estimation in more detail.

10.2 Fitting transition models

As indicated in (7.5.3), the contribution to the likelihood for the *ith* subject can be written as

$$L_i(y_{i1}, \ldots y_{in_i}) = f(y_{i1}) \prod_{j=2}^{n_i} f(y_{ij} \mid H_{ij})$$

in a first order Markov model. In a Markov model of order q, the conditional distribution of Y_{ij} is

$$f(y_{ij} \mid H_{ij}) = f(y_{ij} \mid y_{ij-1}, \ldots, y_{ij-q})$$

so that the likelihood contribution for the *ith* subject becomes

$$f(y_{i1}, \ldots, y_{iq}) \prod_{j=q+1}^{n_i} f(y_{ij} \mid y_{ij-1}, \ldots, y_{ij-q}).$$

The GLM (10.1.1) specifies only the conditional distribution $f(y_{ij} \mid H_{ij})$; the likelihood of the first q observation $f(y_{i1}, \ldots, y_{iq})$ is not specified. In the linear model we assume that Y_{ij} given H_{ij} follows a Gaussian distribution. If $Y_{i1}, \ldots, Y_{iq}$ are also multivariate Gaussian and the covariance structure for the Y_{ij} is weakly stationary, the marginal distribution $f(y_{ij}, \ldots, y_{iq})$ can be fully determined from the conditional distribution model without additional unknown parameters. Hence, full maximum likelihood estimation can be used to fit Gaussian autoregressive models. See Tsay (1984) and references therein for details.

In the logistic and log-linear cases, $f(y_{i1}, \ldots y_{iq})$ is not determined from the GLM assumption about the conditional model, and the full likelihood is unavailable. An alternative is to estimate $\boldsymbol{\beta}$ and $\boldsymbol{\alpha}$ by maximizing the conditional likelihood

$$\prod_{i=1}^{m} f(y_{iq+1}, \ldots, y_{in_i} \mid y_{i1,\ldots,} y_{iq}) = \prod_{i=1}^{m} \prod_{j=q+1}^{n_i} f(y_{ij} \mid H_{ij}). \qquad (10.2.1)$$

When maximizing (10.2.1) there are two distinct cases to consider. In the first, $f_r(H_{ij}; \boldsymbol{\alpha}, \boldsymbol{\beta}) = \alpha_r f_r(H_{ij})$ so that $h(\mu_{ij}^c) = \boldsymbol{x}_{ij}{}'\boldsymbol{\beta} + \sum_{r=1}^{s} \alpha_r f_r(H_{ij})$. Here, $h(\mu_{ij}^c)$ is a linear function of both $\boldsymbol{\beta}$ and $\boldsymbol{\alpha} = (\alpha_1, \ldots, \alpha_s)$ so that estimation proceeds as in GLMs for independent data. We simply regress Y_{ij} on the $(p+s)$-dimensional vector of extended explanatory variables $(\boldsymbol{x}_{ij}, f_1(H_{ij}), \ldots, f_s(H_{ij}))$.

The second case occurs when the functions of past responses include both $\boldsymbol{\alpha}$ and $\boldsymbol{\beta}$. Examples are the linear and log-linear models discussed above. To derive an estimation algorithm for this case, note that the derivative of the log conditional likelihood or conditional score function has the form

$$\boldsymbol{U}^c(\boldsymbol{\delta}) = \sum_{i=1}^{m} \sum_{j=q+1}^{n_i} \frac{\partial \mu_{ij}^c}{\partial \boldsymbol{\delta}} v_{ij}^{c^{-1}} (y_{ij} - \mu_{ij}^c) = 0 \qquad (10.2.2)$$

where $\boldsymbol{\delta} = (\boldsymbol{\beta}, \boldsymbol{\alpha})$. This equation is the conditional analogue of the GLM score equation discussed in Section 4 of the Appendix. The derivative $\partial\mu_{ij}^c/\partial\boldsymbol{\delta}$ is analogous to $\boldsymbol{x}_{ij}$ but it can depend on $\boldsymbol{\alpha}$ and $\boldsymbol{\beta}$. We can still formulate the estimation procedure as an iterative weighted least squares as follows. Let $\boldsymbol{Y}_i$ be the $(n_i - q)$-vector of responses for $j = q+1, \ldots, n_i$ and $\boldsymbol{\mu}_{ij}^c$ its expectation given H_{ij}. Let $\boldsymbol{X}_i^*$ be an $(n_i - q) \times (p+s)$ matrix with kth row $\partial\mu_{iq+k}/\partial\boldsymbol{\delta}$ and $\boldsymbol{W}_i = \text{diag}(1/v_{ik+q}^c,\ k = 1, \ldots, n_i - q)$ an $(n_i - q) \times (n_i - q)$ diagonal weighting matrix. Finally, let $\boldsymbol{Z}_i = \boldsymbol{X}_i^* \hat{\boldsymbol{\delta}} + (\boldsymbol{Y}_i - \hat{\boldsymbol{\mu}}_i^c)$. Then, an updated $\hat{\boldsymbol{\delta}}$ can be obtained by iteratively regressing $\boldsymbol{Z}$ on $\boldsymbol{X}^*$ using weights $\boldsymbol{W}$.

When the correct model is assumed for the conditional mean and variance, the solution $\hat{\boldsymbol{\delta}}$ of (10.2.2) asymptotically, as m goes to infinity, follows a Gaussian distribution with mean equal to the true value, $\boldsymbol{\delta}$, and $(p+s) \times (p+s)$ variance matrix

$$V_{\hat{\boldsymbol{\delta}}} = \left(\sum_{i=1}^{m} X_i^{*\prime} W_i X_i^* \right)^{-1} . \tag{10.2.3}$$

The variance $V_{\hat{\boldsymbol{\delta}}}$ depends on $\boldsymbol{\beta}$ and $\boldsymbol{\alpha}$. A consistent estimate, $\hat{V}_{\hat{\boldsymbol{\delta}}}$, is obtained by replacing $\boldsymbol{\beta}$ and $\boldsymbol{\alpha}$ by their estimates $\hat{\boldsymbol{\beta}}$ and $\hat{\boldsymbol{\alpha}}$. Hence a 95 per cent confidence interval for β_1 is $\hat{\beta}_1 \pm 2\sqrt{\hat{V}_{\hat{\boldsymbol{\delta}}_{11}}}$ where $\hat{V}_{\hat{\boldsymbol{\delta}}_{11}}$ is the element in the first row and column of $\hat{V}_{\hat{\boldsymbol{\delta}}}$.

If the conditional mean is correctly specified but the conditional variance is not, we can still obtain consistent inferences about $\boldsymbol{\delta}$ by using the robust variance from equation (A.6.1) in Appendix A. Here it takes the form

$$V_R = \left(\sum_{i=1}^{m} X_i^{*\prime} W_i X_i^* \right)^{-1} \left(\sum_{i=1}^{m} X_i^{*\prime} W_i V_{Ti} W_i X_i^* \right) \left(\sum_{i=1}^{m} X_i^{*\prime} W_i X_i^* \right)^{-1} . \tag{10.2.4}$$

A consistent estimate $\hat{V}_R$ is obtained by replacing $V_{Ti} = \text{Var}(\boldsymbol{Y}_i \mid H_i)$ in the equation above by its estimate, $(\boldsymbol{Y}_i - \hat{\boldsymbol{\mu}}_i{}^c)(\boldsymbol{Y}_i - \hat{\boldsymbol{\mu}}_i{}^c)'$.

Interestingly, use of the robust variance will often give consistent confidence intervals for $\hat{\boldsymbol{\delta}}$ even when the Markov assumption is violated. However, in that situation, the interpretation of $\hat{\boldsymbol{\delta}}$ is questionable since $\mu_{ij}^c(\hat{\boldsymbol{\delta}})$ is not the conditional mean of Y_{ij} given H_{ij}.

10.3 Transition models for categorical data

This section discusses Markov chain regression models for categorical responses observed at equally spaced intervals. We begin with logistic models for binary responses and then briefly consider extensions to multinominal and ordered categorical outcomes.

As discussed in Section 7.3, a first order binary Markov chain is characterized by the transition matrix

$$\begin{pmatrix} \pi_{00} & \pi_{01} \\ \pi_{10} & \pi_{11} \end{pmatrix}$$

where $\pi_{ab} = \Pr(Y_{ij} = b \mid Y_{ij-1} = a),\ a, b = 0, 1$. For example, π_{01} is the probability that $Y_{ij} = 1$ when the previous response is $Y_{ij-1} = 0$. Note that each row of a transition matrix sums to one since $\Pr(Y_{ij} = 0 \mid Y_{ij-1} = a) + \Pr(Y_{ij} = 1 \mid Y_{ij-1} = a) = 1$. As its name implies, the transition matrix records the probabilities of making each of the possible transitions from one visit to the next.

In the regression setting, we model the transition probabilities as functions of covariates $\boldsymbol{x}_{ij}$. A very general model uses a separate logistic regression for $\Pr(Y_{ij} = 1 \mid Y_{ij-1} = y_{ij})$, $y_{ij} = 0,1$. That is, we assume that

$$\text{logit } \Pr(Y_{ij} = 1 \mid Y_{ij-1} = 0) = \boldsymbol{x}'_{ij}\boldsymbol{\beta}_0$$

and

$$\text{logit } \Pr(Y_{ij} = 1 \mid Y_{ij-1} = 1) = \boldsymbol{x}'_{ij}\boldsymbol{\beta}_1$$

where $\boldsymbol{\beta}_0$ and $\boldsymbol{\beta}_1$ may differ. In words, this model assumes that the effects of explanatory variables will differ depending on the previous response. A more concise form for the same model is

$$\text{logit } \Pr(Y_{ij} = 1 \mid Y_{ij-1} = y_{ij-1}) = \boldsymbol{x}_{ij}'\boldsymbol{\beta}_0 + y_{ij-1}\boldsymbol{x}_{ij}'\boldsymbol{\alpha} \qquad (10.3.1)$$

so that $\boldsymbol{\beta}_1 = \boldsymbol{\beta}_0 + \boldsymbol{\alpha}$. Equation (10.3.1) expresses the two regressions as a single logistic model which includes as predictors the previous response y_{ij-1} as well as the interaction of y_{ij-1} and the explanatory variables. An advantage of the form in (10.3.1) is that we can now easily test whether simpler models fit the data equally well. For example, we can test whether $\boldsymbol{\alpha} = (\alpha_0, 0)$ indicating that the covariates have the same effect on the response probability whether $y_{ij-1} = 0$ or $y_{ij-1} = 1$. Alternately, we can test whether a more limited subset of $\boldsymbol{\alpha}$ is zero indicating that the associated covariates can be dropped from the model. Each of these alternatives is nested within the saturated model so that standard statistical methods for nested models can be applied.

In many problems, a higher order Markov chain may be needed. The second order model has transition matrix

		Y_{ij}	
Y_{ij-2}	Y_{ij-1}	0	1
0	0	π_{000}	π_{001}
0	1	π_{010}	π_{011}
1	0	π_{100}	π_{101}
1	1	π_{110}	π_{111} .

Here, $\pi_{abc} = \Pr(Y_{ij} = c \mid Y_{ij-2} = a, Y_{ij-1} = b)$; for example π_{011} is the probability that $Y_{ij} = 1$ given $Y_{ij-2} = 0$ and $Y_{ij-1} = 1$. By analogy with the regression models for a first order chain, we could now fit four separate logistic regressions, one for each of the four possible histories (Y_{ij-2}, Y_{ij-1}), namely$(0,0)$, $(0,1)$, $(1,0)$, and $(1,1)$ with regression coefficients $\boldsymbol{\beta}_{00}$, $\boldsymbol{\beta}_{01}$, $\boldsymbol{\beta}_{10}$, and $\boldsymbol{\beta}_{11}$ respectively. But is is again more convenient to write a single equation as follows

$$\text{logit } \Pr(Y_{ij} = 1 \mid Y_{ij-2} = y_{ij-2}, Y_{ij-1} = y_{ij-1})$$

$$= \boldsymbol{x}_{ij}{}'\boldsymbol{\beta} + y_{ij-1}\boldsymbol{x}_{ij}{}'\boldsymbol{\alpha}_1 + y_{ij-2}\boldsymbol{x}_{ij}{}'\boldsymbol{\alpha}_2 + y_{ij-1}y_{ij-2}\boldsymbol{x}_{ij}{}'\boldsymbol{\alpha}_3. \tag{10.3.2}$$

By plugging in the different values for y_{ij-2} and y_{ij-1}, we obtain $\boldsymbol{\beta}_{00} = \boldsymbol{\beta}$; $\boldsymbol{\beta}_{01} = \boldsymbol{\beta} + \boldsymbol{\alpha}_1$; $\boldsymbol{\beta}_{10} = \boldsymbol{\beta} + \boldsymbol{\alpha}_2$; and $\boldsymbol{\beta}_{11} = \boldsymbol{\beta} + \boldsymbol{\alpha}_1 + \boldsymbol{\alpha}_2 + \boldsymbol{\alpha}_3$. We would again hope that a more parsimonious model fits the data equally well so that many of the components of the $\boldsymbol{\alpha}_i$ would be zero.

An important special case of (10.3.2) occurs when there are no interactions between the past responses, y_{ij-1} and y_{ij-2}, and the explanatory variables, that is, when all elements of the $\boldsymbol{\alpha}_i$ are zero except the intercept term. In this case, the previous responses affect the probability of a positive outcome but the effects of the explanatory variables are the same regardless of the history. Even in this situation, we must still choose between Markov models of different order. For example, we might start with a third order model which can be written in the form

$$\begin{aligned} \text{logit}\, \Pr(Y_{ij} = 1 \mid Y_{ij-3} = y_{ij-3}, Y_{ij-2} = y_{ij-2}, Y_{ij-1} = y_{ij-1}) \\ = \boldsymbol{x}'_{ij}\boldsymbol{\beta} + \alpha_1 y_{ij-1} + \alpha_2 y_{ij-2} + \alpha_3 y_{ij-3} + \alpha_4 y_{ij-1}y_{ij-2} \\ + \alpha_5 y_{ij-1}y_{ij-3} + \alpha_6 y_{ij-2}y_{ij-3} + \alpha_7 y_{ij-1}y_{ij-2}y_{ij-3}. \end{aligned} \tag{10.3.3}$$

A second order model can be used if the data are consistent with $\alpha_3 = \alpha_5 = \alpha_6 = \alpha_7 = 0$; a first order model is implied if $\alpha_j = 0$ for $j = 2, \ldots, 7$. As with any regression coefficients, the interpretation and value of $\boldsymbol{\beta}$ in (10.3.3) depends on the other explanatory variables in the model, in particular on which previous responses are included. When inferences about $\boldsymbol{\beta}$ are the scientific focus, it is essential to check their sensitivity to the assumed order of the Markov regression model.

As discussed in Section 10.2, when the Markov model is correctly specified, the transition events are uncorrelated so that ordinary logistic regression can be used to estimate regression coefficients and their standard errors. However, there may be circumstances when we choose to model $\Pr(Y_{ij} \mid Y_{ij-1}, \ldots, Y_{ij-q})$ even though it does not equal $\Pr(Y_{ij} \mid H_{ij})$. For example, suppose there is heterogeneity across people in the transition matrix due to unobserved factors, so that a reasonable model is

$$\Pr(Y_{ij} = 1 \mid Y_{ij-1} = y_{ij-1}, U_i) = (\beta_0 + U_i) + \boldsymbol{x}'_{ij}\boldsymbol{\beta} + \alpha y_{ij-1}$$

where $U_i \sim N(0, \sigma^2)$. We may still wish to estimate the population-averaged transition matrix, $\Pr(Y_{ij} \mid Y_{ij-1} = y_{ij-1})$. But here the random intercept U_i makes the transitions for a person correlated. Correct inferences about the population-averaged coefficients can be drawn using the GEE approach described in Section 7.5.

Table 10.1. Number (frequency) of transitions from respiratory disease status Y_{ij-1} at visit $j-1$ to disease status Y_{ij} at visit j for Indonesian Children's Health Study data.

		Y_{ij} 0	Y_{ij} 1	
Y_{ij-1}	0	721	60	781
		(0.923)	(0.077)	(1.0)
	1	64	10	74
		(0.865)	(0.135)	(1.0)
				855

10.3.1 *Indonesian children's study example*

We illustrate the Markov models for binary longitudinal data with the respiratory disease data from our subset of 1200 records from the Indonesian Children's Health Study. Ignoring dependence on covariates for the moment, the observed first order transitions are as presented in Table 10.1.

The table gives the number and frequency of transitions from the infection state at one visit to the next. These rates estimate the transition probabilities $\Pr(Y_{ij} \mid Y_{ij-1})$. Note that there are 855 transitions among the 1200 observations, since we do not observe the child's status prior to visit one. For children who did not have respiratory disease at the prior visit ($Y_{ij-1} = 0$), the frequency of respiratory infection was 7.7 per cent. Among children who did have infection at the prior visit, the proportion was 13.5 per cent, or 1.76 times as high.

An important question is whether vitamin A deficiency, as indicated by the presence of the ocular disease xerophthalmia, is associated with a higher prevalence of respiratory infection. Sommer *et al.* (1984) have demonstrated this relationship in an analysis of the entire Indonesian data set which includes over 20 000 records on 3500 children. In our subset, we can form the cross-tabulation shown in Table 10.2 of xerophthalmia and respiratory infection at a given visit, using the 855 observations in Table 10.1.

The frequency of respiratory infection is $1.49 = .119/.080$ times as high among children who are vitamin A deficient. But there is clearly correlation among repeated respiratory disease outcomes for a given child. We can control for this dependence by examining the effect of vitamin A deficiency separately for transitions starting with $Y_{ij-1} = 0$ or $Y_{ij-1} = 1$ as in Table 10.3. Among children free of infection at the prior visit, the frequency of respiratory disease is $1.44 = 0.108/0.075$ times as high if the child has xerophthalmia. Among those who suffered infection at the prior visit, the xerophthalmia relative risk is $1.54 = 0.200/0.130$. Hence, the xerophthalmia

Table 10.2. Cross-tabulation of respiratory disease Y_{ij} against xerophthalmia status x_{ij} for Indonesian Children's Health Study data from visits 2 to 6.

		Y_{ij}		
		0	1	
x_{ij}	0	748	65	813
		(0.920)	(0.080)	(1.0)
	1	37	5	42
		(0.881)	(0.119)	(1.0)
				855

effect is similar for $Y_{ij-1} = 0$ or $Y_{ij-1} = 1$ even though Y_{ij-1} is a strong predictor of Y_{ij}.

This exploratory analysis suggests a model

$$\text{logit } \Pr(Y_{ij} = 1 \mid Y_{ij-1} = y_{ij-1}) = \boldsymbol{x}_{ij}{}'\boldsymbol{\beta} + \alpha y_{ij-1}.$$

Table 10.4 presents logistic regression results for that model and a number of others. For each set of predictor variables, the table reports the regression coefficient, the standard error from (10.2.3) reported by ordinary logistic regression procedures, and the robust standard error defined by (10.2.4).

The first model predicts the risk of infection using only xerophthalmia status and should reproduce Table 10.2. The frequency of infection among children without xerophthalmia in Table 10.2 is 8.0 per cent which equals $\exp(-2.44)/\{1 + \exp(-2.44)\}$ where -2.44 is the intercept for Model 1 in Table 10.4. The log odds ratio calculated from Table 10.2 is

$$\log\{(0.119/0.881)/(0.080/0.920)\} = 0.44,$$

which is the coefficient for xerophthalmia in Model 1 of Table 10.4. The advantage of the logistic regression formulation is that standard errors are readily calculated.

Model 2 allows the association between xerophthalmia and respiratory disease to differ among children with respiratory disease at the prior visit and reproduces the transition rates from Table 10.3. For children without infection at the prior visit ($Y_{ij-1} = 0$), the log odds ratio for xerophthalmia calculated from Table 10.3 is $\log\{(0.108/0.892)/(0.075/0.925)\} = 0.40$, which equals the coefficient for current xerophthalmia in Model 2. The log odds ratio for children with infection at the prior visit is $0.40 + 0.11 = 0.51$, the sum of the xerophthalmia effect and the xerophthalmia-by-previous-infection interaction.

A comparison of Tables 10.2 and 10.3 indicates that the association of xerophthalmia and respiratory infection is similar for children who did and

Table 10.3. Cross-tabulation of current respiratory disease status Y_{ij} against xerophthalmia x_{ij} and previous respiratory disease status Y_{ij-1} for the Indonesian Children's Study data from visits 2 to 7.

$Y_{ij-1} = 0$

		Y_{ij} = 0	Y_{ij} = 1	
x_{ij}	0	688 (0.925)	56 (0.075)	744 (1.0)
	1	33 (0.892)	4 (0.108)	37 (1.0)
				781

$Y_{ij-1} = 1$

		Y_{ij} = 0	Y_{ij} = 1	
x_{ij}	0	60 (0.870)	9 (0.130)	69 (1.0)
	1	4 (0.800)	1 (0.200)	5 (1.0)
				74

did not have respiratory infection at the previous visit. This is confirmed by the xerophthalmia-by-previous-infection interaction term in Model 2 which is 0.11 with approximate, robust 95 per cent confidence interval $(-2.1, 2.3)$. In Model 3, the interaction has been dropped.

If the first-order Markov assumption is valid, that is, if $\Pr(Y_{ij} \mid H_{ij}) = \Pr(Y_{ij} \mid Y_{ij-1})$, then the standard errors estimated by ordinary logistic regression are valid. The robust standard errors have valid coverage in large samples even when the Markov assumption is incorrect. Hence a simple check of the sensitivity of inferences about a particular coefficient to the Markov assumption is whether the ordinary and robust standard errors are similar. They are similar for both coefficients in Model 1.

Models 1, 2, and 3 illustrate the use of logistic regression to fit simple Markov chains. The strength of this formulation is the ease of adding additional predictors. This is illustrated in Models 4 and 5, where the child's age and a binary indicator of season (1=2nd quarter; 0=other) have been added. In Model 4, we fit all iteractions with Y_{ij-1} which is the same as fitting separate logistic regressions for the cases $Y_{ij-1} = 0$ and 1. None of

Table 10.4. Logistic regression coefficients, standard errors (·) and robust standard errors [·] for several models fit to the 855 respiratory disease transitions in the Indonesian Children's Health Study data.

	Model				
Variable	1	2	3	4	5
Intercept	− 2.44	− 2.51	− 2.51	− 2.85	− 2.81
	(0.13)	(0.14)	(0.14)	(0.19)	(0.18)
	[0.14]	[0.14]	[0.14]	[0.18]	[0.17]
Current	0.44	0.40	0.42	0.79	0.78
xerophthalmia	(0.50)	(0.55)	(0.50)	(0.58)	(0.52)
(1=yes;0=no)	[0.54]	[0.51]	[0.53]	[0.49]	[0.53]
Age-36				− 0.024	− 0.023
(months)				(0.0077)	(0.0073)
				[0.0070]	[0.0065]
Season				1.23	1.11
(1=2nd qtr;				(0.29)	(0.28)
0=other)				[0.29]	[0.27]
Y_{ij-1}		0.61	0.62	0.82	0.62
		(0.39)	(0.37)	(0.47)	(0.38)
		[0.41]	[0.39]	[0.44]	[0.40]
Y_{ij-1} by		0.11		− 0.11	
Xerophthalmia		(1.3)		(1.4)	
		[1.1]		[1.1]	
Y_{ij-1} by Age				0.00063	
				(0.029)	
				(0.024]	
Y_{ij-1} by Season				− 1.24	
				(1.2)	
				[1.1]	

the interactions with prior infection is important so they are dropped in Model 5.

Having controlled for age, season and respiratory infection at the prior visit in Model 5, there is mild evidence in these data for an association between xerophthalmia and respiratory infection; the xerophthalmia coefficient is 0.78 with robust standard error 0.53. As with any regression, it is important to check the sensitivity of the scientific findings to choice of model. With transition models, we must check whether the regression inferences about $\boldsymbol{\beta}$ change with the model for the time dependence. To illustrate, we add Y_{ij-2} as a predictor to Model 5. Because we are using two prior visits as predictors, only data from visits 3 through 7 are relevant giving a total of 591 transitions. Interestingly, the inclusion of Y_{ij-2} reduces the influence of season and Y_{ij-1}, and increases the xerophthalmia coefficient to 1.73. Controlling for respiratory infection at the two prior visits nearly doubles the xerophthalmia coefficient. This example demonstrates an essential feature of transition models: explanatory variables (e.g. xerophthalmia) and previous responses are treated symmetrically as predictors of the current response. Hence, as the time dependence model changes, so might inferences about the explanatory variables. Even when the time dependence is not of primary scientific interest, the sensitivity of the regression inferences must be checked by fitting a variety of time dependence models.

10.3.2 *Ordered categorical data*

The ideas underlying logistic regression models for binary responses carry over to outcomes with more than two categories. In this section, we consider models for ordered categorical data, and then briefly discuss nominal response models. The books by McCullagh and Nelder (1989, Chapter 5) and Agresti (1990, Chapter 9) provide expanded discussions for the case of independent observations.

To formulate a transition model for ordered categorical data, let Y_{ij} indicate a response variable which can take C ordered categorical values, labelled $0, 1, \ldots, C-1$. An example of ordered data is a rating of health status as: poor, fair, good, or excellent. While the outcomes are ordered, any numerical scale that might be assigned would be arbitrary.

The first order transition matrix for Y_{ij} is defined by $\pi_{ab} = \Pr(Y_{ij} = b \mid Y_{ij-1} = a)$ for $a, b = 0, 1, \ldots, C-1$. As with binary data ($C = 2$), a saturated model of the transition matrix can be obtained by fitting a separate regression for each of the C possible values of Y_{ij-1}. That is, we model $\Pr(Y_{ij} = b \mid Y_{ij-1} = a)$ separately for each $a = 0, 1, \ldots, C-1$. With ordered categorical outcomes, we can use a *proportional odds* model (Snell, 1964 and McCullagh, 1980) which we now briefly review.

Since the response categories for ordinal data are usually arbitrary, we would like a regression model whose coefficients have the same interpretation when we combine or split categories. This is achieved by working with the cumulative probabilities $\Pr(Y \leq a)$ rather than the cell probabilities, $\Pr(Y = a)$. Given the cumulative probabilities, we can derive the cell probabilities since $\Pr(Y \leq a) = \Pr(Y \leq a-1) + \Pr(Y = a)$, $a = 1, \ldots, C$.

The proportional odds model for independent observations has the form

Table 10.5. Definition of Y^* variables for proportional odds modelling of ordered categorical data.

Y:	0	1	2
Y_0^*:	1	0	0
Y_1^*:	1	1	0

$$\text{logit}\,\Pr(Y \leq a) = \log \frac{\Pr(Y \leq a)}{\Pr(Y > a)} = \theta_a + \boldsymbol{x}'\boldsymbol{\beta}$$

where $a = 0, 1, \ldots, C-2$. Here and for the remainder of this section, we write the model intercepts as θ_a and do not include an intercept term in $\boldsymbol{x}$. Taking $\boldsymbol{x} = 0$, we see that $\Pr(Y \leq a) = e^{\theta_a}/(1+e^{\theta_a})$. Since $\Pr(Y \leq a)$ is a non-decreasing function of a, we have $\theta_0 \leq \theta_1 \leq \ldots \leq \theta_{C-2}$. If $\theta_a = \theta_{a+1}$, then $\Pr(Y \leq a) = \Pr(Y \leq a+1)$ and categories a and $a+1$ can therefore be collapsed.

The regression parameters $\boldsymbol{\beta}$ have log odds ratio interpretations, since

$$\frac{\Pr(Y \leq a \mid \boldsymbol{x}_1)/\Pr(Y > a \mid \boldsymbol{x}_1)}{\Pr(Y \leq a \mid \boldsymbol{x}_2)/\Pr(Y > a \mid \boldsymbol{x}_2)} = \exp\{(\boldsymbol{x}_1 - \boldsymbol{x}_2)'\boldsymbol{\beta}\}.$$

Following Clayton (1992), it is convenient to introduce the vector of variables $Y^* = (Y_0^*, Y_1^*, \ldots, Y_{C-2}^*)$ defined by $Y_a^* = 1$ if $Y \leq a$ and 0 otherwise. If $C = 3$, $Y^* = (Y_0^*, Y_1^*)$ takes values shown in Table 10.5. The proportional odds model is simply a logistic regression for the Y_a^*, since

$$\log \frac{\Pr(Y \leq a)}{\Pr(Y > a)} = \text{logit}\,\Pr(Y_a^* = 1) = \theta_a + \boldsymbol{x}'\boldsymbol{\beta},\ a = 0, 1, \ldots, C-2.$$

Here, each Y_a^* is allowed to have a different intercept but the proportional odds model requires that covariates have the same effect on each Y_a^*.

Our first application of the proportional odds model to a Markov chain for ordered categorical responses is the saturated first order model without covariates. The transition matrix is $\pi_{ab} = \Pr(Y_{ij} = b \mid Y_{ij-1} = a)$, $a = 0, 1, \ldots, C-1$. We model the cumulative probabilities, $\Pr(Y_{ij} \leq b \mid Y_{ij-1} = a) = \pi_{a0} + \pi_{a1} + \ldots + \pi_{ab}$, assuming that

$$\log \frac{\Pr(Y_{ij} \leq b \mid Y_{ij-1} = a)}{\Pr(Y_{ij} > b \mid Y_{ij-1} = a)} = \theta_{ab}, \qquad (10.3.4)$$

for $a = 0, 1, \ldots, C-1$ and $b = 0, 1, \ldots, C-2$. Now, suppose that covariates $\boldsymbol{x}_{ij}$ have a differential effect on Y_{ij} for each previous state Y_{ij-1}. The model can be written as

$$\log \frac{\Pr(Y_{ij} \leq b \mid Y_{ij-1} = a)}{\Pr(Y_{ij} > b \mid Y_{ij-1} = a)} = \theta_{ab} + \boldsymbol{x}_{ij}{}'\boldsymbol{\beta}_a \tag{10.3.5}$$

for $a = 0, \ldots, C-1$; $b = 0, \ldots, C-2$. As with binary responses, we can rewrite (10.3.5) as a single (although somewhat complicated) regression equation using the interactions between $\boldsymbol{x}_{ij}$ and the vector of derived variables $\boldsymbol{y}^*_{ij-1} = (y^*_{ij-1,0}, \ldots, y^*_{ij-1,C-2})$

$$\log \frac{\Pr(Y_{ij} \leq b \mid \boldsymbol{Y}^*_{ij-1} = \boldsymbol{y}^*_{ij-1})}{\Pr(Y_{ij} > b \mid \boldsymbol{Y}^*_{ij-1} = \boldsymbol{y}^*_{ij-1})} = \theta_b + \sum_{\ell=0}^{C-2} \alpha_{\ell b} y^*_{ij-1\ell} + \boldsymbol{x}'_{ij}\left(\boldsymbol{\beta} + \sum_{\ell=0}^{C-2} \gamma_\ell y^*_{ij-1\ell}\right). \tag{10.3.6}$$

Comparing (10.3.5) and (10.3.6), we see that $\theta_{C-1b} = \theta_b$ and $\alpha_{ab} = \theta_{ab} - \theta_{a+1b}$, $a = 0, 1, \ldots, C-2$; $b = 0, 1, \ldots, C-2$. Similary, $\boldsymbol{\beta}_{C-1} = \boldsymbol{\beta}$ and $\gamma_a = \boldsymbol{\beta}_a - \boldsymbol{\beta}_{a+1}$, $a = 0, 1, \ldots, C-2$.

With this formulation, we can test whether or not the effect of $\boldsymbol{x}_{ij}$ on Y_{ij} is the same for adjacent categories of Y_{ij-1} by testing whether $\gamma_a = 0$. A simple special case of (10.3.6) is to assume that $\gamma_0 = \gamma_1 = \ldots = \gamma_{C-2} = 0$ so the covariates $\boldsymbol{x}_{ij}$ have the same effect on Y_{ij} regardless of Y_{ij-1}. As with binary data we can fit models with different subsets of the interactions between $\boldsymbol{y}^*_{ij-1}$ and $\boldsymbol{x}_{ij}$.

The transition ordinal regression model can be estimated using conditional maximum likelihood, that is, by conditioning on the first observation, Y_{i1}, for each person. Standard algorithms for fitting proportional odds models (McCullagh, 1980) can be used by adding the derived variables $\boldsymbol{y}^*_{ij-1}$ and their interactions with $\boldsymbol{x}_{ij}$ as additional covariates. As an alternative, Clayton (1992) has proposed using GEE to fit simultaneously logistic regressions to $Y^*_{ij0}, \ldots, Y^*_{ijC-2}$, again with $\boldsymbol{y}^*_{ij-1}$ and its interactions with $\boldsymbol{x}_{ij}$ as covariates. For the examples studied by Clayton, the GEE estimates were almost fully efficient.

When the categorical response does not have a natural ordering, we must model the transition probabilities, $\pi_{ab} = \Pr(Y_{ij} = b \mid Y_{ij-1} = a)$, $a, b = 0, 1, \ldots, C-1$. McCullagh and Nelder (1989) and Agresti (1990) discuss a variety of logistic model formulations for independent responses. As we have demonstrated for binary and ordered categorical responses, these can be extended to transition models by using indicator variables for the previous state and their interactions with covariates as additional explanatory variables.

10.4 Log-linear transition models for count data

In this section, we consider extensions of the log-linear model in which the conditional distribution of y_{ij} given the past H_{ij} is Poisson with conditional expectation that depends both on past outcomes $y_{i1}, \ldots, y_{ij-1}$ and on explanatory variables $\boldsymbol{x}_{ij}$. We begin by reviewing three possible models for the conditional mean μ_{ij}^c. In each case, we restrict our attention to a first order Markov chain.

(1) $\mu_{ij}^c = \exp(\boldsymbol{x}_{ij}'\boldsymbol{\beta})\{1 + \exp(-\alpha_0 - \alpha_1 y_{ij-1})\}$, $\alpha_0, \alpha_1 > 0$.

In this model, suggested by Wong (1986), $\boldsymbol{\beta}$ represents the influence of the explanatory variables when the previous response takes the value $y_{ij-1} = 0$. When $y_{ij-1} > 0$, the conditional expectation is decreased from its maximum value, $\exp(\boldsymbol{x}_{ij}'\boldsymbol{\beta})\{1 + \exp(-\alpha_0)\}$, by an amount that depends on α_1. Hence, this model only allows a negative association between the prior and current responses. For a given α_0, the degree of negative correlation increases as α_1 increases. Note that the conditional expectation must vary between $\exp(\boldsymbol{x}_{ij}'\boldsymbol{\beta})$ and twice this value.

(2) $\mu_{ij}^c = \exp(\boldsymbol{x}_{ij}'\boldsymbol{\beta} + \alpha y_{ij-1})$

This model appears sensible by analogy with the logistic model (7.3.1). But it has limited application for count data because when $\alpha > 0$, the conditional expectation grows as an exponential function of time. In fact, when $\exp(\boldsymbol{x}_{ij}'\boldsymbol{\beta}) = \mu$, corresponding to no dependence on covariates, this assumption leads to a stationary process only when $\alpha < 0$. Hence, the model can describe negative association but not positive association without growing exponentially over time. This is a time series analogue of the *auto-Poisson* process discussed for data on a two-dimensional lattice by Besag (1974).

(3) $\mu_{ij}^c = \exp[\boldsymbol{x}_{ij}'\boldsymbol{\beta} + \alpha\{\log(y_{ij-1}^*) - \boldsymbol{x}_{ij-1}'\boldsymbol{\beta}\}]$ where $y_{ij-1}^* = \max(y_{ij-1}, c)$ and $0 < c < 1$.

This is the model introduced by Zeger and Qaqish (1988) and briefly discussed in Section 10.1. When $\alpha = 0$, it reduces to an ordinary log-linear model. When $\alpha < 0$, a prior response greater than its expectation decreases the expectation for the current response and there is negative correlation between y_{ij-1} and y_{ij}. When $\alpha > 0$, there is positive correlation.

For the remainder of this section, we focus on the third Poisson transition model above. It can arise through a simple physical mechanism called a size-dependent branching process. Suppose that $\exp(\boldsymbol{x}_{ij}'\boldsymbol{\beta}) = \mu$. Suppressing the index i for the moment, let y_j represent the number of individuals in a population at generation j. Let $Z_k(y_{j-1})$ be the number of offspring for person k in generation $j-1$. Then for $y_{j-1} > 0$, the total size of the jth generation is

$$y_j = \sum_{k=1}^{y_{j-1}} Z_k(y_{j-1}).$$

If $y_{j-1} = 0$, we assume that the population is restarted with Z_0 individuals. Now, if we asssume that the random variables Z are Poisson with expectation $(\mu/y^*_{j-1})^{1-\alpha}$, then the population size will follow the transition model with $\mu^c_j = \mu(y^*_{j-1}/\mu)^\alpha$. The assumption about the number of offspring per person represents a crowding effect. When the population is large, each individual tends to decrease their number of offspring. This leads to a stationary process.

Figure 10.1 displays five realizations of this transition model for different values of α. When $\alpha < 0$, the sample paths oscillate back and forth about their long-term average level since a large outcome at one time decreases the conditional expectation of the next response. When $\alpha > 0$, the process meanders, staying below the long-term average for extended periods. Notice that the sample paths have sharper peaks and broader valleys. This pattern is in contrast to Gaussian autoregressive sample paths for which the peaks and valleys have the same shape. In the Poisson model, the conditional variance equals the conditional mean. When by chance we get a large observation, the conditional mean and variance of the next value are both large; that is, the process becomes unstable and quickly falls towards the long-term average. After a small outcome, the conditional mean and variance are small, so the process tends to be more stable. Hence, there are broader valleys and sharper peaks.

To illustrate the application of this transition model for counts, we have fitted it to the seizure data. A priori, there is clear evidence that this model is not appropriate for this data set. The correlations among repeated seizure counts for each person do not decay as the time between observations increases. For example, the correlations of the square root transformed seizure number at the first post-treatment period with those at the second, third, and fourth periods are 0.76, 0.70, and 0.78 respectively. The Markov model implies that these should decrease as an approximately exponential function of lag. Nevertheless, we might fit the first-order model if our goal was simply to predict the seizure rate in the next interval, given only the rate in the previous interval. Because there is only a single observation prior to randomization, we have assumed that the pre-randomization means are the same for the two treatment groups. Letting $c = 0.3$, we estimate the treatment effect (treatment-by-time interaction in Tables 8.10 and 9.6) to be -0.10, with a model-based standard error of 0.083. This standard error is not valid because the model does not accurately capture the actual correlation structure in the data. We have also estimated a robust standard error which is 0.30, much larger than 0.083, reflecting the fact that the

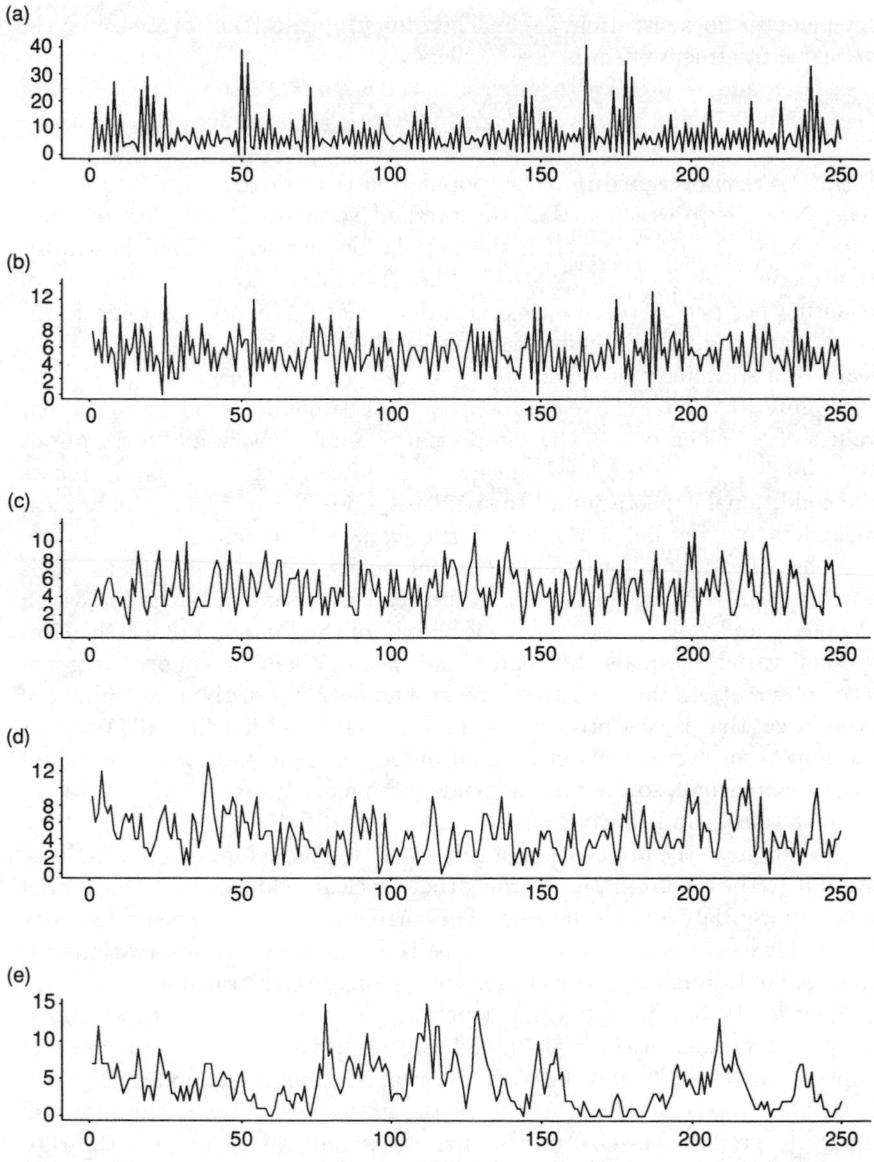

Fig. 10.1. Realisations of the Markov Poisson time series model: (a) $\alpha = -0.8$; (b) $\alpha = -0.4$; (c) $\alpha = 0.0$; (d) $\alpha = 0.4$; (e) $\alpha = 0.8$.

correlation does not decay as expected for the transition model. The estimate of treatment effect is not sensitive to the constant c; it varies between -0.096 and -0.107 as c ranges from 0.1 to 1.0. The estimate of α for $c = 0.3$ is 0.79 and also is not very sensitive to the value of c.

10.5 Further reading

Markov models have been studied by probabilists and mathematical statisticians for several decades. Feller (1968) and Billingsley (1961) are seminal texts. But there has been little theoretical study of Markov regression models except for the linear model (e.g. Tsay, 1984). This is partly because it is difficult to derive marginal distributions given that the conditional distribution is assumed to to follow a GLM regression.

Regression models for binary Markov chains are in common use. Examples of applications can be found in Korn and Whittemore (1979), Stern and Coe (1984), and Zeger *et al.* (1985). The reader is referred to papers by Wong (1986), and Zeger and Qaqish (1988) for methods applicable to count data. There has also been very interesting work on Bayesian Markov regression models with meaurement error. See, for example, the discussion paper by West *et al.* (1985) and the volume by Spall (1988).

11
Missing values in longitudinal data

11.1 Classification of missing values for longitudinal data

Missing values arise in the analysis of longitudinal data whenever one or more of the sequences of measurements from units within the study are incomplete, in the sense that *intended* measurements are not taken, are lost, or are otherwise unavailable. The emphasis is important: if we choose in advance to take measurements every hour on one half of the subjects and every two hours on the other half, the resulting data could also be described as incomplete but there are no missing values in the sense that we use the term; we call such data *unbalanced.* This is not just playing with words. Unbalanced data may raise technical difficulties – we have seen that some methods of analysis can only cope with data for which measurements are made at a common set of times on all units. Missing values raise the same technical difficulties, since of necessity they result in unbalanced data, but also deeper conceptual issues, since we have to ask *why* the values are missing, and more specifically whether their being missing has any bearing on the practical questions posed by the data.

A simple (non-longitudinal) example makes the point explicitly. Suppose that we want to compare the mean concentrations of a hormone in blood samples taken from ten subjects in each of two groups: one a control, the other an experimental treatment intended to suppress production of the hormone. We are presented with ten assayed values of the hormone concentration from the control subjects, eight assayed values from the treated subjects, and are told that the values from the other two treated subjects are 'missing'. If we knew that this was because somebody dropped the test tubes on the way to the assay lab, we would probably be happy to proceed with a two-sample t-test using the 18 non-missing values. If we knew that the missing values were from subjects whose hormone concentrations fell below the sensitivity threshold of the assay, ignoring the missing values would mask the very effect we were looking to detect. Lest this example offend the sophisticated reader, we suggest that more subtle versions of it are not uncommon in real longitudinal studies. For example, it may be that a sequence of atypically low (or high) values on a particular unit foreshadows its removal from the study.

Little and Rubin (1987), give a general treatment of statistical analysis with missing values, which includes a useful hierarchy of missing value mechanisms. Let $\boldsymbol{Y}^*$ denote the complete set of measurements which would have been obtained were there no missing values, and partition this set into $\boldsymbol{Y}^* = (\boldsymbol{Y}^{(o)}, \boldsymbol{Y}^{(m)})$ with $\boldsymbol{Y}^{(o)}$ denoting the measurements actually obtained and $\boldsymbol{Y}^{(m)}$ the measurements which would have been available had they not been missing, for whatever cause. Finally, let $\boldsymbol{R}$ denote a set of indicator random variables, denoting which elements of $\boldsymbol{Y}^*$ fall into $\boldsymbol{Y}^{(o)}$ and which into $\boldsymbol{Y}^{(m)}$. Now, a probability model for the missing value mechanism is a specification of the probability distribution of $\boldsymbol{R}$ conditional on $\boldsymbol{Y}^* = (\boldsymbol{Y}^{(o)}, \boldsymbol{Y}^{(m)})$. Little and Rubin classify the missing value mechanism as:

- *completely random* if $\boldsymbol{R}$ is independent of both $\boldsymbol{Y}^{(o)}$ and $\boldsymbol{Y}^{(m)}$;
- *random* if $\boldsymbol{R}$ is independent of $\boldsymbol{Y}^{(m)}$;
- *informative* if $\boldsymbol{R}$ is dependent on $\boldsymbol{Y}^{(m)}$.

It turns out that for likelihood-based inference, the crucial distinction is between random and informative missing values. To see this, let $f(\boldsymbol{y}^{(o)}, \boldsymbol{y}^{(m)}, \boldsymbol{r})$ denote the joint probability density function of $(\boldsymbol{Y}^{(o)}, \boldsymbol{Y}^{(m)}, \boldsymbol{R})$ and use the standard factorization to express this as

$$f(\boldsymbol{y}^{(o)}, \boldsymbol{y}^{(m)}, \boldsymbol{r}) = f(\boldsymbol{y}^{(o)}, \boldsymbol{y}^{(m)}) f(\boldsymbol{r} \mid \boldsymbol{y}^{(o)}, \boldsymbol{y}^{(m)}). \tag{11.1.1}$$

For a likelihood-based analysis, we need the joint pdf of the observable random variables, $(\boldsymbol{Y}^{(o)}, \boldsymbol{R})$, which we obtain by integrating (11.1.1) to give

$$f(\boldsymbol{y}^{(o)}, \boldsymbol{r}) = \int f(\boldsymbol{y}^{(o)}, \boldsymbol{y}^{(m)}) f(\boldsymbol{r} \mid \boldsymbol{y}^{(o)}, \boldsymbol{y}^{(m)}) d\boldsymbol{y}^{(m)}. \tag{11.1.2}$$

Now, if the missing value mechanism is random, $f(\boldsymbol{r} \mid \boldsymbol{y}^{(o)}, \boldsymbol{y}^{(m)})$ does not depend on $\boldsymbol{y}^{(m)}$ and (11.1.2) becomes

$$\begin{aligned} f(\boldsymbol{y}^{(o)}, \boldsymbol{r}) &= f(\boldsymbol{r} \mid \boldsymbol{y}^{(o)}) \int f(\boldsymbol{y}^{(o)}, \boldsymbol{y}^{(m)}) d\boldsymbol{y}^{(m)} \\ &= f(\boldsymbol{r} \mid \boldsymbol{y}^{(o)}) f(\boldsymbol{y}^{(o)}). \end{aligned} \tag{11.1.3}$$

Finally, taking logarithms in (11.1.3), the log-likelihood function is

$$L = \log f(\boldsymbol{r} \mid \boldsymbol{y}^{(o)}) + \log f(\boldsymbol{y}^{(o)}), \tag{11.1.4}$$

which is maximized by separate maximization of the two terms on the right hand side. Since the first term contains no information about the distribution of $\boldsymbol{Y}^{(o)}$, we can ignore it for the purpose of making inferences about $\boldsymbol{Y}^{(o)}$.

Because of the above result, both completely random and random missing value mechanisms are sometimes referred to without distinction as *ignorable*. However, this can be a misnomer if non-likelihood-based methods of analysis are used. Also, even within a likelihood-based analysis, the argument assumes that $f(\boldsymbol{y}^{(o)})$ and $f(\boldsymbol{r} \mid \boldsymbol{y}^{(o)})$ are separately parameterized, which need not be the case; if there are parameters common to $f(\boldsymbol{y}^{(o)})$ and $f(r \mid \boldsymbol{y}^{(o)})$, ignoring the first term in (11.1.4) leads to a loss of efficiency. Finally, the *marginal* distribution of $\boldsymbol{Y}^{(o)}$ may not be the correct inferential focus. For example, consider a clinical trial concerned with a life-threatening condition in which missing values identify patients who have died before the end of the study and $\boldsymbol{Y}^{(o)}$ measures health status. Then, it may be more sensible to analyse the distribution of time to survival and the *conditional* distribution of $\boldsymbol{Y}^{(o)}$ given survival rather than the marginal distribution of $\boldsymbol{Y}^{(o)}$. For these reasons, we prefer to maintain the distinction between completely random and random missing values.

Another important distinction is whether missing values occur intermittently or as dropouts. Suppose that we intend to take a sequence of measurements, $Y_1, \ldots, Y_n$, on a particular unit. Missing values occur as *dropouts* if whenever Y_j is missing, so are Y_k for all $k \geq j$; otherwise we say that the missing values are *intermittent*.

We know of no well-developed methodology for dealing with informative, intermittent missing values in longitudinal data. When the missing values arise through a known censoring mechanism, for example if all values below a known threshold are missing, the EM algorithm (Dempster *et al.* 1977) provides a possible theoretical framework, but practical implementation for a realistic range of longitudinal data models would seem to be an extremely difficult task (Laird, 1988). When intermittent missing values do not arise from censoring, it may be reasonable to assume that they arise from mechanisms unrelated to the measurement process, and are therefore completely random. In such cases, the resulting data can be analysed by any method which can accommodate unbalanced data. Furthermore, if the method of analysis is likelihood-based, the inferences will be valid under the weaker assumption that the missing value mechanism is random.

Dropouts do not arise as a result of censoring applied to individual measurements, but because some of the units are withdrawn from the study prematurely. Often, a unit's withdrawal is for reasons directly or indirectly connected to the measurement process. For example, the 'dropouts' in the data on protein content of milk samples arose because the cows in question calved after the beginning of the experiment, and there may well be an indirect link between calving date and milk quality. Later in this chapter we shall use the milk protein data to illustrate the analysis of data with dropouts, although this is questionable from a scientific point of view. In fact, the crucial information that in this experiment time was measured relative to date of calving only emerged in response to the analysis reported in section 11.3 below, thereby demonstrating (not for the first time) that a wrong model can yield useful insights! For further details, see Cullis (1994).

An example of dropout directly related to the measurement process arises in clinical trials, where ethical considerations may require a patient to be withdrawn from a trial on the basis of their observed measurement history. Murray and Findlay (1988) discuss this in the context of long-term trials of drugs to reduce blood pressure where 'if a patient's blood pressure is not adequately controlled then there are ethical problems associated with continuing the patient on their study medication'.

These examples may convince the reader, as they convince us, that it would be useful to have methods for analysing data with a view to distinguishing amongst completely random, random and informative dropouts. Fortunately, the additional structure of data with dropouts makes this a more tractable problem than in the case of intermittent missing values.

In the rest of this chapter we concentrate on the analysis of data with dropouts. In Section 11.2 we describe methods for testing whether dropouts are completely random. In Section 11.3 we describe a model-based approach which can accommodate completely random, random or informative dropouts. In Section 11.4 we summarize our conclusions and identify a number of unresolved issues. More research is needed in this area.

11.2 Testing for completely random dropouts

In this section, we assume that a complete set of measurements on a unit would be taken at times t_j, $j = 1, \ldots, n$, but that dropouts occur. Then, the available data on the *ith* of m units are $\boldsymbol{y}_i = (y_{i1}, \ldots, y_{in_i})$, with $n_i \leq n$ and y_{ij} taken at time t_j. As always, the units are allocated into a number of different treatment groups. Our objective is to test the hypothesis that the dropouts are completely random, that is, that the probability that a unit drops out at time t_j is independent of the observed sequence of measurements on that unit at times $t_1, \ldots, t_{j-1}$. We view such a test as a preliminary screening device, and therefore wish to avoid any parametric assumptions about the process generating the measurement data. Note that our definition of completely random dropouts makes no reference to the possible influence of explanatory variables on the dropout process. For example, suppose that in a study comparing two groups with different mean response profiles, the dropout rate is higher in the group with the higher mean response. If we ignored the group structure, the dropout process would appear to be informative in the undifferentiated data, even if it were completely random within each group. We therefore recommend that for preliminary screening, the data should first be divided into homogeneous sub-groups.

Let p_{ij} denote the probability that the ith unit drops out at time t_j. Under the assumption of completely random dropouts, the p_{ij} may depend on time, treatment, or other explanatory variables but cannot depend on the observed measurements, $\boldsymbol{y}_i$. The method developed in Diggle (1989) to test this assumption consists of applying separate tests at each time within each treatment group and analysing the resulting sample of p-values for departure from the uniform distribution on (0,1). Combination of the separate p-values is necessary if the procedure is to have any practical value, because the individual tests are typically based on very few cases and therefore have low power.

The individual tests are constructed as follows. For each k, define a function, $h_k(y_1, \ldots, y_k)$. We will discuss the choice of $h_k(\cdot)$ shortly. Now, within each group and for each time-point t_k, $k = 1, \ldots, n-1$, identify the R_k units which have $n_i \geq k$ and compute the set of scores $h_{ik} = h_k(y_{i1}, \ldots, y_{ik})$, for $i = 1, \ldots, R_k$. Within this set of R_k scores, identify those r_k scores which correspond to units with $n_i = k$, that is, units which are about to drop out. If $1 \leq r_k < R_k$, test the hypothesis that the r_k scores so identified are a random sample from the 'population' of R_k scores previously identified. Finally, investigate whether the complete set of p-values observed by applying this procedure to each time within each treatment group behaves like a random sample from the uniform distribution on (0,1). The implicit assumption that the separate p-values are mutually independent is valid precisely because once a unit drops out it never returns.

Our first decision in implementing this procedure is how to choose the functions $h_k(\cdot)$. Our aim is to choose these so that extreme values of the scores, h_{ik}, constitute evidence against completely random dropouts. A sensible choice is a weighted sum,

$$h_k(y_1, \ldots, y_k) = \Sigma_{j=1}^{k} w_j y_j. \qquad (11.2.1)$$

As with any *ad hoc* procedure, the success of this will be influenced by the investigator's ability to pick a set of weights which reflect the actual dependence of the dropout probabilities on the observed measurement history. If dropout is suspected to be an immediate consequence of an abnormally low measurement we should choose $w_k = 1$ and all other $w_j = 0$. If it is suspected to be the culmination of a sustained sequence of low measurements we would do better using equal weights, $w_j = 1$ for all j.

The next decision is how to convert the scores, h_{ik}, into a test statistic and p-value. A natural test statistic is $\bar{h}_k$, the mean of the r_k scores corresponding to those units with $n_i = k$ which are about to drop out. If dropouts occur completely at random, the approximate sampling distribution of $\bar{h}_k$ is Gaussian, with mean $\bar{H}_k = R_k^{-1} \Sigma_{i=1}^{R_k} h_{ik}$, and variance $S_k^2(R_k - r_k)/(r_k R_k)$, where $S_k^2 = (R_k - 1)^{-1} \Sigma_{i=1}^{R_k} (h_{ik} - \bar{H}_k)^2$. This is a standard result from elementary sampling theory. See, for example, Cochran

(1977). However, in the present context some of the r_k and R_k will be small and the Gaussian approximation may then be poor. For an exact test, we can evaluate the complete randomization distribution of each $\bar{h}_k$ under the null hypothesis of random sampling. If $\binom{R_k}{r_k}$ is too large for this to be practical, a feasible alternative procedure is to sample from the randomization distribution. We recompute $\bar{h}_k$ after each of $s-1$ independent random selections of r_k scores chosen without replacement from the set h_{ik}, $i = 1, \ldots, R_k$, and let x denote the rank of the original $\bar{h}_k$ amongst the recomputed values. Then, $p = x/s$ is the p-value of an exact, Monte Carlo test (Barnard, 1963).

The final stage consists of analysing the resulting set of p-values. Informal graphical analyses can be very revealing; for example, a plot of the empirical distribution function of the p-values with different plotting symbols to identify the different treatment groups. Diggle (1989) also suggests a formal test of departure from uniformity using the Kolmogorov–Smirnov statistic; in practice, we will usually have arranged that a preponderance of *small* p-values will lead us to reject the hypothesis of completely random dropouts, and the appropriate form of the Kolmogorov–Smirnov statistic is its one-sided version, $D_+ = sup\{\hat{F}(p) - p\}$. Another technical problem now arises because each p-value derives from a *discrete* randomization distribution whereas the Kolmogorov–Smirnov statistic tests for departure from a continuous uniform distribution on the interval $0 \leq p \leq 1$. We would not want to over-emphasize the formal testing aspect of this procedure, preferring to regard it as a method of exploratory data analysis. Nevertheless, if we do want a formal test we can again use Barnard's Monte Carlo testing idea to give us an exact statement of significance. We simply rank the observed value of D_+ amongst simulated values based on the appropriate set of discrete uniform distributions which hold under the null hypothesis.

Example 11.1. Dropouts in the milk protein data

Recall that these data consist of up to 19 weekly measurements of protein content in milk samples taken from each of 79 cows. Also, the cows were allocated amongst three different treatments, representing different diets, in a completely randomized design. Of the 79 cows, 38 dropped out during the study. There were also 11 intermittent missing values. In Section 5.4 we fitted a model to these data in which, amongst other things, the mean response in each treatment group was assumed to be constant after an initial–settling–in period. We noted that an apparent rise in the *observed* mean response near the end of the experiment was not supported by testing against an enlarged model, and speculated that this might be somehow connected to the dropout process. As a first stage in pursuing this question, we now test whether the dropouts are completely random.

The dropouts are confined to four of the last five weeks of the study, and

Table 11.1. Attained significance levels of tests for completely random dropouts in the milk protein data.

	Treatment (diet)		
Dropout time (wks)	Barley	Mixed	Lupins
15	0.001	0.001	0.012
16	0.016	0.001	0.011
17	0.022	0.053	0.254
19	0.032	0.133	0.206

occur in 12 distinct treatment-by-time combinations. To construct the 12 test statistics we use $h_k(y_1, \ldots, y_k) = y_k$, the latest observed measurement, and implement a Monte Carlo test using s = 999 random selections from the randomization distribution of $\bar{h}_k$ for each treatment-by-time combination. The resulting p-values range from 0.001 to 0.254. On the basis of these results, we reject firmly the hypothesis of completely random dropouts. For example, if we use the Kolmogorov–Smirnov statistic to test for uniformity of the empirical distribution of the p-values, again via a Monte Carlo implementation with s = 999 to take account of the discreteness of the problem, we obtain a p-value of 0.001. Incidentally, note that 0.001 is the smallest possible p-value for a Monte Carlo test with s = 999.

Table 11.1 shows the 12 p-values cross-classified by dropout time and treatment. It is noticeable that the larger p-values predominate in the third treatment group and at the later dropout times, although the second of these features may simply be a result of the reduced power of the tests as the earlier dropouts lead to smaller sample sizes at later times.

We obtained very similar results for the milk protein data when we used $h_k(y_1, \ldots, y_k) = r^{-1}\Sigma_{j=k-r+1}^{k} y_j$, for each of $r = 2, 3, 4, 5$. The implication is that dropouts predominate amongst cows whose protein measurements in the preceding weeks are below average. Now recall the likelihood-based analysis of these data reported in Section 5.4. There, we accepted a constant mean response model against an alternative which allowed for a rising trend in the mean response, whereas the empirical mean responses suggested a rise towards the end of the experiment. We now have a possible explanation. The likelihood-based analysis is estimating $\mu_1(t)$, the mean response which would apply to a population with no dropouts, whereas the empirical means are estimating $\mu_2(t)$, the mean response of a unit conditional on its not having dropped out by time t. Under completely random dropouts, or under random dropouts with independent measurements at different times, $\mu_1(t) = \mu_2(t)$, whereas under random dropouts with serially correlated measurements, $\mu_1(t) \neq \mu_2(t)$. This underlines the danger of regarding random dropouts as ignorable in the colloquial sense.

Ridout (1991) points out a connection between Diggle's (1989) procedure and logistic regression analysis. At each time-point, we could use the function $h_k(y_1, \ldots, y_k)$ as an explanatory variable in a logistic regression model for the probability of dropout. Thus, if p_k is the probability that a unit will drop out at time t_{k+1}, we assume that

$$\log\{p_k/(1-p_k)\} = \alpha + \beta h_k. \qquad (11.2.2)$$

Then, conditional on the observed values, h_{ik}, for all the units who have not dropped out previously, the mean, $\bar{h}$, of those about to drop out is the appropriate statistic to test the hypothesis that $\beta = 0$ (Cox and Snell, 1989, Chapter 2).

Clearly, it is possible to fit the logistic model (11.2.2), or extensions of it, using standard methodology for generalized linear models as suggested by Aitkin *et al.* (1989, Chapter 3). For example, we can introduce an explicit dependence of p_k on time and/or the experimental treatment administered.

Example 11.2. Protein content of milk samples (continued)
We now give the parametric version of the analysis described in Example 11.1. Let p_{gk} denote the probability that a unit in the gth treatment group drops out at the kth time. We assume that

$$\log\{p_{gk}/(1-p_{gk})\} = \alpha_{gk} + \beta_{gk} h_k, \qquad (11.2.3)$$

with $h_k(y_1, \ldots, y_k) = y_k$. In this analysis, we consider only those four of the 19 time-points at which dropouts actually occurred, giving a total of 24 parameters. The residual deviances from this 24-parameter model and various simpler models are given in Table 11.2. Note that the analysis is based on 234 binary responses. These consist of 79 responses in week 15, the first occasion on which any units drop out, 59 in week 11 from the units which did not drop out in week 15, and similarly 50 and 46 from weeks 17 and 19; there were no dropouts in week 18.

From lines 1 to 5 in Table 11.2, we conclude firstly that there is a strong dependence of the dropout probability on the most recently observed measurement; for example, the log-likelihood ratio statistic to test model 5 within model 4 is 197.66 – 119.32 = 78.34 on 1 degree of freedom. We also conclude that the nature of the dependence does not vary between treatments or times; none of lines 1 to 3 gives a significant reduction in the residual deviance by comparison with line 4. The dependence of the dropout rate on treatment or time is investigated in lines 6 to 8 of the table where we test different assumptions about the α_{gk} parameters in the model. We now have some evidence that the dropout rate depends on treatment. Comparing lines 7 and 8 the log-likelihood ratio statistic is 7.76 on 2 degrees of freedom, corresponding to a p-value of 0.021. The

Table 11.2. Analysis of deviance for a logistic regression analysis of dropouts in the milk protein data.

Model for $\log\{p_{gk}/(1-p_{gk})\}$	Residual deviance	df
1. $\alpha_{gk} + \beta_{gk}h_k$	111.97	210
2. $\alpha_{gk} + \beta_g h_k$	116.33	219
3. $\alpha_{gk} + \beta_k h_k$	118.63	218
4. $\alpha_{gk} + \beta h_k$	119.32	221
5. α_{gk}	197.66	222
6. $\alpha_g + \alpha_k + \beta h_k$	124.16	227
7. $\alpha_g + \beta h_k$	131.28	230
8. $\alpha + \beta h_k$	139.04	232

evidence for a dependence on time is rather weak; from lines 6 and 7 the log likelihood ratio statistic is 7.12 on 3 degrees of freedom, $p = 0.089$.

The parametric analysis in Example 11.2 allows a more detailed description of departures from completely random dropouts than was possible from the non-parametric analysis in Example 11.1. As always, the relative merits of non-parametric and parametric approaches involve a balance between the robustness and flexibility of the non-parametric approach and the greater sensitivity of the parametric approach when the assumed model is approximately correct. Note that in both approaches the implied alternative hypothesis is of random dropouts. In the next section we develop the parametric approach to allow for the possibility of informative dropouts.

11.3 Modelling the dropout process

In this section we describe recent work by Diggle and Kenward (1994). This work proposes a modelling framework for longitudinal data with informative dropouts, in which random or completely random dropouts are included as explicit sub-models. The aim is to highlight the practical implications of the distinctions between completely random, random and informative dropouts and to provide a possible framework for routine analysis of longitudinal data with dropouts. The formal development is in terms of an underlying linear model for a continuous response variable. The ideas, and the issues of interpretation which are raised by informative dropouts, apply equally well to discrete responses, but a working implementation has yet to be developed.

To convey the essential ideas we first develop the model for a single sequence of measurements on one unit. Let $\boldsymbol{Y}^* = (Y_1^*, \ldots, Y_n^*)$ denote the complete vector of intended measurements and $\boldsymbol{t} = (t_1, \ldots, t_n)$ the corresponding set of times at which measurements are taken. Let $\boldsymbol{Y} = (Y_1, \ldots, Y_n)$ denote the vector of observed measurements, with missing values coded as zero. The crucial assumption in the model for the dropout process is that $\boldsymbol{Y}$ and $\boldsymbol{Y}^*$ coincide as long as the unit remains in the study. This implies the relationship

$$Y_k = \begin{cases} Y_k^*, & k = 1, 2, \ldots, D-1 \\ 0 & k \geq D, \end{cases} \tag{11.3.1}$$

where D is a random variable such that $2 \leq D \leq n$ identifies the dropout time and $D = n+1$ identifies no dropout.

In the spirit of the models used in Chapter 5, we assume that the joint distribution of $\boldsymbol{Y}^*$ is multivariate Gaussian with pdf $f^*(\boldsymbol{y}; \boldsymbol{\beta}, \boldsymbol{\alpha})$, where $\boldsymbol{\beta}$ and $\boldsymbol{\alpha}$ respectively parametrize the mean and covariance structure of $\boldsymbol{Y}^*$. Our most general model for the dropout process assumes that the probability of dropout at time t_d depends on the history of the measurement process up to and including time t_d. For each k, write $H_k = (y_1, \ldots, y_{k-1})$ for the observed history up to time t_{k-1}. Then we specify the model for the dropout process in the form

$$\Pr(D = d \mid \text{history}) = p_d(H_d, y_d^*; \boldsymbol{\phi}). \tag{11.3.2}$$

Equation (11.3.2) allows the dropout probability to depend on both the *observed* measurement history H_d and the *unobserved* value y_d^*, as well as on a set of parameters $\boldsymbol{\phi}$. Using (11.3.1), (11.3.2) and the assumed multivariate Gaussian distribution of $\boldsymbol{Y}^*$, we can now derive the joint distribution of the observable random vector, $\boldsymbol{Y}$, via the sequence of conditional distributions for Y_k given H_k. Let $f_k^*(y \mid H_k; \boldsymbol{\beta}, \boldsymbol{\alpha})$ denote the conditional univariate Gaussian pdf of $\boldsymbol{Y}_k^*$ given H_k and $f_k(y \mid H_k; \boldsymbol{\beta}, \boldsymbol{\alpha}, \boldsymbol{\phi})$ the conditional pdf of Y_k given H_k. Then,

$$\Pr(Y_k = 0 \mid H_k, Y_{k-1} = 0) = 1, \tag{11.3.3}$$

$$\Pr(Y_k = 0 \mid H_k, Y_{k-1}, \neq 0) = \int p_k(H_k, y; \boldsymbol{\phi}) f_k^*(y \mid H_k; \boldsymbol{\beta}, \boldsymbol{\alpha}) dy \tag{11.3.4}$$

and for $y \neq 0$,

$$f_k(y \mid H_k; \boldsymbol{\beta}, \boldsymbol{\alpha}, \boldsymbol{\phi}) = \{1 - p_k(H_k, y; \boldsymbol{\phi})\} f_k^*(y \mid H_k; \boldsymbol{\beta}, \boldsymbol{\alpha}). \tag{11.3.5}$$

Equations (11.3.3), (11.3.4), and (11.3.5) determine the joint distribution of $\boldsymbol{Y}$, and hence the likelihood function for $\boldsymbol{\beta}, \boldsymbol{\alpha}$, and $\boldsymbol{\phi}$. Suppressing the dependence on the parameters, the joint pdf of a complete sequence, $\boldsymbol{Y}$, is

$$f(\boldsymbol{y}) = f_1^*(y_1) \prod_{k=2}^{n} f_k(y_k \mid H_k) \tag{11.3.6}$$

$$= f^*(\boldsymbol{y}) \left[\prod_{k=2}^{n} \{1 - p_k(H_k, y_k)\} \right],$$

whilst for an incomplete sequence with dropout at the dth time-point,

$$f(\boldsymbol{y}) = \left\{ f_1^*(y_1) \prod_{k=2}^{d-1} f_k(y_k \mid H_k) \right\} \Pr(y_d = 0 \mid H_d, Y_{d-1} \neq 0) \qquad (11.3.7)$$

$$= f_{d-1}^*(\boldsymbol{y}) \left[\prod_{k=2}^{d-1} \{1 - p_k(H_k, y_k)\} \right] \Pr(Y_d = 0 \mid H_d, Y_{d-1} \neq 0),$$

where $f_{d-1}^*(\boldsymbol{y})$ denotes the joint pdf of the first $d-1$ elements of $\boldsymbol{Y}$ and the product term within square brackets is absent if $d = 2$.

Following Little and Rubin's terminology, we adopt the following three-way classification of the dropout process:

(1) *informative dropouts* (ID): $p_k(\cdot)$ depends on y_k^*;

(2) *random dropouts* (RD): $p_k(\cdot)$ depends on H_k but not on y_k^*;

(3) *completely random dropouts* (CRD): $p_k(\cdot)$ depends neither on H_k nor on y_k^*.

Note that under either CRD or RD, (11.3.4) reduces to

$$\Pr(Y_k = 0 \mid H_{k-1}, Y_{k-1} \neq 0) = p_k(H_k; \boldsymbol{\phi}).$$

Together with (11.3.5), (11.3.6) and (11.3.7), this shows that the likelihood function then separates into two components, one for $(\boldsymbol{\beta}, \boldsymbol{\alpha})$ and one for $\boldsymbol{\phi}$. The practical implication of this is that we can ignore the dropout process for making inferences about $(\boldsymbol{\beta}, \boldsymbol{\alpha})$, and it is in this sense that CRD and RD are sometimes collectively called ignorable. However, as we discussed in Section 11.1, these may not be the relevant inferences for the practical questions posed by the data. In general, it seems important to be able to make inferences about the enlarged parameter set, $(\boldsymbol{\beta}, \boldsymbol{\alpha}, \boldsymbol{\phi})$ in both the ID and RD cases, which is what the model provides.

We now consider the form of the likelihood function for a set of data consisting of m units, in which $\boldsymbol{y}_i = \{y_{ij},\ j = 1, \ldots, d_i - 1\}$ represents the sequence of observed measurements on the ith unit, where $d_i = n+1$ if the unit does not drop out and d_i identifies the dropout time otherwise. Then, the log-likelihood for $(\boldsymbol{\beta}, \boldsymbol{\alpha}, \boldsymbol{\phi})$ can be partitioned as

$$L(\boldsymbol{\beta}, \boldsymbol{\alpha}, \boldsymbol{\phi}) = L_1(\boldsymbol{\beta}, \boldsymbol{\alpha}) + L_2(\boldsymbol{\phi}) + L_3(\boldsymbol{\beta}, \boldsymbol{\alpha}, \boldsymbol{\phi}),$$

where

$$L_1(\boldsymbol{\beta}, \boldsymbol{\alpha}) = \sum_{i=1}^{m} \log\{f_{d_i-1}^*(\boldsymbol{y}_i)\},$$

$$L_2(\boldsymbol{\phi}) = \sum_{i=1}^{m} \sum_{k=1}^{d_i - 1} \log\{1 - p_k(H_k, y_{ik})\}$$

and

$$L_3(\boldsymbol{\beta}, \boldsymbol{\alpha}, \boldsymbol{\phi}) = \sum_{i:d_i \leq n} \log\{\Pr(D = d_i \mid \boldsymbol{y}_i)\}.$$

Recall that under RD, $L_3(\boldsymbol{\beta}, \boldsymbol{\alpha}, \boldsymbol{\phi})$ depends only on $\boldsymbol{\phi}$, and can therefore be absorbed into $L_2(\boldsymbol{\phi})$.

Diggle and Kenward (1994) develop the implementation of this model with a probit or linear logistic specification for $p_k(H_k, y; \boldsymbol{\phi})$. The logistic version is

$$\log\{p_k/(1 - p_k)\} = \phi_0 + \phi_1 y_k + \Sigma_{j=2}^{k} \phi_j y_{k+1-j}. \tag{11.3.8}$$

One motivation for (11.3.8) is that the probability of dropout at time t in the underlying continuous-time process, $Y(t)$, might depend on a discounted integral of the form

$$\int_0^t w(s) y(t - s) ds. \tag{11.3.9}$$

Then, the expression $\phi_1 y_k + \Sigma_{j=2}^{k} \phi_j y_{k+1-j}$ can be interpreted as a quadrature formula for (11.3.9) in which the weight function, $w(\cdot)$, has been chosen empirically to give the best fit to the data. An important extension is to allow the parameter ϕ_0 to depend on external covariates such as time or the experimental treatment applied. Another, not unrelated, is to replace the measurements y_k by the deviations of the y_k from their fitted means. Note that this would lose the separation between the likelihoods for $(\boldsymbol{\beta}, \boldsymbol{\alpha})$ and for $\boldsymbol{\phi}$ in the RD case.

Example 11.3. Protein content of milk samples (concluded)

We now fit the Diggle and Kenward model to the milk protein data, our objectives being to establish whether the dropout process is informative, and if so to find out if this affects our earlier conclusions about the mean response profiles.

For the mean response profiles we use a simple extension of the model fitted in Section 5.4, where we implicitly assumed random dropouts. Note that this model allows the possibility of an increase in mean response towards the end of the experiment. If $\mu_g(t)$ denotes the mean response at time t under diet g, we assume that

$$\mu_g(t) = \begin{cases} \beta_{0g} + \beta_1 t\,, & t \leq 3 \\ \beta_{0g} + 3\beta_1 + \beta_2(t - 3) + \beta_3(t - 3)^2\,, & t > 3 \end{cases}$$

In Section 5.4, Example 5.1, we used a model for the covariance structure of the complete measurement process, $Y^*(t)$, which included three distinct

Table 11.3. Likelihood analysis of dropout mechanisms for the milk protein data.

Dropout mechanism	$2L_{max}$
ID	2381.63
RD ($\beta_1 = 0$)	2369.32
CRD ($\beta_1 = \beta_2 = 0$)	2299.06

components of variation: a random intercept component between animals, a serially correlated component within animals, and an uncorrelated measurement error component. However, the estimated component of variation between animals was very small, and in what follows we have chosen to set this component to zero. Finally, for the dropout process we note that all of the dropouts occur in the last five weeks of the experiment. Writing p_k for the probability of dropout at time k, we therefore assume that $p_k = 0$ for $k \leq 14$, whereas for $k \geq 15$,

$$\log\{p_k/(1-p_k)\} = \phi_0 + \phi_1 y_k + \phi_2 y_{k-1}.$$

Table 11.3 gives values of twice the maximized log-likelihood for the full model and for the reduced models with $\phi_1 = 0$ (random dropouts), and with $\phi_1 = \phi_2 = 0$ (completely random dropouts). Comparison of the RD and CRD lines in Table 11.3 confirms our earlier conclusion that dropouts are not completely random. More interestingly, there is overwhelming evidence in favour of informative dropouts: from the ID and RD lines, the likelihood ratio statistic to test the RD assumption is 12.31 on one degree of freedom, corresponding to a p-value of 0.0005.

In principle, rejection of the RD assumption forces us to re-assess our model for the underlying measurement process, $Y^*(t)$. However, it turns out that the maximum likelihood estimates of the parameters are virtually the same as under the RD assumption. This is not surprising, as most of the information about these parameters is contained in the 14 weeks of dropout-free data. With regard to the possibility of an increase in the mean response towards the end of the experiment, the maximum likelihood estimates of β_2 and β_3 are both close to zero, and the likelihood ratio statistic to test $\beta_2 = \beta_3 = 0$ is 1.70 on two degrees of freedom, corresponding to a p-value of 0.43.

In the case of the milk protein data, our re-assessment of the dropout process has not led to any substantive changes in our inferences concerning the mean response profiles for the underlying dropout-free process $Y^*(t)$. This is not always so. See Diggle and Kenward (1994) for examples.

11.4 Discussion

We are not yet able to deal in complete generality with missing values in longitudinal data. Further research is needed, especially with regard to non-linear models for discrete response variables. Further research is also needed to deal with the specific problems raised by intermittent missing values, which are rather different in character from dropouts.

Intermittent missing values can arise through explicitly stated censoring rules. For example, values outside a stated range may be simply unreliable because of the limitations of the measuring techniques in use – this is a feature of many bioassay techniques. Methods for handling censored data are now well established in survival analysis (e.g. Cox and Oakes, 1984; Lawless, 1982), but are less well developed to cope with the types of correlated data structures which arise with longitudinal data (Laird, 1988).

When censoring is not an issue, it will often be reasonable to assume that intermittent missing values are either completely random or random, in which case a likelihood-based analysis ignoring the missing values should give the relevant inferences. An exception would be in longitudinal clinical trials with voluntary compliance, where a patient may miss an appointment because they are feeling particularly unwell on the day.

We feel that the issues raised by a consideration of dropouts in longitudinal data are somewhat more complex than in the case of intermittent values. Firstly, an assumption of completely random dropouts will often be implausible. Secondly, if dropouts are *not* completely random we need to think carefully what are the relevant inferences. Do we want to know about the dropout process; or about the conditional distribution of the measurements given that the unit has not dropped out; or about the distribution that the measurements would follow in the absence of dropouts? Suppose, for example, that a longitudinal process has a mean response which is constant over time, but that dropouts occur at random with the dropout probability directly dependent on the observed sequence of measurements. Then, the empirical mean response would show a trend over time as the higher yielding units drop out of the study, whereas a likelihood-based analysis would tend to support the correct, constant mean model. Put the other way round, data which show a trend in the empirical mean response may be equally compatible with a model incorporating a trend in the population mean and completely random dropouts, or one incorporating a constant population mean and random dropouts. Note that this ambiguity of interpretation arises only because of serial correlation in the data. Figure 11.1 gives an example of the problem. The two sets of data were simulated from a model in which the mean response was constant over time, and the random variation within units followed the uniform correlation model described in Section 4.2.1. Ten responses, at unit time-intervals, were simulated for each of 100 subjects. Dropouts occurred at random, according to the logistic model (11.3.8) with $\beta_0 = -1$, $\beta_1 = -2$ and all other $\beta_j = 0$. In Fig. 11.1(a), the correlation between any two responses on the same unit was $\rho = 0.9$, and the empirical mean responses, calculated at each time-point from those subjects who had not yet dropped out, show a steadily rising trend. Nevertheless, a likelihood-based analysis ignoring the dropout process leads, correctly, to the conclusion that there is no time-trend in the underlying mean response. In Fig. 11.1(b), the within-unit correlation was $\rho = 0$, and both the empirical means and a likelihood-based inference tell the same, correct, story – that there is no time-trend in the mean response.

Diggle and Kenward (1994) use their model to investigate some of the consequences of informative dropouts. Using simulated data, they establish that wrongly treating informative dropouts as random dropouts introduces bias into parameter estimates, and that likelihood-based methods can be used to identify informative dropout processes from data sets of a realistic size. A much more difficult problem is to identify a unique model for any real dataset, where all aspects of the model are unknown *a priori* and, as we have suggested above, random or informative dropout processes may be partially confounded with time-trends in the mean response. It seems inevitable that a model which attempts to describe the influence of *unobserved* measurements will be hard to validate from observed data. In this context, it is perhaps worth emphasizing that the Diggle and Kenward model represents but one of a number of possible ways of describing informative dropouts. Wu and Carroll (1988) and Wu and Bailey (1989) develop a somewhat different approach, based on random effects models. They also use a binary regression model for the dropout probability, but assume that dropout is driven by the realized values of the underlying random effects, which are of course unobservable, rather than by the measurement process itself. Other suggestions for modelling informative dropouts can be found amongst the contributions to the published discussion of Diggle and Kenward (1994).

Finally, an important assumption in all of the model-based work described here is that we can postulate a single model for all of the units, whether or not they eventually drop out of the study. This is stated most explicitly in (11.3.1), but is implicit elsewhere. If, in fact, the potential dropouts are characterized by atypical measurement histories prior to the actual dropout, we are liable to make wrong inferences from the start unless we have both a very precise statement of the objectives of the study and a richer class of models than is currently available.

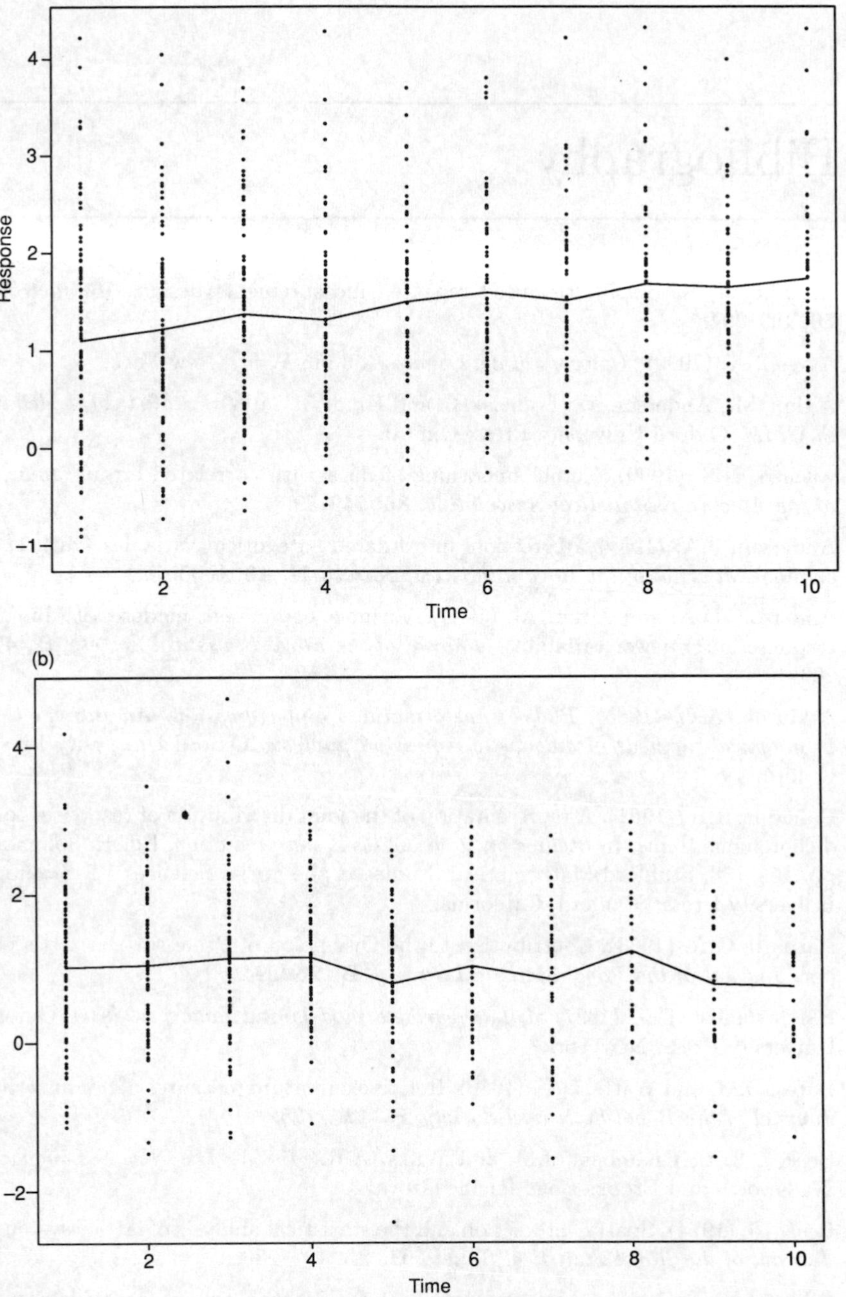

Fig. 11.1. Simulated realizations of a model with a constant mean response, uniform correlation and random dropouts: (a) within-unit correlation $\rho = 0.9$; (b) within-unit correlation $\rho = 0.0$. ... : data; —— : empirical mean response.

Bibliography

Afsarinejad, K. (1983). Balanced repeated measurements designs. *Biometrika*, **70**, 199–204.

Agresti, A. (1990). *Categorical data analysis.* John Wiley, New York.

Aitkin, M., Anderson, D., Francis, B., and Hinde, J. (1989). *Statistical modelling in GLIM.* Oxford University Press, Oxford.

Altman, N.S. (1990). Kernel smoothing of data with correlated errors. *Journal of the American Statistical Association*, **85**, 749–59.

Anderson, J.A. (1984). Regression and ordered categorical variables (with Discussion). *Journal of the Royal Statistical Society*, **B, 46**, 1–30.

Anderson, D.A. and Aitkin, M. (1985). Variance component models with binary response: interviewer variability. *Journal of the Royal Statistical Society*, **B, 47**, 203–10.

Atkinson, A.C. (1985). *Plots, transformations and regression. An introduction to graphical methods of diagnostic regression analysis.* Oxford University Press, Oxford.

Bahadur, R.R. (1961). A representation of the joint distribution of responses to n dichotomous items. In *Studies on item analysis and prediction*, Ed. H. Solomon, pp. 158–168. Stanford Mathematical Studies in the Social Sciences VI, Stanford University Press, Stanford, California.

Barnard, G.A. (1963). Contribution to the Discussion of Professor Bartlett's paper. *Journal of the Royal Statistical Society*, **B, 25**, 294.

Bartholomew, D.J. (1987). *Latent variable models and factor analysis.* Oxford University Press, New York.

Bates, D.M. and Watts, D.G. (1980). Relative curvature measures of nonlinearity. *Journal of the Royal Statistical Society*, **B, 42**, 1–25.

Becker, R.A., Chambers, J.M. and Wilks, A.R. (1988). *The New S Language.* Wadsworth and Brooks-Cole, Pacific Grove.

Besag, J. (1974). Spatial interaction and the statistical analysis of lattice systems. *Journal of the Royal Statistical Society*, **B, 36**, 192–236.

Billingsley, P. (1961). *Statistical inference for Markov processes.* University of Chicago Press, Chicago, Illinois.

Bishop, S.H. and Jones, B. (1984). A review of higher-order cross-over designs. *Journal of Applied Statistics*, **11**, 29–50.

Bishop, Y.M.M., Fienberg, S.E. and Holland, P.W. (1975). *Discrete multivariate analysis: theory and practice.* MIT Press, Cambridge, Massachussetts.

Bloomfield, P. and Watson, G.S. (1975). The inefficiency of least squares. *Biometrika*, **62**, 121-128

Box, G.P. and Jenkins, G.M. (1970). *Time series analysis - forecasting and control* (Revised edition). Holden-Day, San Francisco, California.

Breslow, N.E. (1984). Extra-Poisson variation in log linear models. *Applied Statistics*, **33**, 38–44.

Breslow, N.E. and Clayton, D.G. (1993). Approximate inference in generalized linear mixed models. *Journal of the American Statistical Association*, **88**, 9–25.

Breslow, N.E. and Day, N.E. (1980). *Statistical methods in cancer research, Volume I.* IARC Scientific Publications No. 32. Lyon.

Carey, V.C. (1992). Regression analysis for large binary clusters. Unpublished PhD thesis, Department of Biostatistics, The Johns Hopkins University, Baltimore, Maryland.

Carey, V.C., Zeger, S.L., and Diggle, P.J. (1993). Modelling multivariate binary data with alternating logistic regressions. *Biometrika*, **80**, 517–26.

Chambers, J.M. and Hastie, T.J. (1992). *Statistical Models in S.* Wadsworth and Brooks-Cole, Pacific Grove.

Chambers, J.M., Cleveland, W.S., Kleiner, B., and Tukey, P.A. (1983). *Graphical methods for data analysis.* Wadsworth, Belmont, California.

Clayton, D.G. (1974). Some odds ratio statistics for the analysis of ordered categorical data. *Biometrika*, **61**, 525–31.

Clayton, D.G. (1992). Repeated ordinal measurements: a generalised estimating equation approach. *Medical Research Council Biostatistics Unit Technical Report* Cambridge, England.

Cleveland, W.S. (1979). Robust locally weighted regression and smoothing scatterplots. *Journal of the American Statistical Association*, **74**, 829–36.

Cochran, W.G. (1977). *Sampling techniques.* John Wiley, New York.

Conaway, M.R. (1990). A random effects model for binary data. *Biometrics*, **46**, 317–28.

Cook, D. and Weisberg, S. (1982). *Residuals and influence in regression.* Chapman and Hall, London.

Courcier. Reissued with a supplement, 1806. Second supplement published 1820. A portion of the appendix was translated, 1929, pp. 576–579 in *A Source Book in Mathematics*, D.E. Smith, ed., trans. by H.A. Ruger and H.M. Walker, McGraw Hill, New York; reprinted 1959 in 2 volumes, Dover, New York.

Cox, D.R. (1970). *The analysis of binary data.* Chapman and Hall, London.

Cox, D.R. (1972). Regression models and life tables (with Discussion). *Journal of the Royal Statistical Society*, **B, 74**, 187–200.

Cox, D.R. and Miller, H.D. (1965). *The theory of stochastic processes.* John Wiley, New York.

Cox, D.R. and Oakes, D. (1984). *Analysis of survival data.* Chapman and Hall, London.

Cox, D.R. and Snell, E.J. (1989). *Analysis of binary data.* Chapman and Hall, London.

Crouch, A.C. and Spiegelman, E. (1990). The evalulation of integrals of the form $\int f(t)\exp(-t^2)dt$: application to logistic-normal models. *Journal of the American Statistical Association*, **85**, 464–469.

Cullis, B.R. (1994). Contribution to the Discussion of the paper by Diggle and Kenward. *Applied Statistics*, **43**, 79–80.

Cullis, B.R. and McGilchrist, C.A. (1990). A model for the analysis of growth data from designed experiments. *Biometrics*, **46**, 131–42.

Dale, J.R. (1986). Global cross-ratio models for bivariate discrete ordered responses. *Biometrics*, **42**, 909–917.

Davidian, M. and Gallant, A.R. (1992). The nonlinear mixed effects model with a smooth random effects density. *Department of Statistics Technical Report, North Carolina State University*, Campus Box 8203, Raleigh, North Carolina 27695.

Dempster, A.P., Laird, N.M., and Rubin, D.B. (1977). Maximum likelihood from incomplete data via the EM algorithm (with Discussion). *Journal of the Royal Statistical Society*, **B**, **39**, 1–38.

Diggle, P.J. (1988). An approach to the analysis of repeated measures. *Biometrics*, **44**, 959–71.

Diggle, P.J. (1989). Testing for random dropouts in repeated measurement data. *Biometrics*, **45**, 1255–58.

Diggle, P.J. (1990). *Time series: a biostatistical introduction.* Oxford University Press, Oxford.

Diggle, P.J. and Kenward, M.G. (1994). Informative dropout in longitudinal data analysis (with Discussion). *Applied Statistics*, **43**, 49–93.

Draper, N. and Smith, H. (1981). *Applied regression analysis (second edition).* Wiley, New York.

Evans, J.L. and Roberts, E.A. (1979). Analysis of sequential observations with applications to experiments on grazing animals and perennial plants. *Biometrics*, **35**, 687–93.

Fahrmeir, L. (1992). Posterior mode estimation by extended Kalman filtering for multivariate dynamic generalized linear models. *Journal of the American Statistical Association*, **87**, 501–509.

Fearn, T. (1977). A two-stage model for growth curves which leads to Rao's covariance-adjusted estimates. *Biometrika*, **64**, 141–43.

Feller, W. (1968). *An introduction to probability theory and its applications* (3rd edn). John Wiley, New York.

Fitzmaurice, G.M., Laird, N.M., and Rotnitsky, A.G. (1993). Regression models for discrete longitudinal responses (with Discussion). *Statistical Science*, **8**, 284–309.

Gabriel, K.R. (1962). Ante dependence analysis of an ordered set of variables. *Annals of Mathematical Statistics*, **33**, 201–12.

Gauss, C.F. (1809). *Theoria motus corporum celestium.* Hamburg: Perthes et Besser. Translated, 1857, as *Theory of motion of the heavenly bodies moving about the sun in conic sections*, trans. C. H. Davis. Little, Brown, Boston. Reprinted, 1963; Dover, New York. French translation of the portion on least squares, pp. 11-134 in Gauss, 1855.

Gilmour, A.R., Anderson, R.D. and Rae, A.L. (1985). The analysis of binomial data by a generalized linear mixed model. *Biometrika*, **72**, 593–99.

Glasbey, C.A.(1980). Nonlinear regression with autoregressive time series errors. *Biometrics*, **36**, 135–40.

Glasbey, C.A.(1986). Conservative estimates of the variances of regression parameter estimators for classes of error model. *Biometrika*, **73**, 746–50.

Glasbey, C.A. (1988). Examples of regression with serially correlated errors. *The Statistician*, **37**, 277-92.

Godambe, V.P. (1960). An optimum property of regular maximum likelihood estimation. *Annals of Mathematical Statistics*, **31**, 1208–12.

Goldfarb, N. (1960). *An introduction to longitudinal statistical analysis – the method of repeated observations from a fixed sample.* Free Press of Glencoe, Illinois.

Goldstein, H.(1979). *The Design and analysis of longitudinal studies: their role in the measurement of change.* Academic Press, London.

Gourieroux, C., Monfort, A., and Trognon, A. (1984). Psuedo-maximum likelihood methods: theory. *Econometrica*, **52**, 681–700.

Graybill, F. (1976). *Theory and application of the linear model.* Wadsworth, California.

Greenwood M. and Yule, G.U. (1920). An enquiry into the nature of frequency distributions to the occurrence of multiple attacks of disease or of repeated accidents. *Journal of the Royal Statistical Society*, **A**, **83**, 255–279.

Griffiths, D.A. (1973). Maximum likelihood estimation for the beta-binomial distribution and an application to the household distribution of the total number of cases of a disease. *Biometrics*, **29**, 637–48.

Härdle, W. (1990). *Applied nonparametric regression.* Cambridge University Press, New York.

Hart, J.D. (1991). Kernel regression estimation with time series errors. *Journal of the Royal Statistical Society*, **B 53**, 173–87.

Hart, J.D. and Wehrly, T.E. (1986). Kernel regression estimation using repeated measurements data. *Journal of the American Statistical Association*, **81**, 1080–88.

Harville, D. (1974). Bayesian inference for variance components using only error contrasts. *Biometrika*, **61**, 383–85.

Harville, D. (1977). Maximum likelihood estimation of variance components and related problems. *Journal of the American Statistical Association*, **72**, 320–40.

Hastie, T.J. and Tibshirani, R.J. (1990). *Generalized additive models*. Chapman and Hall, New York.

Hausman, J.A. (1978). Specification tests in econometrics. *Econometrica*, **46**, 1251–72.

Heckman, J.J. and Singer, B. (1985). *Longitudinal analysis of labour market data*. Cambridge University Press, Cambridge.

Hedayat, A. and Afsarinejad, K. (1975). Repeated measures designs, I. In *A Survey of Statistical Design and Linear Models* (ed. J.N. Srivastava), pp.229–42. North-Holland, Amsterdam.

Hedayat, A. and Afsarinejad, K. (1978). Repeated measures designs, II. *Annals of Statistics*, **6**, 619–28.

Huber, P.J. (1967). The behaviour of maximum likelihood estimators under non-standard conditions. *Proceedings of the Fifth Berkeley Symposium on Mathematical Statistics and Probability*, **1**, LeCam, L.M. and Neyman, J. editors, University of California Press, pp. 221–33.

Jones, B. and Kenward, M.G. (1987). Modelling binary data from a three-point cross-over trial. *Statistics in Medicine*, **6**, 555–64.

Jones, B. and Kenward, M.G. (1989). *Design and analysis of cross-over trials*. Chapman and Hall, London.

Jones, M.C. and Rice, J.A. (1992). Displaying the important features of large collections of similar curves. *The American Statistician*, **46**, 140–45.

Jones, R.M. (1993). *Longitudinal data with serial correlation: a state-space approach*. Chapman and Hall, London.

Jones, R.H. and Ackerson, L.M. (1990). Serial correlation in unequally spaced longitudinal data. *Biometrika*, **77**, 721–731.

Jones, R.H. and Boadi-Boteng, F. (1991). Unequally spaced longitudinal data with serial correlation. *Biometrics*, **47**, 161–75.

Journel, A.G. and Huijbregts C.J. *Mining geostatistics*. Academic Press, New York.

Jowett, G.H. (1952). The accuracy of systematic sampling from conveyor belts. *Applied Statistics*, **1**, 50–59.

Kalbfleisch, J.D. and Prentice, R.L. (1980). *The statistical analysis of failure time data*. John Wiley, New York.

Kalman, R.E. (1960). A new approach to linear filtering and prediction problems. *Journal of Basic Engineering*, **82**, 34–45.

Karim, M.R. (1991). Generalized linear models with random effects; a Gibbs sampling approach. Unpublished PhD thesis. The Johns Hopkins University Department of Biostatistics, Baltimore, Maryland.

Kaslow, R.A., Ostrow, D.G., Detels, R. *et al.* (1987). The Multicenter AIDS Cohort Study: rationale, organization and selected characteristics of the participants. *American Journal of Epidemiology*, **126**, 310–18.

Kaufmann, H. (1987). Regression models for nonstationary categorical time series: asymptotic estimation theory. *Annals of Statistics*, **15**, 863–71.

Kenward, M.G. (1987). A method for comparing profiles of repeated measurements. *Applied Statistics*, **36**, 296-308.

Korn, E.L. and Whittemore, A.S. (1979). Methods for analyzing panel studies of acute health effects of air pollution. *Biometrics*, **35**, 795–802.

Laird, N.M. (1988). Missing data in longitudinal studies. *Statistics in Medicine*, **7**, 305–15.

Laird, N.M. and Ware, J.H. (1982). Random-effects models for longitudinal data. *Biometrics*, **38**, 963–74.

Lange, N. and Ryan, L. (1989). Assessing normality in random effects models. *Annals of Statistics*, **17**, 624–42.

Lawless, J.F. (1982). *Statistical models and methods for lifetime data.* John Wiley, New York.

Legendre, A.M. (1805). *Nouvelles méthodes pour la detérmination des orbites des comètes.* John Wiley, New York.

Lepper, A.W.D. (1989). Effects of altered dietary iron intake in *Mycobacterium paratuberculosis*-infected dairy cattle: sequential observations on growth, iron and copper metabolism and development of paratuberculosis. *Research in Veterinary Science* **46**, 289–96.

Liang, K.-Y. and McCullagh, P. (1993). Case studies in binary dispersion. *Biometrics*, **49**, 623–30.

Liang, K.-Y. and Zeger, S.L. (1986). Longitudinal data analysis using generalized linear models. *Biometrika*, **73**, 13–22.

Liang, K.-Y., Zeger, S.L., and Qaqish, B. (1992). Multivariate regression analyses for categorical data (with Discussion). *Journal of the Royal Statistical Society*, **B 54**, 3–40.

Lindstrom, M.J. and Bates, D.M. (1990). Nonlinear mixed effects models for repeated measures data. *Biometrics*, **46**, 673–87.

Lipsitz, S. (1989). Generalized estimating equations for correlated binary data: using the odds ratio as a measure of association. Technical Report, Department of Biostatistics, Harvard School of Public Health.

Lipsitz, S., Laird, N., and Harrington, D. (1991). Generalized estimating equations for correlated binary data: using odds ratios as a measure of association. *Biometrika*, **78**, 153–60.

Little, R.J.A. and Rubin, D.B. (1987). *Statistical analysis with missing data.* John Wiley, New York.

Mason, W.B. and Fienberg, S.E. (eds) (1985). *Cohort analysis in social research: beyond the identification problem.* Springer-Verlag, New York.

McCullagh, P. (1980). Regression models for ordinal data (with Discussion). *Journal of the Royal Statistical Society*, **B**, **42**, 109–42.

McCullagh, P. (1983). Quasi-likelihood functions. *Annals of Statistics*, **11**, 59–67.

McCullagh, P. and Nelder, J.A. (1989). *Generalized linear models.* Chapman and Hall, New York.

Mead, R. and Curnow, R.N. (1983). *Statistical methods in agriculture and experimental biology.* Chapman and Hall, London.

Morton, R. (1987). A generalized linear model with nested strata of extra-Poisson variation. *Biometrika*, **74**, 247–57.

Mosteller, F. and Tukey, J.W. (1977). *Data analysis and regression – a second course in statistics.* Addison-Wesley, Reading, Massachusetts.

Moyeed, R.A. and Diggle, P.J. (1994). Rates of convergence in semi-parametric modelling of longitudinal data. *Australian Journal of Statistics*, **36**.

Müller, M.G. (1988). *Nonparametric regression analysis of longitudinal data.* Lecture Notes in Statistics, **41**. Springer-Verlag, Berlin.

Munoz, A., Carey, V., Schouten, J.P., Segal, M., and Rosner, B. (1992). A parametric family of correlation structures for the analysis of longitudinal data. *Biometrics*, **48**, 733–42.

Murray, G.D. and Findlay, J.G. (1988). Correcting for the bias caused by dropouts in hypertension trials. *Statistics in Medicine*, **7**, 941–46.

Nelder, J.A. and Mead, R. (1965). A simplex method for function minimisation. *The Computer Journal*, **7**, 303–13.

Neuhaus, J.M. and Jewell, N.P. (1990). Some comments on Rosner's multiple logistic model for clustered data. *Biometrics*, **46**, 523–534.

Neuhaus, J.M., Kalbfleisch, J.D. and Hauck, W.W. (1991). A comparison of cluster-specific and population averaged approaches for analyzing correlated binary data. *International Statistical Review*, **59**, 25–36.

Neyman, J. and Scott, E.L. (1948). Consistent estimates based on partially consistent observations. *Econometrica*, **16**, 1–32.

Paik, M.C. (1992). Parametric variance function estimation for non-normal repeated measurement data. *Biometrics*, **48**, 18–30.

Pantula, S.G. and Pollock, K.H.(1985). Nested analysis of variance with autocorrelated errors. *Biometrics*, **41**, 909–20.

Patterson, H.D. (1951). Change-over trials. *Journal of the Royal Statistical Society*, **B 13**, 256–71.

Patterson, H.D. and Thompson, R. (1971). Recovery of inter-block information when block sizes are unequal. *Biometrika*, **58**, 545–54.

Pierce, D.A. and Sands, B.R. (1975). Extra-Bernoulli variation in binary data. Technical Report 46, Department of Statistics, Oregon State University.

Plewis, I. (1985) *Analysing change. Measurement and explanation using longitudinal data.* John Wiley, New York.

Poisson, S.D. (1837). *Recherches sur la probabilité des jugements en matière criminelle et en matière civile, précédés des règles générales du calcul des probabilités.* Bachelier, Paris.

Prentice, R.L. (1986). Binary regression using an extended beta-binomial distribution, with discussion of correlation induced by covariate measurement errors. *Journal of the American Statistical Association*, **81**, 321–27.

Prentice, R.L. (1988). Correlated binary regression with covariates specific to each binary observation. *Biometrics*, **44**, 1033–48.

Prentice, R.L. and Zhao, L.P. (1991). Estimating equations for parameters in means and covariances of multivariate discrete and continuous responses. *Biometrics*, **47**, 825–39.

Priestley, M.B. and Chao, M.T. (1972). Non-parametric function fitting. *Journal of the Royal Statistical Society*, **B**, **34**, 384–92.

Rao, C.R. (1965). The theory of least squares when the parameters are stochastic and its application to the analysis of growth curves. *Biometrika* **52**, 447–58.

Rao, C.R. (1973). *Linear statistical inference and its applications* (Second Edition). John Wiley, New York.

Ratkowsky, D.A. (1983). *Non-linear regression modelling.* Marcel Dekker, New York.

Rice, J.A. and Silverman, B.W. (1991). Estimating the mean and covariance structure nonparametrically when the data are curves. *Journal of the Royal Statistical Society*, **B**, **53**, 233-43.

Ridout, M. (1991). Testing for random dropouts in repeated measurement data. *Biometrics*, **47**, 1617–21.

Rosner, B. (1984). Multivariate methods in ophthalmology with application to other paired-data situations. *Biometrics*, **40**, 1025–35.

Rosner, B. (1989). Multivariate methods for clustered binary data with more than one level of nesting. *Journal of the American Statistical Association*, **84**, 373–80.

Rowell, J.G. and Walters, D.E. (1976). Analysing data with repeated observations on each experimental unit. *Journal of Agricultural Science*, **87**, 423–32.

Royall, R.M. (1986). Model robust inference using maximum likelihood estimators. *International Statistical Review*, **54**, 221–26.

Sandland, R.L. and McGilchrist, C.A. (1979). Stochastic growth curve analysis. *Biometrics*, **35**, 255–71.

Schall, R. (1991). Estimation in generalized linear models with random effects. *Biometrika*, **78**, 719–27.

Seber, G.A.F. (1977). *Linear regression analysis.* John Wiley, New York.

Senn, S.J. (1992). *Crossover trials in clinical research.* John Wiley, Chichester.

Silverman, B.W. (1984). Spline smoothing: the equivalent variable kernel method. *Annals of Statistics*, **12**, 898–916.

Silverman, B.W. (1985). Some aspects of the spline smoothing approach to non-parametric regression curve fitting (with Discussion). *Journal of the Royal Statistical Society*, **B**, **47**, 1–52.

Skellam, J.G. (1948). A probability distribution derived from the binomial distribution by regarding the probability of success as variable between the sets of trials. *Journal of the Royal Statistical Society*, **B**, **10**, 257–61.

Snedecor, G.W. and Cochran, W.G. (1989). *Statistical methods* (Eighth edition). Iowa State University Press, Ames, Iowa.

Snell, E.J. (1964). A scaling procedure for ordered categorical data. *Biometrics*, **40**, 592–607.

Sommer, A. (1982). *Nutritional blindness.* Oxford University Press, New York.

Sommer, A., Katz, J., and Tarwotjo, I. (1984). Increased risk of respiratory infection and diarrhea in children with pre-existing mild vitamin A deficiency. *American Journal of Clinical Nutrition*, **40**, 1090–95.

Spall, J.C. (1988). *Bayesian analysis of time series and dynamic models.* Marcell Dekker, New York.

Stefanski, L.A. and Carroll, R.J. (1985). Covariate measurement error in logistic regression. *Annals of Statistics*, **13**, 1335–51.

Stern, R.D. and Coe, R. (1984). A model fitting analysis of daily rainfall data (with Discussion). *Journal of the Royal Statistical Society*, **A**, **147**, 1–34.

Stiratelli, R., Laird, N. and Ware, J.H. (1984). Random effects models for serial observations with binary responses. *Biometrics*, **40**, 961–71.

Stram, D.O., Wei, L.J. and Ware, J.H. (1988). Analysis of repeated ordered categorical outcomes with possibly missing observations and time-dependent covariates. *Journal of the American Statistical Association*, **83**, 631–37.

Thall, P.F. and Vail, S.C. (1990). Some covariance models for longitudinal count data with overdispersion. *Biometrics*, **46**, 657–71.

Tsay, R. (1984). Regression models with time series errors. *Journal of the American Statistical Association*, **79**, 118–124.

Tufte, E.R. (1983). *The visual display of quantitative information.* Graphics Press, Cheshire, Connecticut.

Tufte, E.R. (1990). *Envisioning information*. Graphics Press, Cheshire, Connecticut.

Tukey, J.W. (1977). *Exploratory data analysis*. Addison-Wesley, Reading, Massachusetts.

Tunnicliffe-Wilson, G. (1989). On the use of marginal likelihood in time series model estimation. *Journal of the Royal Statistical Society*, **B**, **51**, 15–27.

Velleman, P.F. and Hoaglin, D.C. (1981). *Applications, basics, and computing of exploratory data analysis*. Duxbury Press, Boston, Massachusetts.

Verbyla, A.P.(1986). Conditioning in the growth curve model. *Biometrika*, **73**, 475–83.

Verbyla, A.P. and Cullis, B.R. (1990). Modelling in repeated measures experiments. *Applied Statistics*, **39**, 341–56.

Verbyla, A.P. and Venables, W.N.(1988). An extension of the growth curve model. *Biometrika*, **75**, 129–38.

Volberding, P.A., Lagakos, S.W., Koch, M.A. *et al.* (1990). Zidovudine in asymptomatic human immunodeficiency virus infection. *The New England Journal of Medicine*, **322**, 941–49.

Waclawiw, M.A. and Liang, K.-Y. (1993). Prediction of random effects in the generalized linear model. *Journal of the American Statistical Association*, **88**, 171–78.

Ware, J.H. (1985). Linear models for the analysis of longitudinal studies. *The American Statistician*, **39**, 95–101.

Ware, J.H., Lipsitz, S., and Speizer, F.E. (1988). Issues in the analysis of repeated categorical outcomes. *Statistics in Medicine*, **7**, 95–107.

Ware, J.H., Dockery, D., Louis, T.A. *et al.* (1990). Longitudinal and cross-sectional estimates of pulmonary function decline in never-smoking adults. *American Journal of Epidemiology*, **32**, 685–700.

Wedderburn, R.W.M. (1974). Quasi-likelihood functions, generalized linear models and the Gaussian method. *Biometrika*, **61**, 439–47.

West, M., Harrison, P.J., and Migon, H.S. (1985). Dynamic generalized linear models and Bayesian forecasting (with Discussion). *Journal of the American Statistical Association*, **80**, 73–97.

White, H. (1982). Maximum likelihood estimation of misspecified models. *Econometrics*, **50**, 1–25.

Whittaker, J.C. (1990). *Graphical models in applied multivariate statistics*. John Wiley, New York.

Williams, E.J. (1949). Experimental designs balanced for the estimation of residual effects of treatments. *Australian Journal of Scientific Research*, **2**, 149–68.

Williams, D.A. (1975). The analysis of binary responses from toxicological experiments involving reproduction and teratogenicity. *Biometrics*, **31**, 949–52.

Williams, D.A. (1982). Extra-binomial variation in logistic linear models. *Applied Statistics*, **31**, 144–48.

Winer, B.J. (1971). *Statistical principles in experimental design* (second edition). McGraw-Hill, New York.

Wishart, J. (1938). Growth-rate determinations in nutrition studies with the bacon pig, and their analysis. *Biometrika*, **30**, 16–28.

Wong, W.H. (1986). Theory of partial likelihood. *Annals of Statistics*, **14**, 88–123.

Wu, M.C. and Bailey, K.R. (1989). Estimation and comparison of changes in the presence of informative right censoring: conditional linear model. *Biometrics*, **45**, 939–55.

Wu, M.C. and Carroll, R.J. (1988). Estimation and comparison of changes in the presence of right censoring by modeling the censoring process. *Biometrics*, **44**, 175–88.

Yule, G.U. (1927). On a method of investigating periodicities in disturbed series with special reference to Wolfer's sunspot numbers. *Philosophical Transactions of the Royal Society of London*, A **226**, 267–98.

Zeger, S.L. and Diggle, P.J. (1994). Semi-parametric models for longitudinal data with application to CD4 cell numbers in HIV seroconverters. *Biometrics*, **50**.

Zeger, S.L. and Karim, M.R. (1991). Generalized linear models with random effects: a Gibbs sampling approach. *Journal of the American Statistical Association*, **86**, 79–86.

Zeger, S.L. and Liang, K.-Y. (1986). Longitudinal data analysis for discrete and continuous outcomes. *Biometrics*, **42**, 121–30.

Zeger, S.L. and Liang, K.-Y. (1992). An overview of methods for the analysis of longitudinal data. *Statistics in Medicine*, **11**, 1825–39.

Zeger, S.L., Liang, K.-Y., and Albert, P.S. (1988). Models for longitudinal data: a generalized estimating equation approach. *Biometrics*, **44**, 1049–60.

Zeger, S.L., Liang, K.-Y. and Self, S.G. (1985). The analysis of binary longitudinal data with time-indpendent covariates. *Biometrika*, **72**, 31–8.

Zeger, S.L. and Qaqish, B. (1988). Markov regression models for time series: a quasi-likelihood approach. *Biometrics*, **44**, 1019–31.

Zhao, L.P. and Prentice, R.L. (1990). Correlated binary regression using a generalized quadratic model. *Biometrika*, **77**, 642–48.

Appendix A
Statistical Background

A.1 Introduction

This appendix provides a brief review of some basic statistical concepts used throughout the book. Readers should find sections A.2 and A.3 useful for the material that is presented in Chapters 4, 5, 6 and 11 which deal with methods for analysing data with a continuous response variable. These four chapters also make extensive use of the method of maximum likelihood, which is the subject of section A.4. Sections A.5 and A.6 outline the basic concepts of generalized linear models, which provide a unified methodology for the analysis of data with continuous or discrete responses. This material is most relevant for Chapters 7 to 10.

A.2 The linear model and the method of least squares

In many scientific studies, the main goal is to predict or describe an *outcome* or *response* variable, in terms of other variables which we will refer to as *predictors, covariates* or *explanatory variables.* The explanatory variables can either be fixed in advance, such as the treatment assignment in an experiment, or uncontrolled, such as smoking status in an observational study. Regression analysis has been used since the early nineteenth century to describe the relationship between the expectation of a response variable, Y, and a set of explanatory variables, $x_j : j = 1, ..., p$. The *linear regression model* assumes that the response variable and the explanatory variables are related through

$$Y_i = \beta_1 x_{i1} + \beta_2 x_{i2} + \cdots + \beta_p x_{ip} + \epsilon_i = \boldsymbol{x}_i'\boldsymbol{\beta} + \epsilon_i, \quad i = 1, \ldots, m,$$

where Y_i is the response for the ith of m subjects and x_{ij} is the value of the jth explanatory variable for the ith subject. Usually, $x_{i1} = 1$ for every subject, so that β_1 is the *intercept* of this regression model. The ϵ_i are random variables which are assumed to be uncorrelated with each other, and to have $\mathrm{E}(\epsilon_i) = 0$ and $\mathrm{Var}(\epsilon_i) = \sigma^2$. This implies that the first two moments of Y_i are $\mathrm{E}(Y_i) = \boldsymbol{x}_i'\boldsymbol{\beta}$ and $\mathrm{Var}(Y_i) = \sigma^2$, where $\boldsymbol{x}_i$ and $\boldsymbol{\beta}$ are p-element vectors. Note that these statements involve no distributional assumption about Y_i, although a common assumption is that the joint distribution of $Y_1, ..., Y_m$ is multivariate Gaussian. Models of this kind have been widely used in both experimental and observational studies. In fact, linear models include as special cases

- the analysis of variance, in which the x_j are dummy variables, used to indicate the allocation of experimental units to treatments,
- multiple regression, in which the x_j are quantitative variables,
- the analysis of covariance, in which the x_js are a mixture of continuous and dummy variables.

Each regression coefficient , β_j, describes the change in the expected value of the response variable, Y, per unit change of its corresponding explanatory variable, x_j, all other variables held fixed.

The method of *least squares* (Legendre, 1805; Gauss, 1809) is a long-standing method for estimating the vector of regression coefficients, $\boldsymbol{\beta}$. The idea is to find an estimate, $\hat{\boldsymbol{\beta}}$ say, which minimizes the sum of squares

$$\text{RSS} = \sum_{i=1}^{m} (Y_i - \boldsymbol{x}_i'\boldsymbol{\beta})^2.$$

This procedure is formally equivalent to solving

$$\frac{\partial \text{RSS}}{\partial \boldsymbol{\beta}} = -2 \sum_{i=1}^{m} \boldsymbol{x}_i (Y_i - \boldsymbol{x}_i'\boldsymbol{\beta}) = 0,$$

which gives rise to

$$\hat{\boldsymbol{\beta}} = \left(\sum_{i=1}^{m} \boldsymbol{x}_i \boldsymbol{x}_i' \right)^{-1} \sum_{i=1}^{m} \boldsymbol{x}_i Y_i.$$

An alternative, and more familiar, form of the least squares estimate $\hat{\boldsymbol{\beta}}$ is obtained by defining an m-element vector $\boldsymbol{Y} = (Y_1, ..., Y_m)$ and an m by p matrix X with ij^{th} element x_{ij}. Then,

$$\hat{\boldsymbol{\beta}} = (X'X)^{-1}X'\boldsymbol{Y}.$$

The least squares estimate, $\hat{\boldsymbol{\beta}}$, enjoys many desirable statistical properties. Firstly, it is an unbiased estimator of $\boldsymbol{\beta}$, i.e. $\text{E}(\hat{\boldsymbol{\beta}}) = \boldsymbol{\beta}$. Its variance matrix is

$$\text{Var}(\hat{\boldsymbol{\beta}}) = \sigma^2 (X'X)^{-1}.$$

Secondly, for any vector $\boldsymbol{a}$ of known coefficients, if we let $\phi = \boldsymbol{a}'\boldsymbol{\beta}$ then $\hat{\phi} = \boldsymbol{a}'\hat{\boldsymbol{\beta}}$ has the smallest possible variance amongst all unbiased estimators for ϕ which are linear combinations of the Y_i. This optimality property of least squares estimation is known as the Gauss-Markov Theorem.

The constant variance, σ^2, of the ϵ_i is usually unknown but can be estimated by

$$\hat{\sigma}^2 = \sum_{i=1}^{m} (Y_i - \boldsymbol{x}_i'\hat{\boldsymbol{\beta}})^2/(m-p).$$

Many books, including Seber (1977) and Draper and Smith (1981), give more detailed discussions of least squares estimation.

A.3 Multivariate Gaussian theory

This section reviews some of the important results for multivariate Gaussian observations. For more detailed treatment of multivariate Gaussian theory see, for example, Graybill (1976) or Rao (1973).

A random vector $\boldsymbol{Y} = (Y_1, \ldots Y_n)$ is said to follow a multivariate Gaussian distribution if its probability density is of the form

$$f(\boldsymbol{y}; \boldsymbol{\mu}, V) = (2\pi)^{-n/2} \mid V \mid^{-1/2} \exp\{-(\boldsymbol{y} - \boldsymbol{\mu})'V^{-1}(\boldsymbol{y} - \boldsymbol{\mu})/2\},$$

where $-\infty < y_j < \infty$, $j = 1, \ldots, n$. As in the univariate case, this distribution is fully specified by its first two moments, $\boldsymbol{\mu} = \mathrm{E}(\boldsymbol{Y})$ and $V = \mathrm{Var}(\boldsymbol{Y})$. A convenient shorthand notation is $\boldsymbol{Y} \sim MVN(\boldsymbol{\mu}, V)$.

The following properties of the multivariate Gaussian distribution are used extensively in the book:

- Each Y_j has a univariate Gaussian distribution.
- More generally, if $\boldsymbol{Z}_1 = (Y_1, \ldots, Y_{n_1})$ with $n_1 < n$, then $\boldsymbol{Z}_1$ also follows a multivariate Gaussian distribution with mean $(\mu_1, \ldots, \mu_{n_1})$ and covariance matrix V_1 which is the upper left n_1 by n_1 submatrix of V.
- If, additionally, $\boldsymbol{Z}_2 = (Y_{n_1+1}, \ldots, Y_n)$, then the conditional distribution of $\boldsymbol{Z}_1$ given $\boldsymbol{Z}_2 = \boldsymbol{z}_2$ is multivariate Gaussian. Its conditional mean vector is

  $$\begin{pmatrix} \mu_1 \\ \vdots \\ \mu_{n_1} \end{pmatrix} + V_{12}{V_{22}}^{-1} \left(\boldsymbol{z}_2 - \begin{pmatrix} \mu_{n_1+1} \\ \vdots \\ \mu_n \end{pmatrix} \right),$$

  and its conditional variance matrix is

  $$V_{11} - V_{12}{V_{22}}^{-1}{V_{12}}',$$

  where

  $$V = \begin{pmatrix} V_{11} & V_{12} \\ {V_{12}}' & V_{22} \end{pmatrix}.$$

- If B is an $m \times n$ matrix of rank $m \leq n$, then BY is also distributed as a multivariate Gaussian, with mean vector $B\boldsymbol{\mu}$ and variance matrix BVB'.

- The random variable $U = (\boldsymbol{Y} - \boldsymbol{\mu})'V^{-1}(\boldsymbol{Y} - \boldsymbol{\mu})$ has a *chi-squared distribution* with n degrees of freedom, which we write as $U \sim \chi^2_n$.

A.4 Likelihood inference

Likelihood inference is based on a specification of the probability or probability density for the observed data, $\boldsymbol{y}$. This expression, $f(\boldsymbol{y};\boldsymbol{\theta})$, is indexed by a vector of unknown parameters, $\boldsymbol{\theta}$. Once the data are observed, the only quantities in $f(\cdot)$ that are unknown to the investigators are $\boldsymbol{\theta}$. Then, the *likelihood function* for $\boldsymbol{\theta}$ is the function

$$L(\boldsymbol{\theta} \mid \boldsymbol{y}) = f(\boldsymbol{y};\boldsymbol{\theta}).$$

Note that the likelihood is interpreted as a function of $\boldsymbol{\theta}$, with $\boldsymbol{y}$ held fixed at its observed value.

The *maximum likelihood estimate* of $\boldsymbol{\theta}$ is the value, $\hat{\boldsymbol{\theta}}$, which maximizes the likelihood function or, equivalently, its logarithm. That is, for any value of $\boldsymbol{\theta}$,

$$L(\boldsymbol{\theta} \mid \boldsymbol{y}) \leq L(\hat{\boldsymbol{\theta}} \mid \boldsymbol{y}).$$

According to the likelihood principle, $\hat{\boldsymbol{\theta}}$ is then regarded as the value of $\boldsymbol{\theta}$ which is most strongly supported by the observed data. In practice, $\hat{\boldsymbol{\theta}}$ is obtained either by direct maximisation of $\log L$, or by solving the set of equations

$$\boldsymbol{S}(\boldsymbol{\theta}) = \partial \mathrm{log} L / \partial \boldsymbol{\theta} = 0. \qquad \text{(A.4.1)}$$

The function $\boldsymbol{S}(\boldsymbol{\theta})$ is known as the *score function* for $\boldsymbol{\theta}$. Very often, numerical methods are required to evaluate the maximum likelihood estimate. Popular methods include the Nelder and Mead (1965) simplex algorithm for direct maximisation of $\log L$, or Newton-Raphson iteration for solution of the score equations (A.4.1).

The maximum likelihood estimate is known to enjoy many optimality properties in large samples. In particular, under mild regularity conditions, $\hat{\boldsymbol{\theta}}$ is asymptotically unbiased, and asymptotically efficient in the sense that the elements of $\hat{\boldsymbol{\theta}}$ are estimated with the smallest possible asymptotic variances of any asymptotically unbiased estimators. The asymptotic covariance matrix of $\hat{\boldsymbol{\theta}}$ is given by the expression

$$V = \{-E(\partial^2 \mathrm{log} L / \partial \boldsymbol{\theta}^2)\}^{-1}.$$

The matrix V^{-1} is also known as the *Fisher information matrix* for $\boldsymbol{\theta}$.

Example A.1. Consider Y_1 and Y_2 to be two independent binomial observations with sample sizes and probabilities (n_1, p_1) and (n_2, p_2), respectively. A typical example for which this setting is appropriate is a clinical trial where two treatments are being compared. Here, Y_i denotes the number of patients responding negatively to the treatment i to which n_i subjects were assigned, and p_i is the corresponding probability of negative response for $i = 1, 2$. It is convenient to transform the parameters (p_1, p_2) to (θ_1, θ_2), where

$$\theta_1 = \log\left(\frac{p_1(1-p_2)}{p_2(1-p_1)}\right), \quad \theta_2 = \log\left(\frac{p_2}{1-p_2}\right).$$

This leads to a likelihood function for $\boldsymbol{\theta} = (\theta_1, \theta_2)$ of the form

$$\begin{aligned} L(\boldsymbol{\theta} \mid y_1, y_2) &\propto p_1^{y_1}(1-p_1)^{n_1-y_1}\, p_2^{y_2}(1-p_2)^{n_2-y_2} \\ &= \left(\frac{p_1}{1-p_1}\right)^{y_1}\left(\frac{p_2}{1-p_2}\right)^{y_2}(1-p_1)^{n_1}(1-p_2)^{n_2} \\ &= \exp\{\theta_1 y_1 + \theta_2 y_1 + \theta_2 y_2 - n_1 \log\left(1+e^{\theta_1+\theta_2}\right) - n_2 \log\left(1+e^{\theta_2}\right)\}. \end{aligned}$$

The parameter θ_1 is called the *log odds ratio*. A zero value for θ_1 denotes equality of p_1 and p_2. The maximum likelihood estimate, $\hat{\boldsymbol{\theta}}$, can be derived as the solution of the pair of equations,

$$y_1 \quad \frac{n_1 \exp(\theta_1+\theta_2)}{1+\exp(\theta_1+\theta_2)} = y_1 - n_1 p_1 = 0$$

and

$$y_1 + y_2 - \frac{n_1 \exp(\theta_1+\theta_2)}{1+\exp(\theta_1+\theta_2)} \quad \frac{n_2 \exp(\theta_2)}{1+\exp(\theta_2)} = y_1 + y_2 - n_1 p_1 - n_2 p_2 = 0.$$

This gives

$$\hat{\theta}_1 = \log\left(\frac{y_1(n_2-y_2)}{y_2(n_1-y_1)}\right), \quad \hat{\theta}_2 = \log\left(\frac{y_2}{n_2-y_2}\right).$$

Fisher's information matrix for θ can be obtained by straightforward algebra, and is

$$V^{-1} = \begin{pmatrix} \frac{n_1 \exp(\theta_1+\theta_2)}{\{1+\exp(\theta_1+\theta_2)\}^2} & \frac{n_1 \exp(\theta_1+\theta_2)}{\{1+\exp(\theta_1+\theta_2)\}^2} \\ \frac{n_1 \exp(\theta_1+\theta_2)}{\{1+\exp(\theta_1+\theta_2)\}^2} & \frac{n_1 \exp(\theta_1+\theta_2)}{\{1+\exp(\theta_1+\theta_2)\}^2} + \frac{n_2 \exp(\theta_2)}{\{1+\exp(\theta_2)\}^2} \end{pmatrix}.$$

The asymptotic variance of $\hat{\theta}_1$ is the upper left entry of V, namely

$[n_1 \exp(\theta_1+\theta_2)/\{1+\exp(\theta_1+\theta_2)\}^2]^{-1} + [n_2 \exp(\theta_2)/\{1+\exp(\theta_2)\}^2]^{-1}$

which can be estimated consistently by

$$\frac{1}{y_1} + \frac{1}{n_1 - y_1} + \frac{1}{y_2} + \frac{1}{n_2 - y_2}.$$

The word 'asymptotic' in this example means that both n_1 and n_2 are large.

Likelihood inference proceeds by fitting a series of sub-models which are *nested.* This means that each sub-model in the sequence is contained within the previous one. In Example A.1, an interesting hypothesis, or sub-model, to test is that $\theta_1 = 0$, corresponding to equality of p_1 and p_2. The difference between this sub-model and the full model with no restriction on θ_1 can be examined by calculating the *likelihood ratio test statistic*, which is defined as

$$G = -2\{\log L(\hat{\boldsymbol{\theta}}_0 \mid y) - \log L(\hat{\boldsymbol{\theta}} \mid y)\},$$

where $\hat{\boldsymbol{\theta}}_0$ and $\hat{\boldsymbol{\theta}}$ are the maximum likelihood estimates of $\boldsymbol{\theta}$ under the null hypothesis or sub-model, and the unrestricted model, respectively. Assuming that the sub-model is correct, the sampling distribution of G is approximately chi-squared, with number of degrees of freedom equal to the difference between the numbers of parameters specified under the sub-model and the unrestricted model.

An alternative testing procedure is to examine the *score statistic*, $\boldsymbol{S}(\boldsymbol{\theta})$, as in (A.4.1). The score test statistic is

$$\boldsymbol{S}(\hat{\boldsymbol{\theta}}_0)' V(\hat{\boldsymbol{\theta}}_0) \boldsymbol{S}(\hat{\boldsymbol{\theta}}_0),$$

whose null sampling distribution is also chi-squared, with the same degrees of freedom as for the likelihood ratio test statistic. In either case, the sub-model is rejected in favour of the unrestricted model if the test statistic is too large.

Example A.1 (continued). Suppose that we want to test whether the probabilities, p_1 and p_2, from two treatments are identical. This is equivalent to testing the sub-model with $\theta_1 = 0$. Note that the value of θ_2 is unspecified by the null hypothesis and therefore has to be estimated. The algebraic form of the likelihood ratio test statistic G is complicated, and we do not give it here, although in applications it can easily be evaluated numerically. The score statistic has the simple form

$$\frac{(y_1 - E_1)^2}{E_1} + \frac{(y_2 - E_2)^2}{E_2}$$

where $E_i = n_i(y_1 + y_2)/(n_1 + n_2)$ is the expected value for Y_i under the null model that the two groups are the same. This statistic is also known as the

Pearson χ^2 test statistic. The number of degrees of freedom in this example is one, because the sub-model has one parameter, whereas the unrestricted model has two.

A.5 Generalized linear models

Regression models for independent, discrete and continuous responses have been unified under the class of *generalized linear models*, or GLMs (McCullagh and Nelder, 1989), thus providing a common body of statistical methodology for different types of response. Here, we review the salient features of this class of models.

We begin by considering two particular GLMs, logistic and Poisson regression models, and then discuss the general class. Because GLMs apply to independent responses, we focus on the cross-sectional situation, as in Section A.2, with a single response Y_i and a vector $\boldsymbol{x}_i$ of p explanatory variables associated with each of m experimental units. The objective is to describe the dependence of the mean response, $\mu_i = \mathrm{E}(Y_i)$, on the explanatory variables.

A.5.1 *Logistic regression*

This model has been used extensively for dichotomous response variables such as the presence or absence of a disease. The logistic model assumes that the logarithm of the odds of a positive response is a linear function of explanatory variables, so that

$$\log \frac{\Pr(Y_i = 1)}{\Pr(Y_i = 0)} = \log \frac{\mu_i}{1 - \mu_i} = \boldsymbol{x}_i' \boldsymbol{\beta}.$$

Figure A.1 shows plots of $\Pr(Y = 1)$ against a single explanatory variable x, for several values of β. A major distinction between the logistic regression model and the linear model in Section A.2 is that the linearity applies to a transformation of the expectation of Y_i, in this case the log odds transformation, rather than to the expectation itself. Thus, the regression coefficients, $\boldsymbol{\beta}$, represent the change of the log odds of the response variable per unit change of x. Another feature of the dichotomous response variable is that the variance of Y_i is completely determined by its mean, μ_i. Specifically,

$$\mathrm{Var}(Y_i) = \mathrm{E}(Y_i)\{1 - \mathrm{E}(Y_i)\} = \exp(\boldsymbol{x}_i'\boldsymbol{\beta})/\{1 + \exp(\boldsymbol{x}_i'\boldsymbol{\beta})\}^2.$$

This is to be contrasted with the linear model, where $\mathrm{Var}(Y_i)$ is usually assumed to be a constant, σ^2, which is independent of the mean.

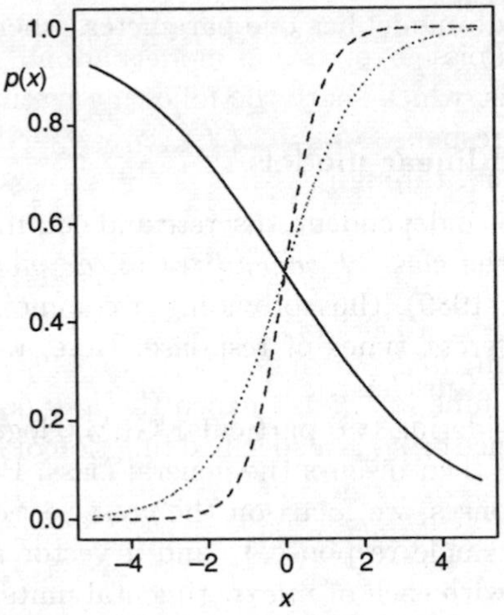

Fig. A.1. The logistic model, $p(x) = \exp(\beta x)/\{1 + \exp(\beta x)\}$; ——— : $\beta = -0.5$; ········ : $\beta = 1$; – – – : $\beta = 2$.

A.5.2 *Poisson regression*

Poisson regression, or *log-linear*, models are applicable in problems in which the response variable represents the number of events occurring in a fixed period of time. One instance is the number of seizures in a given time-interval, as in Example 1.6. Because of the discrete and non-negative nature of count data, a reasonable assumption is that the *logarithm* of the expected count is a linear function of explanatory variables, so that

$$\log \mathrm{E}(Y_i) = \boldsymbol{x}_i'\boldsymbol{\beta}.$$

Here, the regression coefficient for a particular explanatory variable can be interpreted as the logarithm of the ratio of expected counts before and after a one unit increase in that explanatory variable, with all other explanatory variables held constant. The term 'Poisson' refers to the distribution for counts derived by Poisson (1837),

$$p(y) = \exp(-\mu)\mu^y/y!\ ,\quad y = 0, 1, \ldots$$

As in logistic regression, the assumption that Y_i follows a Poisson distribution implies that the variance of Y_i is determined by its mean. In this case, the mean and variance are the same,

$$\mathrm{Var}(Y_i) = \mathrm{E}(Y_i) = \exp(\boldsymbol{x}_i'\boldsymbol{\beta}).$$

A.5.3 *The general class*

Linear, logistic and Poisson regression models are all special cases of generalized linear models, which share the following features.

First, the mean response, $\mu_i = \mathrm{E}(Y_i)$, is assumed to be related to a vector of covariates, $\boldsymbol{x}_i$, through

$$h(\mu_i) = \boldsymbol{x}_i'\boldsymbol{\beta}.$$

For logistic regression, $h(\mu_i) = \log\{\mu_i/(1-\mu_i)\}$; in Poisson regression, $h(\mu_i) = \log(\mu_i)$. The function $h(\cdot)$ is called the *link function*.

Second, the variance of Y_i, is a specified function of its mean, μ_i, namely

$$\mathrm{Var}(Y_i) = v_i = \phi v(\mu_i).$$

In this expression, the known function $v(.)$ is referred to as the *variance function*; the scaling factor, ϕ, is a known constant for some members of the GLM family, whereas in others it is an additional parameter to be estimated.

Third, each class of generalized linear models corresponds to a member of the exponential family of distributions, with a likelihood function of the form

$$f(y_i) = \exp[\{y_i\theta_i - \psi(\theta_i)\}/\phi + c(y_i, \phi)]. \qquad \text{(A.5.1)}$$

The parameter θ_i is known as the *natural parameter*, and is related to μ_i through $\mu_i = \partial\psi(\theta_i)/\partial\theta_i$. For example, the Poisson distribution is a special case of the exponential family, with

$$\theta_i = \log\mu_i, \quad \psi(\theta_i) = \exp(\theta_i), \quad c(y_i, \phi) = -\log(y_i!), \quad \phi = 1.$$

Other distributions within this family include the Gaussian or Normal distribution, the binomial distribution and the two-parameter gamma distribution.

In any generalized linear model the regression coefficients, $\boldsymbol{\beta}$, can be estimated by solving the same estimating equation,

$$\boldsymbol{S}(\boldsymbol{\beta}) = \sum_{i=1}^{m} \left(\frac{\partial\mu_i}{\partial\boldsymbol{\beta}}\right)' v_i^{-1}\{Y_i - \mu_i(\boldsymbol{\beta})\} = 0, \qquad \text{(A.5.2)}$$

where $v_i = \mathrm{Var}(Y_i)$. Note that $\boldsymbol{S}(\boldsymbol{\beta})$ is the derivative of the logarithm of the likelihood function. The solution $\hat{\boldsymbol{\beta}}$, which is the maximum likelihood estimate, can be obtained by iteratively reweighted least squares; see

McCullagh and Nelder (1989) for a detailed discussion. Finally, in large samples $\hat{\boldsymbol{\beta}}$ follows a Gaussian distribution with mean $\boldsymbol{\beta}$ and variance

$$V = \left(\sum_{i=1}^{m} \left(\frac{\partial \mu_i}{\partial \boldsymbol{\beta}} \right)' v_i^{-1} \frac{\partial \mu_i}{\partial \boldsymbol{\beta}} \right)^{-1}. \tag{A.5.3}$$

This variance can be estimated by $\hat{V}$ which is obtained by replacing $\boldsymbol{\beta}$ with $\hat{\boldsymbol{\beta}}$ in the expression (A.5.3).

A.6 Quasi-likelihood

One important property of the GLM family is that the score function, $\boldsymbol{S}(\boldsymbol{\beta})$, depends only on the mean and the variance of the Y_i. Wedderburn (1974) was the first to point out that the estimating equation (A.5.2) can therefore be used to estimate the regression coefficients for any choices of link and variance functions, whether or not they correspond to a particular member of the exponential family. The name *quasi-score function* was coined for $\boldsymbol{S}(\boldsymbol{\beta})$ in (A.5.2), since its integral with respect to $\boldsymbol{\beta}$ can be thought of as a 'quasi-likelihood' even if it does not constitute a proper likelihood function. This suggests an approach to statistical modelling in which we make assumptions about the link and variance functions without attempting to specify the entire distribution of Y_i. This is desirable, since we often do not understand the precise details of the probabilistic mechanisms by which data were generated. McCullagh (1983) showed that the solution, $\hat{\boldsymbol{\beta}}$, of the quasi-score function has a sampling distribution which, in large samples, is approximately Gaussian with mean $\boldsymbol{\beta}$ and variance given by equation (A.5.3).

Example A.2. Let $Y_1, \ldots, Y_m$ be independent counts whose expectations are modelled as

$$\log \mathrm{E}(Y_i) = \boldsymbol{x}_i' \boldsymbol{\beta}, \quad i = 1, \ldots, m.$$

In biomedical studies, frequently the variance of Y_i is greater than $\mathrm{E}(Y_i)$, the variance expression induced by the Poisson assumption. This phenomenon is known as *over-dispersion*. One way to account for this is to assume that $\mathrm{Var}(Y_i) = \phi \mathrm{E}(Y_i)$ where ϕ is a non-negative scalar parameter. Note that for the Poisson distribution, $\phi = 1$; if we allow $\phi > 1$, we no longer have a distribution from the exponential family. However, if we define $\hat{\boldsymbol{\beta}}$ as the solution to

$$\sum_{i=1}^{m} \boldsymbol{x}_i \{ Y_i - \exp(\boldsymbol{x}_i' \boldsymbol{\beta}) \} = 0,$$

then a simple calculation gives the asymptotic variance matrix of $\hat{\boldsymbol{\beta}}$ as

$$\phi \left(\sum_{i=1}^{m} \boldsymbol{x}_i \exp(x_i'\boldsymbol{\beta})\boldsymbol{x}_i' \right)^{-1}.$$

Thus, by comparison with (A.5.3) the variance of $\hat{\boldsymbol{\beta}}$ is inflated by a factor of ϕ. Clearly, ignoring over-dispersion in the analysis would lead to under-estimation of standard errors, and consequent over-statement of significance in hypothesis testing.

In the above example, $\phi \mathrm{E}(Y_i)$ is but one of many possible choices for the variance formula which would take account of over-dispersion in count data. Fortunately, the solution, $\hat{\boldsymbol{\beta}}$, is a consistent estimate of $\boldsymbol{\beta}$ as long as $h(\mu_i) = \boldsymbol{x}_i'\boldsymbol{\beta}$, whether or not the variance function is correctly specified. This robustness property holds because the expectation of $\boldsymbol{S}(\boldsymbol{\beta})$ remains zero so long as $\mathrm{E}(Y_i) = \mu_i(\boldsymbol{\beta})$. However, the asymptotic variance matrix of $\hat{\boldsymbol{\beta}}$ has the form

$$V_2 = V \left(\sum_{i=1}^{m} \left(\frac{\partial \mu_i}{\partial \boldsymbol{\beta}} \right)' v_i^{-1} \mathrm{Var}(Y_i) v_i^{-1} \frac{\partial \mu_i}{\partial \boldsymbol{\beta}} \right) V.$$

Note that V_2 is identical to V in (A.5.3) only if $\mathrm{Var}(Y_i) = v_i$. When this assumption is in doubt, confidence limits for $\boldsymbol{\beta}$ can be based on the estimated variance matrix

$$\hat{V}_2 = \hat{V} \left(\sum_{i=1}^{m} \left(\frac{\partial \mu_i}{\partial \boldsymbol{\beta}} \right)' v_i^{-1} \{Y_i - \mu_i(\boldsymbol{\beta})\}^2 v_i^{-1} \frac{\partial \mu_i}{\partial \boldsymbol{\beta}} \right) \hat{V}, \qquad \text{(A.6.1)}$$

evaluated at $\hat{\boldsymbol{\beta}}$. We call $\hat{V}$ a *model-based* variance estimate of $\hat{\boldsymbol{\beta}}$ and $\hat{V}_2$ a *robust* variance estimate in that $\hat{V}_2$ is consistent regardless of whether the specification of $\mathrm{Var}(Y_i)$ is correct.

Example A.2 (continued). An alternative to the variance function

$$v_i = \phi \mu_i$$

is the form induced by the Poisson-gamma distribution (Breslow, 1984),

$$v_i = \mu_i(1 + \mu_i \phi).$$

With a limited amount of data available, it is difficult to choose empirically between these two variance functions (Liang and McCullagh, 1993). The availability of the robust variance estimate, $\hat{V}_2$, helps to alleviate the concern regarding the choice of variance formula in larger samples. It is interesting to note that in the special case where $\mu_i = \mu$ and hence $\log \mu_i = \beta_0$, the estimate $\hat{V}_2$ reduces to

$$\sum_{i=1}^{m} (Y_i - \bar{Y})^2 / m^2,$$

the sampling variance of $\hat{\beta}_0 = \log \bar{Y}$ (Royall, 1986).

Index

Note: Figures are shown by italic page numbers, Tables by numbers in square brackets

adaptive kernel estimator 107–8
age effects 1, 161
alternating logistic regressions (ALRs) 152
  see also generalized estimating equations (GEEs)
analysis-of-variance (ANOVA) methods 117–30
  limitations 117, 129
  main advantage 129
  repeated-measures ANOVA 125–9
  split-plot ANOVA 127–8, 129–30
    example of use 128–9
  time-by-time ANOVA 118–22, 129
    disadvantages 118–19
    example of use 119, 122
ante-dependence models 85–6, 118
autocorrelation function 49–51
  estimation from sample variogram 51
  for exponential model *57*
  further reading recommended 53
  relationship to variogram 51
auto-Poisson process 204

back-fitting algorithm 111
Bahadur parametrization (of log-linear model) 149
beta-binomial distribution 178–9
  applications 178
  extension to allow covariates to vary within clusters 179
bias 23–5
bibliography 224–34
binary responses
  logistic regression for
    conditional likelihood approach used 175–8
    generalized estimating equations used 151–2
  log-linear model used 147–8
    with marginal parameters 148–51
  marginal models used 147–52
    examples 153–62
  random effects models used 178–80
    examples 180–3
  sample size calculations 31–2
boxcar window technique 43
boxplots, epileptic seizure data *15*

calves intestinal parasites trial 119
  data [120–1]
  time-by-time ANOVA analysis applied 119, 122
carry-over effect
  experimental assessment 13, 156
  ignored in cerebrovascular deficiency/treatment trial 154
categorical data
  transition models for 194–203
  see also ordered categorical data
CD4+ cell numbers data 3–4
  correlation structure explored 47–50, 51–2, *53*
  curve-smoothing methods used *41–6*
  effect of depressed state of person 39–40
  graphical presentations *3*, *37–40*
  individual-trajectory estimation 115, *116*
  population-mean estimation 112–13
  random effects model applied 135
  variogram constructed 51–2, *53*
censoring of data 210, 221
cerebrovascular deficiency/treatment trial 153
  marginal model used 153–4, [155], [181]
  random effects model used 180–1
  random intercept model fitted 176–7
chi-squared distribution 238
chi-squared test statistic *see* Pearson chi-squared test statistic
cohort effects 1
complete data score function 173
completely random dropouts 211, 218
  testing for 211–16
completely random missing values 209
conditional likelihood 143, 170, 171–2
  advantages 177–8

applications
random intercept logistic model 175–8
random intercept log-linear model for count data 183–5
transition ordinal regression model 203
maximization in transition models 193
confidence regions 94
confirmatory analysis 33
connected-line graphs 34
examples *36*, *38*
avoidance of clutter 37–8, *39*, *40*
continuous responses, sample size calculations 29–31
correlated errors, general linear models with 55–8
correlation (in longitudinal data set) 47–52
correlation matrix 49
count data
conditional likelihood estimation of random intercept model 183–5
examples 162
generalized estimating equations used 165
log-linear transition models for 203–7
marginal models fitted in examples 165–7
parametric modelling for 162–5
random effects models for 185–7
counted responses 162–7, 183–5
covariance structure
modelling of 77–116
parametric models 78–106
cow-weight data 100–1, [100], [101]
parametric model 102–6
crossover trials 153
conditional likelihood approach 175–7
examples 13, 153–8
further reading recommended 32
ordinary least-squares estimation used 61–2
relative efficiency [62]
see also cerebrovascular...; epileptic...
cross-sectional studies
contrasted with longitudinal studies 1, 17, 23, 161
longitudinal data marginal mean modelled as in 19
cross-validation 47
curve-smoothing methods 41–7
see also kernel estimation; lowess; splines

data score functions 173
derived variables 18, 122–6, 129
examples of use 123–6
design considerations 23–32
diagnostic checking (of model) 95
Diggle–Kenward model (of dropout process) 219
application to milk-protein data 219–20
informative dropouts studied 219–20, 222
dropout process
classification of 218
modelling of 216–20
discussion on 221–2
example 219–20
dropouts 210
completely random dropouts 211, 218
testing for 211–16
informative dropouts 218
random dropouts 218
reasons for occurrence 210–11
see also missing values
dysmenorrheal pain treatment study 13, 155
data [13], [156]
marginal models fitted 154–8
random intercept model fitted 177

efficiency, longitudinal studies 25–7
EM algorithm 173, 210
empirical Bayes estimates 115, *116*
epileptic seizure/treatment study 13–14
boxplots of data *15*
conditional likelihood method used 184–5
data [14], [163]
log-linear transition model fitted 205, *206*, 207
marginal model fitted 165–7, [168]
Poisson–Gaussian random effects models fitted 187–8
explanatory variables 235
exploratory data analysis (EDA) 33
further reading recommended 52
exponential correlation function, compared with Gaussian correlation function 82
exponential correlation model 57–8, 81, 86
variogram of model 81–2, *83*
extra-binomial variation 178

factorial effects, cow-weight data 102
first-order autoregressive processes 85
models for 58, 85
see also ante-dependence...; exponential correlation model
Fisher information matrix 238, 239
F-statistics 118, 124, 126, 128, 129

Gaussian assumptions
general linear model 56
maximum likelihood estimation 63–4
Gaussian correlation function 82
compared with exponential correlation function 82
Gaussian correlation model, variograms of model *84*
Gaussian kernel 43, 107
Gaussian random effects, examples of logistic models with 180–3
Gauss–Markov Theorem 237
generalized estimating equations (GEEs) 143–5
applications
to count data 165
to logistic regression model 151–2
further reading recommended 144, 145, 167
generalized linear mixed models, estimation for 171–5
generalized linear models (GLMs) 131–45, 241–4
general class 243–4
logistic regression model 241, *242*
marginal models 131–3, 146–68
Poisson regression model 242
random effects models 133–5, 169–89
transition models 135–6, 190–207
general linear models 55–77
with correlated errors 55–8
exponential correlation model 57–8
uniform correlation model 56
geostatistics 51
graphical presentation of data 34–41
further reading recommended 52
growth curve model 89

holly leaves, effect of pH on drying rate 124–6, [127]
hypothesis tests 94

ignorable dropout processes 218
ignorable missing value mechanisms 210
individual trajectories
estimation of 113–15
example using CD4+ cell data 115, *116*
Indonesian Children's Health Study (ICHS) 4–5
data [4], [159]
marginal model used 132–3, 140–1, 146, 158–62
random effects model used 134, 135, 139–40, 181–3
transition model used 135–6, 197–201
infant-growth linear-regression model
random effects model considered 133–4, 137–8
transition model used 138–9
inference 142–5
informative dropouts 218
Diggle–Kenward model used to investigate 219–20, 222
informative missing values 209
intermittent missing values 210, 221

kernel estimation 42–4
approximation of smoothing spline 45
compared with other curve-smoothing techniques *43*
effect of kernel bandwidth *45*
effect of outliers 46
visualization *44*
kernel function 107
see also Gaussian kernel
Kolmogorov–Smirnov statistic 213, 214

lag 49
latent variable model 135
least-squares estimator 24, 25, 236
least-squares method 236–7
likelihood functions 171–2, 173, 238
see also conditional...; maximum likelihood...
likelihood inference 238–41
likelihood ratio testing 94–5
likelihood ratio test statistic 220, 240
linear models 235–6
least-squares method used 236–7
marginal modelling approach 137
parametric modelling of treatment contrasts using 111
random effects modelling approach 137–8
transition modelling approach 138–9
see also general...; generalized linear models
linear regression model 235
link functions 191–2, 243
logistic regression model 241, *242*
fitting in examples 154, [155], [157]
marginal-modelling approach 140–2, 151–2
random effects modelling approach 139–40, 169, 175–80
examples 180–3
log-linear models 147–8, 242
canonical parameters in 147, 148, 158
equivalence of random effects and marginal parameters 142
log-linear transition models for count data 203–7
for marginal means 148–51

offset introduced 165
parametrization using marginal means 149
log odds ratio 239
longitudinal data
example data sets 3–16
CD4+ cell numbers data 3–4
dysmenorrheal pain treatment study 13
epileptic seizure/treatment study 13–14, *15*
Indonesian children's health study 4–5
milk protein study 5, [9–11], *12*
pig weight data [35]
Sitka spruce growth data 5, [6–7], *8*
exploring correlation structure 47–52
general linear models for 55–77
with correlated errors 55–8
graphical presentation 34–41
guidelines for presentation 33
marginal mean modelled as in cross-sectional study 19
smoothing methods used 41–7
longitudinal data analysis
approaches 18–21
marginal analysis 19
random effects model 19
transition model 19–20
two-stage/derived variable analysis 18
confirmatory analysis 33
consequences of ignoring correlation in data 20
classification of problems 20–1
exploratory analysis 33–54
see also marginal...; random effects...; transition models
longitudinal studies
advantages 1, 17–18, 23
contrasted with cross-sectional studies 1, 17, 23, 161
defining characteristic 1, 2
efficiency 25–7
sample size calculations 27–32
lowess smoothing 34, 37, 41–2, 45
compared with other curve-smoothing techniques *43*, *46*
effect of outliers 46
examples *3*, *37*, *41–3*

marginal analysis 19
marginal expectation 131, 142
marginal likelihood 68
see also restricted maximum likelihood (REML) estimation
marginal mean response 18
marginal means, log-linear models for 148–51
marginal models 131–3, 146–68
assumptions 131–2
examples of use 153–62
further reading recommended 167
likelihood of data determined 143
linear models 137–8
logistic models 140–1
marginal odds ratio 150
Markov models 85, 191
example illustrating 197–201
further reading recommended 207
likelihood contribution in 192
see also transition models
Markov Poisson time series model, realizations *206*
maximum likelihood estimation 63–4, 238, 239, 240
compared with REML estimation 68, 92, 93
hat notation used 59, 68
for random effects models 172–5
for transition models 143
see also restricted maximum likelihood estimation
mean response
generalized linear models 243
non-parametric modelling of 106–13
mean response profiles
calves intestinal parasites trial *119*
cow-weight data 102, *105*
definition in ANOVA 117
milk-protein data 96, [97], 99–100, *99*
measurement error
with random effects 88–90
with serial correlation 87
and random intercept 87
as source of random variation 80
measurement variation 29
micro/macro graph design strategy 37–8
milk protein data 5, [9–11], *12*
dropouts in data 210–11, 213–14, 215–16
parametric model 96–100
variogram 52, *54*, 96
missing values 208–23
classification of 208–11
effects 77
meaning of term 208
see also dropouts
model-based variance 245
model-fitting process 90–5
diagnostics 95
estimation 92–3
formulation 91
inference 93–5

models *see* ante-dependence...; generalized linear...; general linear...; linear...; logistic...; log-linear...; marginal...; Markov...; random effects...; saturated...; serial correlation...; transition models
Monte Carlo techniques, testing for completely random dropouts using 213, 214
Multicenter AIDS Cohort Study (MACS) 3
  CESD scores 40
  objective 4
  see also CD4+ cell numbers data
multivariate Gaussian theory 237–8

natural parameter 243
negative-binomial distribution 163–4, 186
Nelder–Mead simplex algorithm 93, 238
nested sub-models 240
Newton–Raphson iteration 238
non-parametric curve estimation techniquess 42–6
  see also kernel estimation; lowess; smoothing spline
non-parametric modelling of mean response 106–13
notation 16–17
number of observations per person 29

odds ratios 132, 133, 150
  see also log odds ratio
ordered categorical data, transition models for 201–3
ordering statistics, data presentation using 38–9
ordinary least-squares (OLS) estimation
  effect of ignoring correlation in data 20, *21*
  naive approach 62
    errors arising 62
  relative efficiency 59, 60
    in crossover example [62]
    in linear regression example [61]
  robust estimation of errors using 69
  small-sample behaviour in nonlinear regression modelling 123
  variogram plotted for milk protein data 52, *54*
outliers, curve-smoothing affected by 46
over-dispersed count data, models for 163–4
over-dispersion 164, 186, 244

panel studies 2
parametric models 79–90
  for count data 162–5
  examples of application 95–106
    cow-weight data 100–6
    milk-protein data 96–100
  fitting of model to data 90–5
  further reading recommended 115
  notation 78–9
  sources of random variation 79–80
Pearson chi-squared test statistic 185, 241
pig weight data [35]
Poisson distribution 162–3, 185, 242, 243
Poisson–Gaussian random effects models 187–8
Poisson regression models 242
  see also log-linear models
Poisson transition models 204–7
power (of statistical test) 29
predictive squared error (PSE) 46–7
principal components analysis 39
proportional odds model 201–2
  application to Markov chain for ordered categorical data 202–3

quasi-likelihood estimation approach 186, 244–6
quasi-Newton algorithm 93
quasi-score function 244

random effects models 19, 89, 133–5, 169–89
  further reading recommended 189
  likelihood of data estimated 142–3
  linear models 138
  logistic models 139–40
  most-useful applications 135
random intercept models
  approaches to inference 170
  conditional likelihood approach 175–8
    examples 176–7
random missing values 209
random selection, limited data subset 38, *39*
random variation (in data), sources 79–80
reading ability/age example 1, *2*
references listed 224–34
regression equation, matrix notation 16–17
relative efficiency, longitudinal vs cross-sectional studies 26, *27*
relative growth rate (RGR) 89–90
repeated-measures ANOVA 125–9
repeated observations
  correlation among 29
  number per person 29
restricted maximum likelihood (REML) estimation 64–8
  application to milk-protein data 96–7
  compared with maximum likelihood estimation 68, 92, 93

robust estimation of standard errors using 69–70
Rice–Silverman prescription (for choosing *h*) 108–9
robust estimation of standard errors 68–77
example 72–5, *76*
factors affecting 77
robust smoothing methods 45
see also lowess
robust variance 194, 245
roughness penalty 44

sample size calculations 27–32
binary responses 31–2
continuous responses 29–31
sample variograms 51
estimation of autocorrelation function from 51
examples 51–2, *53–4*, *104*
Sandland–McGilchrist RGR model 89–90
saturated models 52, 64
scatterplot 33
examples *36–46*
scatterplot matrix, example *48*
score equations 173
score function 238
score test statistic 240
semi-parametric models 79, 111
serial correlation 79–80
with measurement error 87
and random intercept 87
pure serial correlation 81–2, 85–6
Sitka spruce growth study 5, [6–7], *8*
derived variables used 123–4
robust estimation of standard errors 72–5, *76*
split-plot ANOVA used 128–9
size-dependent branching process 204
smallest meaningful difference 29
smoothing methods 41–7
further reading recommended 52–3
see also kernel estimation; lowess; splines
smoothing splines 44–5
approximated by kernel estimate 45
compared with other curve-smoothing techniques *43*
specification tests 170
splines 44
see also smoothing splines
split-plot analysis/model 56, 88–9, 127
split-plot ANOVA 127–8, 129–30
example of use 128–9
standard errors
robust estimation of 68–77
example 72–5, *76*
statistical background 235–46
stochastic process, variogram defined 51, 79
survival analysis 2

time-by-time ANOVA 118–22, 129
disadvantages 118–19
example of use 119, 122
time series analysis, and longitudinal studies 2, 88
tracking 34
trajectory *see* individual trajectories
transition matrix 194
first-order Markov chain 194
second-order Markov chain 195
transition models 19–20, 135–6, 190–207
for categorical data 194–203
example 197–201
ordered categorical data 201–3
fitting of models 192–4
likelihood of data determined 143
with linear link function 191
linear models 138–9
with logit link function 191
log-linear transition models for count data 203–7
with log-link function 192
see also Markov models
transition ordinal regression model 203
tree growth study 5
see also Sitka spruce growth study
t-tests 119, [122], 124
two-stage variable analysis 18
type I error rate 28

unbalanced data 208
uniform correlation model 56, 88

variance function 243
variograms 51
estimation of autocorrelation function from 51
estimation of covariance structure from 111
examples 51–2, *53–4*
for exponential correlation model 81–2, *83*
for Gaussian correlation model *84*
parametric models
cow-weight data 102, *104*
milk-protein data 98, *99*
for random intercepts/effects model *89*, *90*
for serial correlation models 81–2, *83–4*, 87, *88*, *89*
for stationary process 79
for stochastic process 51, 79
see also sample variogram

vitamin A deficiency study 4–5, 158
see also Indonesian Children's Health Study (ICHS)

weakly stationary process 47
weighted least-squares estimation 58–62, 77
Wong transition model 204
working covariance matrix 69, 75

Zeger–Qaqish transition model 204
realizations *206*